AB = 0.2042

BEL ET UTILE
THE WORK OF THE ROBERT DE VAUGONDY FAMILY OF MAPMAKERS

Published in a limited edition of 750 copies

this copy No.

BEL ET UTILE

THE WORK OF THE ROBERT DE VAUGONDY FAMILY OF MAPMAKERS

Mary Sponberg Pedley

MAP COLLECTOR PUBLICATIONS 1992

For John

First published in Great Britain by Map Collector Publications Ltd., 1992
48 High Street, Tring, Herts HP23 5BH, England

Book design and typography by Roy Walker

Printed by BPCC Wheatons Ltd, Exeter

British Library Cataloguing in Publication Data

Pedley, Mary

Bel et Utile: The Work of the Robert de Vaugondy Family of Mapmakers

ISBN 0 906430 12 7

A Note on Abbreviations and Terms Used

In the references, these abbreviations are used:
AN = Archives Nationales, Paris
BN = Bibliothèque Nationale, Paris
BL = British Library, London
LC = Library of Congress, Washington, D.C.
WLCL = William L. Clements Library, University of Michigan

Endpapers: *Manuscript celestial globe gores by Gobin, for Didier Robert de Vaugondy. [Cat. 3] (Courtesy of the Bibliothèque Nationale, Paris)*

Frontispiece: *Didier Robert de Vaugondy, terrestrial globe of 1773, 490 mm (Photo courtesy of the Bibliothèque Nationale, Paris)*

CONTENTS

ATLAS
UNIVERSEL
DRESSÉ
SUR LES MEILLEURES CARTES
MODERNES
1776
À VENISE
Chez P. Santini rue Ste. Justine.

FOREWORD

The eighteenth century in France has already generated important studies in the field of the history of cartography, particularly those of Père de Dainville on the cartography of the provinces[1], that of Numa Broc[2], which has the virtue of not separating geography and cartography, and the more scientific and technical work of Josef Konvitz[3] which has examined how and why cartography has advanced. But numerous areas still remain to be explored. The work of Mary Pedley has the distinction of helping us to discover the conditions under which maps were created before being broadly diffused, thanks to the efforts of one family which dedicated itself to compiling maps in the scientific and critical atmosphere of the eighteenth century.

As is known, cartography during the century of the Enlightenment benefited from the methodological advances which were accomplished in the second half of the seventeenth century by the Academy of Sciences, under the impetus of Colbert, founder of the Academy, and of Jean Picard and Jean-Dominique Cassini, members of the Academy. These advances were themselves preceded by the birth of truly French cartographic publications which had tried to challenge the powerful Dutch publishing houses, especially that of Blaeu. The Sanson family, whose role has been analysed by Mireille Pastoureau[4], participated actively in this movement.

Thus eighteenth century French geographers were heirs to the Sansons by virtue of their involvement at the start of original French cartography. They were also heirs to the work by members of the Academy which they could not ignore without being accused of neglecting progress. The Vaugondys did not escape this double heritage, especially since Gilles Robert de Vaugondy literally inherited his stock from Pierre Moullart-Sanson, the grandson of the great Nicolas Sanson.

One of the characteristics of the work of Nicolas Sanson, which made it different from Dutch publications, had been the separation of the map from descriptive texts, whether historical or geographical. The first French atlas of the world, published in 1658, could not therefore entirely rival the geography of Blaeu. In the following century, the text was restored to the map but lost the dominant position that the Dutch had given to it. Didier Robert de Vaugondy was thus led to participate as an illustrator in the publications of the *Encyclopédie* brought out by Panckoucke, the *Histoire naturelle* of Buffon, the *Histoire des navigations aux Terres Australes* by Président de Brosses, and *l'Esprit des lois* by Montesquieu. These collaborations were the source for many of Vaugondy's ideas and were reflected in his own work, such as polar explorations (Buffon and de Brosses) and human geography (Montesquieu).

The Vaugondys complemented the work of the Sansons by strengthening a rather weak branch of French cartography, that of globes, an activity which linked two groups: geographers and scientific instrument makers. The eighteenth century brought a new public to cartography which would be attracted by Vaugondy publications. As the production of cartographic documents grew, so did the demand for them. The climate was favourable for great enterprises like the manufacture of luxurious globes or magnificent world atlases, but also for less costly publications like school atlases or guide-plans of Paris and *environs*.

The Vaugondys' *Atlas universel* placed them at the heart of the disputes in eighteenth century cartography, a cartography under intense scrutiny. If it was inadvisable simply to copy one's predecessors, it was equally important to pay attention to the accuracy of maps and to use, in order to accomplish this, the calculations of the members of the Academy. Such ambient perfectionism naturally led to corrections on engraved plates, which complicated relations with map engravers. Yet the quarrels between geographers have the virtue of lifting the veil on their working methods. The criticisms directed against the maps of the *Atlas universel* show how the Vaugondys adapted themselves to methods illustrated by Guillaume Delisle and Jean-Baptiste Bourguignon d'Anville. The new *cartographie de cabinet*, just like the *cartographie de terrain*, progressed spectacularly during the eighteenth

---

Left *The frontispiece from the* Atlas Universel, *a particularly fine example of mid-eighteenth-century design and engraving. The depiction of the four continents, Asia, Africa, Europe, and the Americas, the globe with putti armed with pen and dividers, the sun around which the continents and globe revolve in Copernican fashion all emphasize the Vaugondy goals of beauty and usefulness.*

century. The sources used by cartographers grew more and more numerous, and the quality of the maps was the consequence of choices made and analyses performed on the documents selected. Among these sources were recent astronomical determinations which permitted a cartographer to map more accurately the information, whether textual or cartographic, drawn from accounts of voyages. In spite of obvious progress, the *cartographes de cabinet* were aware of the limits of their science, and like the Vaugondys, they could not help but foresee inevitable corrections. In becoming more precise, the map acquired a greater political weight. The cartographer had to be prudent and not contradict the official line taken by the authorities of his nation.

In spite of advances, eighteenth century cartography could not cover the globe without reference to some of the hypotheses which were feeding famous disputes, such as the hypothetical discoveries of Admiral de la Fuente on the northwest coast of North America. These particular theories concerned the Northeast and Northwest passages from Europe to Asia and the land link between Asia and America, their economic implications and even their religious significance. From them, Didier Robert de Vaugondy came to construct a geographic model: he did not intend to represent reality but to create a synthesis of different hypotheses which he rendered complementary but which exposed him to new criticisms.

The Vaugondys' place in French society could be measured by their commercial success. Yet in spite of an abundant and diversified production, Didier Robert was far from ending his life as a wealthy man. The fate of the Sansons earlier had not been much better. Such economic difficulties made geographers seek titles which simultaneously brought them scientific legitimacy and financial support. Amongst these were the titles of *premier géographe du roi* and *géographe adjoint* of the Academy of Sciences. If the quest for the Academy title revived personal squabbles, the existence of the former bore witness to the contacts between the King and geographers. These connections did not necessarily coincide with those established between the State and cartography, intensified in the eighteenth century thanks to the work of the engineers of the Ponts et Chaussées (Bridges and Roads) and the work on the *Carte de Cassini* (the Map of France directed by Cassini).

The personal role of Louis XV, and then Louis XVI, was a powerful force in the development of cartography and knowledge of the world. The visit to the King reported by Mary Pedley is thus very significant: after having seen Louis XV, Vaugondy and de Croÿ went to present their work to the young princes. In the end, Vaugondy could not obtain the titles he wanted; the nephew of Buache, younger but with better connections, obtained all d'Anville's spoils after the death of the great geographer in 1782 to the detriment of Didier Robert de Vaugondy whose situation became all the more critical.

Would geographers all be bad businessmen? In fact, Charles-François Delamarche, a lawyer by training, who came into possession of the Vaugondy stock, succeeded very well, thanks notably to the sale of school materials (maps and globes).

Mary Pedley's study introduces us to two different but complementary worlds, that of the *géographes reconnus*, whose works did more than feed the debates in the Academy of Sciences, and that of commercial cartographers, who in order to survive had to reconcile two apparently contradictory objectives: the selling of old stock (which itself had not been easily assembled) and the constant up-dating of stock with work reflecting recent movements of men and ideas.

But geography was also an affair of State, a State which could scarcely function without it, a monarchy which since the sixteenth century had made this discipline a "*labeur royal*," before putting it under the control of the Académie des Sciences in the second half of the seventeenth century.

The study of geography during the Enlightenment constitutes a vast and exciting subject, and Mary Pedley must be thanked for having revealed a part of it which is not limited to the destiny of one family but which adds to a larger discussion.

Monique Pelletier
(*Conservateur en chef, Département des Cartes et Plans, Bibliothèque Nationale, Paris*)

References

1 See the studies by P. de Dainville on the map of the Guyenne, the maps of the Languedoc, the map of Bourgogne, and the old maps of the Church in France, appearing between 1956 and 1966.
2 Broc, Numa, *La géographie des philosophes* Paris, 1974.
3 Konvitz, Josef, *Cartography in France*, Chicago, 1987.
4 Sanson, Nicolas, *Atlas du monde*, 1665, presented by Mireille Pastoureau, Paris, 1988.

PREFACE

This book examines the lives and work of Gilles and Didier Robert de Vaugondy, father and son, whose map business flourished in Paris in the middle of the eighteenth century (1731-1778), a period particularly important for the history of geography and cartography. Two scholars have laid the foundation of any study of French cartography in this period. Numa Broc, in *La Géographie des Philosophes*, analyzes in rich, close detail the impact of exploration and the philosophy of the Enlightenment on the development of geographical thought. Josef Konvitz explores the relationship between cartography and the emergent state in *Cartography in France: 1660-1848*. That the French were at the forefront of developments in cartography in Europe is undisputed, as Konvitz points out: "French scientists, engineers, and public officials were responsible for the most important and distinctive innovations in cartography in eighteenth century Europe." (p. xviii) The accurate determination of longitude, the triangulation mapping of France, and the voyages of discovery in the Pacific were among the century's contributions to mapmaking.

The Robert de Vaugondys were among the geographers who disseminated the results of these developments. They produced a wide range of geographical works: atlases, maps, globes, cartographic illustration for books, articles for the *Encyclopédie*, innovative projections and new mountings for globes. They were also interested in the broader issues of cartography: the problems of scale, the distortions of projections, the reliability of sources, and the achievement of accuracy. In exploring these questions they showed themselves to be children of the Enlightenment and its 'positive geography,' a guiding principle that held that the physical world could be mapped by means of measurement and the rational analysis of all sources, be they literary, mathematical, or cartographic.

The contribution of the Vaugondy family was not just intellectual but also commercial. They formed part of a map-selling business that lasted nearly two hundred years. They inherited their stock from the family of Nicolas Sanson, who had established himself in Paris, helping it to become an important cartographic centre in the seventeenth century. They continued publishing Sanson maps along with their own during the eighteenth century. By the time Didier Robert de Vaugondy died in 1786, Charles Delamarche had acquired the Sanson-Vaugondy stock; he carried their names well into the nineteenth century. Vaugondy maps were sold and copied throughout Europe, even after their geographic content was outdated. As maps are a means by which the world is known and visualized, the imprint of Vaugondy maps on a European consciousness was significant.

One aspect of this study of the Vaugondy family and their *oeuvre* remains elusive: that is whether the Vaugondys were typical or unique, whether their contributions were surprising or were part of the larger fabric of geographical thought woven during the years of the Enlightenment. These questions remain unanswered for lack of comparative material: the biographies and studies of Guillaume Delisle, Philippe Buache, and Jean Baptiste Bourguignon d'Anville have yet to be written. They were the geographers who had the most significant impact of the course of cartography in the eighteenth century and in whose shadow the Vaugondys worked. Until we know more about the complete work of these men and about their lives and the conditions in which they produced their maps, we cannot adequately assess the lives and contributions of the Vaugondys in the context of their peers of the eighteenth century.

Instead we must judge the Vaugondys on their own terms. Gilles and Didier Robert de Vaugondy were clear about their cartographic goals, as Didier explains in the preface to his *Usage des Globes* (1751): "My principal aim has been to satisfy the mind with precision and exactitude, but since it is also necessary to please the eyes, I have taken every care possible to insure that the execution of my work corresponds in its beauty to its underlying design." (p. xii) Such a statement of purpose has changed little for cartography since 1751. Compare the Michelin Company's description of their 1/200,000 map of France from 1961: "We will summarize the characteristics of this map by saying that it should be *pleasant to the eye, practical, useful, and interesting.*" ['Histoire et évolution des publications cartographiques Michelin', *International Yearbook of Cartography*, (1963), p. 173.] This short biography of Gilles and Didier Robert de Vaugondy and the annotated catalogues of maps, atlases, and globes, attempt to show how their philosophy of *bel et utile* — beautiful and useful — permeated their work and set a standard for cartography which remains unaltered today.

Acknowledgments

My study of the Robert de Vaugondy family began in a modest way when D. W. Marshall, former curator of maps at the William L. Clements Library, University of Michigan, suggested that the subscription list of the *Atlas Universel* deserved analysis. This led to a further study on the history of the publication of that atlas, and soon work on the Vaugondys turned into a thesis, thanks to the energetic encouragement and supervision of Professor Eila Campbell, Birkbeck College, University of London. Along the way, advice and suggestions have come from many quarters. On matters of history and geography I am indebted to David Bien, Hugh Clout, Michael Conzen, Elizabeth Eisenstein, Robert Darnton, John Davis, George Kish, Josef Konvitz, P.D.A. Harvey, James McClellan, Isabelle Raynaud-Nguyen, Judith Siegel, and Robert Ward. Map curators Chris Baruth, David Bosse, Ed Dahl, Robert Karrow, Lucie Lagarde, Richard Stephenson, and Ken Winch have helped me untangle many bibliographical problems. Map dealers, who have often seen the least well known maps and have remarkable memories, made helpful contributions to both catalogue and text: Richard Arkway, Dudley Barnes, Tony Campbell (now of the British Library), Darrell Leeson, J.L. Monckton and Trevor Waterman. My appreciation also goes to the private collectors who have shared their treasures: Russell Morrison, Mrs. L.C. Smith (whose collection is now in the University of Southern Maine), and Robert Ward. In the early stages of my research in Paris, I was fortunate enough to meet the late Marcel Destombes, whose kindness and guidance made a lasting impression. Mireille Pastoureau of the Département des Cartes et Plans at the Bibliothèque nationale in Paris has provided inspiration from the beginning, encouraging this work as a continuation of her own on the Sanson family and sharing unstintingly her time and scholarship. The time and patient talents of William Ingram are responsible for whatever computer literacy I have, without which this work would still be *in embryo*. The Central Research Fund of the University of London and the National Endowment for the Humanities of Washington, D. C. helped to make research in France possible.

Any cartographic study owes much to the rich resources of libraries and archives in many countries and I give particular thanks to those people who make research in their map rooms such a pleasure: Monique Pelletier (Bibliothèque Nationale, Paris), Helen Wallis (formerly of the British Library, London), Betty Fathers (Bodleian Library, Oxford), Roger Fairclough (Cambridge University Library), Francis Herbert (Royal Geographical Society), John Wolters (Library of Congress, Washington, D.C.), Robert Karrow and Jim Akerman (The Newberry Library, Chicago), Susan Danforth (The John Carter Brown Library, Providence, Rhode Island).

To John Dann, director of the William L. Clements Library in Ann Arbor, Michigan, I owe very special thanks. He has aided all my research on French cartography by providing a home for me in the Map Division of that library. In addition, the Library has generously provided most of the illustrations in this book, photographed by Paul Jaronski and Donny Goings of the University of Michigan Photographic Services. The University further supported this book with aid from the Horace H. Rackham School of Graduate Studies.

The editing skills of Valerie Scott and Jo French of Map Collector Publications have worked wonders on my convoluted prose and the complexities of the catalogue, and Tessa Campbell aided greatly with the illustrations. In earlier stages, with sharp eye and even sharper pencil, David Bosse and Elizabeth Ingram did their best to clarify the obscure and streamline the prolix. I owe a great deal not only to their critical faculties but to their motivating friendship. I dedicate this book to my husband, whose unwavering support made its existence possible. He has accepted the Vaugondys into the family without a murmur and never once jibbed at a trip to Paris. To friends, family, and colleagues, I give every thanks, attributing to them what is good in this work and accepting the errors as my own.

PROLOGUE

The Vaugondy Family and Geography in Eighteenth-Century France

No century has been as touched by the science, culture, and language of France as the eighteenth. French influence weighed heavily in every cultural domain from the fine arts to the sciences. In the field of cartography, the name of a French geographer on a map was the hallmark of both a high standard of scientific precision and elegant engraving on copperplate. The maps of Gilles and Didier Robert de Vaugondy, father and son, were no exception, and their work exemplifies that period in history known as the Enlightenment. France and its intellectual leaders or *philosophes* were central to the development of a philosophy that held that the useful application of scientific knowledge would lead to limitless progress. The legacy of the Enlightenment was this positivist view of the world: that nature was knowable, observable and could be objectively analyzed, that enquiry could be conducted without theology.

As the *philosophe* writers and scientists like Voltaire, Rousseau, Buffon, Montesquieu, Diderot and d'Alembert were making their mark, so French geographers (*ipso facto* cartographers — for the term was not used until the nineteenth century) were having a significant impact on the development of cartography in Europe. The century saw great changes in ideas about the nature of the Earth and its relationship to the Sun: by mid-century there was general acceptance of a Copernican solar system, and Newtonian laws of gravity had changed perceptions of the shape of the Earth. Most importantly for geography, it had also become feasible to determine longitude and locate places on the globe accurately. The French names most commonly linked to the development of geography as a discipline and to the 'accurate' mapping of the world were the Cassini family, Guillaume Delisle, Philippe Buache, and Jean Baptiste Bourguignon d'Anville. Gilles and Didier Robert de Vaugondy, though perhaps not in the vanguard, performed the invaluable service of making such geography and mapping widely available. Their work included maps, atlases, globes, guide books, wall maps, treatises, a history of cartography, astronomical charts, maps as book illustrations, and maps of ideas. Theirs was an approach which contributed to the dissemination of a new image of the world.

Gilles Robert Vaugondy (1696-1766) was a professor of mathematics who by luck of inheritance found himself in possession of one of the most extensive and prestigious geographical businesses in Paris: that of Nicolas Sanson and his sons. Gilles, together with his son, Didier Robert de Vaugondy, turned the bequest to good account, establishing themselves as major figures in the world of eighteenth-century geography. So entwined were the works of both Vaugondys that maps produced by one are still easily mistaken for those by the other. Only the geographers' names are a clue, for the father Gilles habitually signed his work as "M[onsieur]. Robert," inexplicably leaving off his last name, while the son Didier signed his "Robert de Vaugondy," adding the aristocratic "de". Their maps are not easily distinguished by style, though the decorative elements on Gilles' maps tend to be more baroque than the restrained cartouches on maps by Didier. The corpus of their maps displays the finest of eighteenth-century French engraving on copper plate with fine linework and lettering and cleverly designed cartouches which often represent the areas portrayed on the map. The combination of beautiful engraving and carefully compiled geographical information demonstrated the Vaugondy principle that maps, atlases, globes, and texts should always be *bel et utile*: beautiful and useful.

The Vaugondy reputation was such that Voltaire included them in his list of geographers to be feared if they were to notice a topographical mistake he had made in a couplet: "In order not to sadden M. d'Anville, Messrs. Delisle, Buache and Vaugondy, I have changed the two verses."[1] Vaugondy maps were sold throughout Europe, copied in England and Italy, and published well into the nineteenth century. Their work also illustrated several of the most important publications of the French Enlightenment: Buffon's *Histoire Naturelle*, Montesquieu's *De l'Esprit des Lois*, and Diderot and d'Alembert's *Encyclopédie*.

Josué ordonna
qu'on choisisse trois
hommes de chaque Tribu pour faire
le Tour et la Description de la Terre
Pole Arctique Septentrion
TERRES ARCTIQUES
Cercle du Pole Arctique
GROENLAND
Detroit de Davis
Hemisphere Occidental
CANADA
AMERIQUE SEPTENTRIONALE
LOUISIANE
Mariland
Virginie
Caroline
CALIFORNIE
NOUVEAU MEXIQUE
GOLFE DU MEXIQUE
Bermudes
Lucayes
Cuba
ANTILLES
Banc de Terreneuve ou Grand Banc
ISLES AÇORES
ISLES DU CAP VERD
Tropique du Cancer
Ecliptique
GRANDE MER DU SUD ou MER PACIFIQUE
Equateur ou Ligne Equinoctiale nomée communement la Ligne
Galapes
Occident
CASTILLE DOR
TERRE FERME
GUIANE
PAYS DES AMAZONES
TAPOUYES
AMERIQUE MERIDIONALE
PEROU
BRESIL
Baye de Tous les Saints
PARAGUAY
Chaco
CHILI
TERRE MAGELLANIQUE
I. SALOMON
les Marquises Mendose
Tropique du Capricorne
NOUVELLE ZELANDE
Baye des Assasins
Detroit de Magellan
TERRES ANTARCTIQUES AUSTRALES MAGELLANIQUES
Cercle Polaire Antarctique
TERRES INCONNUES qui pouront former un Continent
Pole Antarctique Midi

Ils partirent la reconnurent avec soin
en sept parts quils Ecrivirent dans un Livre. Josue ch. 18 v. 4. 8. 9.
NDE
PTION
ERRESTRE
general le rapport
t avec les
lles
NSON
Roi
TERRES ARCTIQUES
MER GLACIALE
MER DU NORD
Cercle Polaire
NOUVELLE ZEMBLE
SAMOGEDIE
KAMTSCHATKA
TERRE D'YESSO
EUROPE
MOSCOVIE
POLOGNE
ROYME DE SIBERIE
GRANDE TARTARIE
MEDITERRANÉE
BARBARIE
SAARA ou DESERT de BARBARIE
NIGRITIE
AFRIQUE
HAUTE GUINÉE
ASIE
PETIT THIBET
GRAND THIBET
ETATS DU MOGOL
PERSE
ARABIE
CHINE
MER DE LA CHINE
Tropique du Cancer
Isles des Larrons ou
Isles Marianes
MER DU SUD
ORIENTAL
ISLES PHILIPPINES
Nouvelles Philippines
OCEAN
MER D'ARABIE
GOLFE de BENGALE
I. de Ceylan
I. MALDIVES
quateur ou
Ligne Equinoctiale
N. Bretagne
Ecliptique
Detroit de la Sonde
Gilolo
Borneo
Celebes
ISLES MOLUQUES
Terre de Diemens
Carpentarie
Terre du St. Esprit
NOUVELLE HOLLANDE
Landt d'Entracht ou
Terre de Concorde
Terre de Nuits
Terre de Leuwen
MER DES INDES
Tropique de Capricorne
Cap de bonne Esperance
Route d'Abel Tasman en 1642
TERRE DE DIEMENS
TERRES ANTARCTIQUES AUSTRALES
Cercle Polaire Antarctique
Premier Meridien
BERT
du Roi
rloge
grave par le Tellier

While Gilles and Didier Robert de Vaugondy rode the intellectual currents of the Enlightenment, they were buffeted by the scientific debates and material uncertainties of the period. The difficulties they encountered were as linked to economic competition and defence of intellectual turf as to a developing science, and their story highlights the mixed fortunes of the map trade and the unhappy consequences of academic dispute. When Gilles died in 1766, he was probably confident that his son would carry on a thriving concern. By this date they had both achieved a great deal: three atlases, numerous presentations to the Royal Academy of Sciences, globes and spheres of all sizes. But personal tragedy, bad business, and bad luck dogged Didier from his father's death until his own, in poverty, in 1786. Nonetheless their work, achievements, and troubles were woven into the larger geographical questions which faced mapmakers during the Enlightenment.

There is perhaps no scientific field so inextricably linked with politics and state as cartography. National surveys demanded by increasingly centralized governments, atlases required by establishment schools, and maps created for military purposes are aspects of cartography which flower with government support. In the France of Louis XIV and his chief minister Colbert in the seventeenth century cartography flourished; no previous French king had supported and encouraged the sciences to such a degree. His founding of the Royal Academy of Sciences in 1666 provided an impetus to geography and cartography that was felt until the French Revolution and beyond.[2] The establishment of the Academy in Paris and the support given by the Sun King to geographers like Nicolas Sanson, Nicolas de Fer, and Guillaume Delisle allowed France to overtake the markets of the foreign houses of Ortelius, Mercator, and Blaeu. Not even England with its own Royal Society (est. 1662), could compete cartographically with the vast output of the Paris geographers. Maps of the known world changed rapidly with the increased discovery and exploration of the seventeenth century, during which most of the eastern seaboard of North America was colonised and the great trading depots in the Far East established. Along with prints, maps became the means by which the known world was visualized.

The expansion of literacy insured the dominance of the printed word and image. Ideas about the world and the nature of the universe were spread by means of printing: books, of course, but also the more affordable broadsheets, pamphlets, gazettes, newspapers, and intellectual journals, many of which were illustrated by maps. The educational curriculum, particularly in the hands of the Jesuits, contained a strong dose of geography, and to learn geography was "to know the map."[3] In France, as literacy increased,

Portrait of Nicolas Sanson (1600-1667), the 'father of French cartography'. (Photo courtesy of the British Library)

Preceding page: *A Mappemonde [Map of the World] by Gilles Robert, 1743 [Cat. 4]. It is a re-working of a Sanson map, "reviewed, augmented and adjusted to astronomical observations." (Photo courtesy of the William L. Clements Library, University of Michigan)*

Japan (1750) from the **Atlas Universel** *[Cat. 410]*

personal library building became both fashionable and an index of status. Maps formed part of these private libraries, and some very large private map collections were built at the end of the seventeenth and early eighteenth centuries: Jean-Nicolas de Tralage left 5400 maps in 1698 to the Abbaye St. Victor; the collection of the abbé Michel-Antoine Baudraud left to the Abbaye St. Germain des Près inspired the young geographer d'Anville to develop his own of nearly 10,000 items, left to the Bibliothèque Nationale after his death.

By 1752, the military geographer Roch-Joseph Julien was confident enough of a general market for maps that he opened a map store which sold the work of a wide variety of European cartographers, particularly that of the German house of Homann Heirs. The diversity of his stock was the first instance in Paris of a geographer selling more than his own work.

By the eighteenth century, it was assumed that the literate public had a knowledge of geography. Voltaire, for example, relied on the familiarity of his readers with the literature of voyages to appreciate the travels of Candide across Europe, in South America, and finally to Constantinople. The century also saw many proposals designed to improve the general public's taste for and the young person's understanding of geography. A certain De Bouys envisaged a children's geographical garden, the *parterre géographique*, wherein coloured obelisks, footpaths and flower beds would be named for different parts of the world. Similarly, the abbé Teisserence proposed making Paris a geographical town by renaming all its streets with those of cities and regions in France, placed in their respective positions.[4] Rousseau went even further by suggesting that geography be taught literally "in the field," allowing the student to wander about unguided in a forest to the point of nearly becoming lost until the pangs of hunger and some knowledge of the position of the sun would lead him home for lunch.[5]

A taste for maps and competence in geography extended to the monarchy as well. Both Louis XV (1710-1774) and XVI (1774-1793) were tutored in geography, the former by Guillaume Delisle, the latter by Philippe Buache and his nephew Jean Nicolas Buache de la Neuville. Like their luminous forebear, Louis XIV, both kings encouraged more accurate mapmaking. Their reigns saw the completion of the first edition of Cassini's "Carte de France," the first national survey by triangulation on a uniform scale. They also supported Pacific explorers like La Pérouse towards the end of the century.

Eighteenth-century France was ready for change not only in the content of maps but also in their form. A more innovative approach introduced the school wall map, maps printed on silk for travellers and army officers, pocket globes, and variations on the armillary sphere. The 'look' of the eighteenth-century map was altered, distinguishing itself from its predecessors by a great reduction in decorative detail; by the end of the century, the ships and sea monsters, vignettes of faraway places, cartouches extending well into the geographic area are completely gone. Instead, blank spaces increased, as the eighteenth-century geographer no longer felt a *horror vacui*, that fear of the empty space.

Titles also changed, proclaiming the map to be "adjusted to the latest astronomical observations." This desire to appear scientific, to strip the map of any decoration which might imply a point of view, reflected the positivism of the Enlightenment. The separation of facts from values, the elimination of theology and metaphysics from discourse by working only from experiment and objective analysis became the basis of what Didier Robert de Vaugondy called "the positive geography." It was pursued by the *géographe de cabinet* who "gathers, without leaving his study, all the details he can acquire, combines and compares them with authentic relations and determines them with the help of astronomical observations."[6]

The "Positive" Geography: Longitude and the Shape of the Earth

Geography in the eighteenth century saw three great changes in ideas about the position, shape and measurement of the earth: first, the final acceptance of the Copernican solar system; second, the shape of the earth as an oblate spheroid based on Newtonian laws of gravity; and third, an accurate way to determine longitude and thereby the correct location of places. To locate a place accurately on the surface of the earth, exact measurements of latitude and longitude are necessary. While measuring latitude had gone on with increasing accuracy since the time of Ptolemy, the problem of longitude remained until the works of Isaac Newton in England and Christian Huygens initiated its solution in France in the seventeenth century.

The determination of longitude requires accurate time pieces and some standard observable phenomenon by which to measure time. As longitude lines converge at the poles, they reduce the linear space between them; a linear degree of longitude at the equator is not equal to that 45 degrees north of the equator. The difference in degrees of longitude is best measured in terms of time. By observing a phenomenon visible at the same time in two or more places and noting the differences in time, the degrees of longitude between these places may be ascertained. Two discoveries in the seventeenth century made such measurements possible. The Italian astronomer Jean-Dominique Cassini, invited to Paris to the Academy of Sciences, proposed that the regular and frequent eclipses of the moons of Jupiter be used as the phenomenon visible from any point on Earth. Then Christian Huygens, a Dutch astronomer working in Paris, invented a pendulum clock (1657), which depended on the constant pull of gravity to keep it from gaining or losing time. So by the turn of the century, the French scientific community was poised to make the accurate determination of longitude.

However, the use of Huygens' pendulum clock raised a further problem: the shape of the earth. Current calculations of both latitude and longitude assumed the earth to be a perfect sphere. But when the abbé Jean Picard used the pendulum clock on a scientific expedition in Cayenne in 1672, the pendulum lost time compared to its movement in Paris. To explain this, Newton and Huygens postulated that such variation could only be accounted for by the centrifugal force of the earth's revolution around the sun, causing a bulge at the equator. This suggested that the shape of the earth was an oblate spheroid, flattened at thc poles. As a result, a degree of latitude would be longer at the poles than at the equator. To verify these conclusions, the Royal Academy of Sciences sponsored two expeditions: one led by de La Condamine to the equator (Peru) in 1735, the other by Maupertuis to the arctic circle (Lapland) in 1736. By 1744, the results were in: the earth was flattened at the poles and the accuracy of locational measurements had been refined to a degree hitherto unknown.

The Positive Geographers

The presence of the Cassinis at the Royal Academy of Sciences and that institution's continuous support for geographic research meant that Paris, by the early eighteenth century, was becoming the cartographic centre of Europe in its state support of map projects and in its commercial endeavours. The private market was dominated by three families whose scions would perpetuate the family names until the French Revolution: Sanson, Delisle, and Jaillot. Of these Nicolas Sanson commanded respect as the "father of French cartography." His publication of an atlas of the world in 1658, produced on a uniform projection of his own devising (and which bears his name) and on a consistent scale, was a first for French cartography. His geographical tables were innovations which guaranteed his fame. These were vertical tables that organized, classified, and ranked names of places, regions, governments, ecclesiastical authorities, and natural

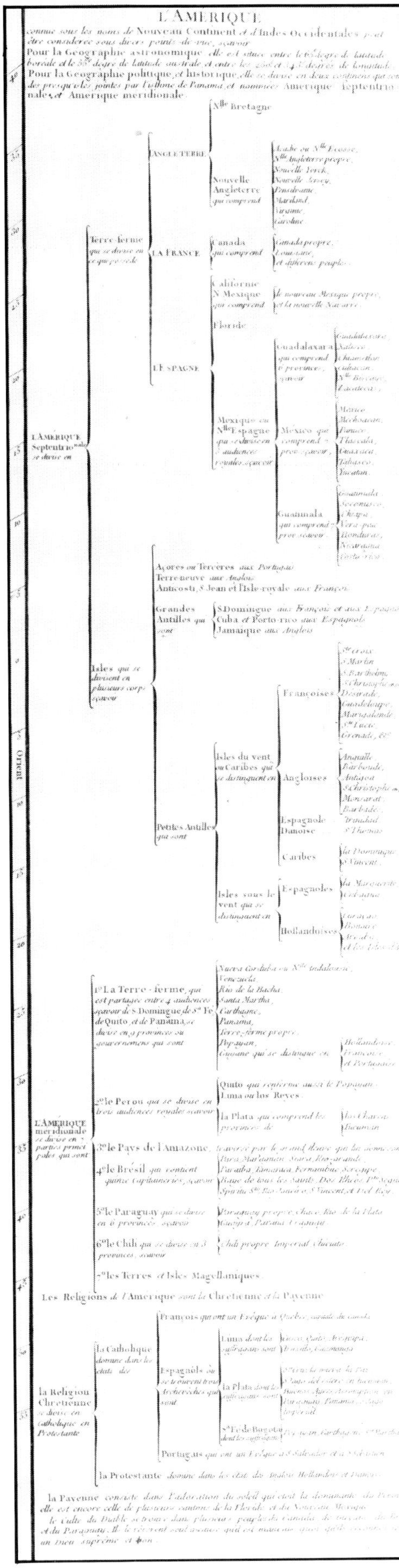

The geographical table on Gilles Robert's map of the western hemisphere of 1749, 'L'Amérique Septentrionale et Méridionale' [Cat. 19]. The table is adapted from those published by Nicolas Sanson in the seventeenth century. (Photo courtesy of the William L. Clements Library, University of Michigan)

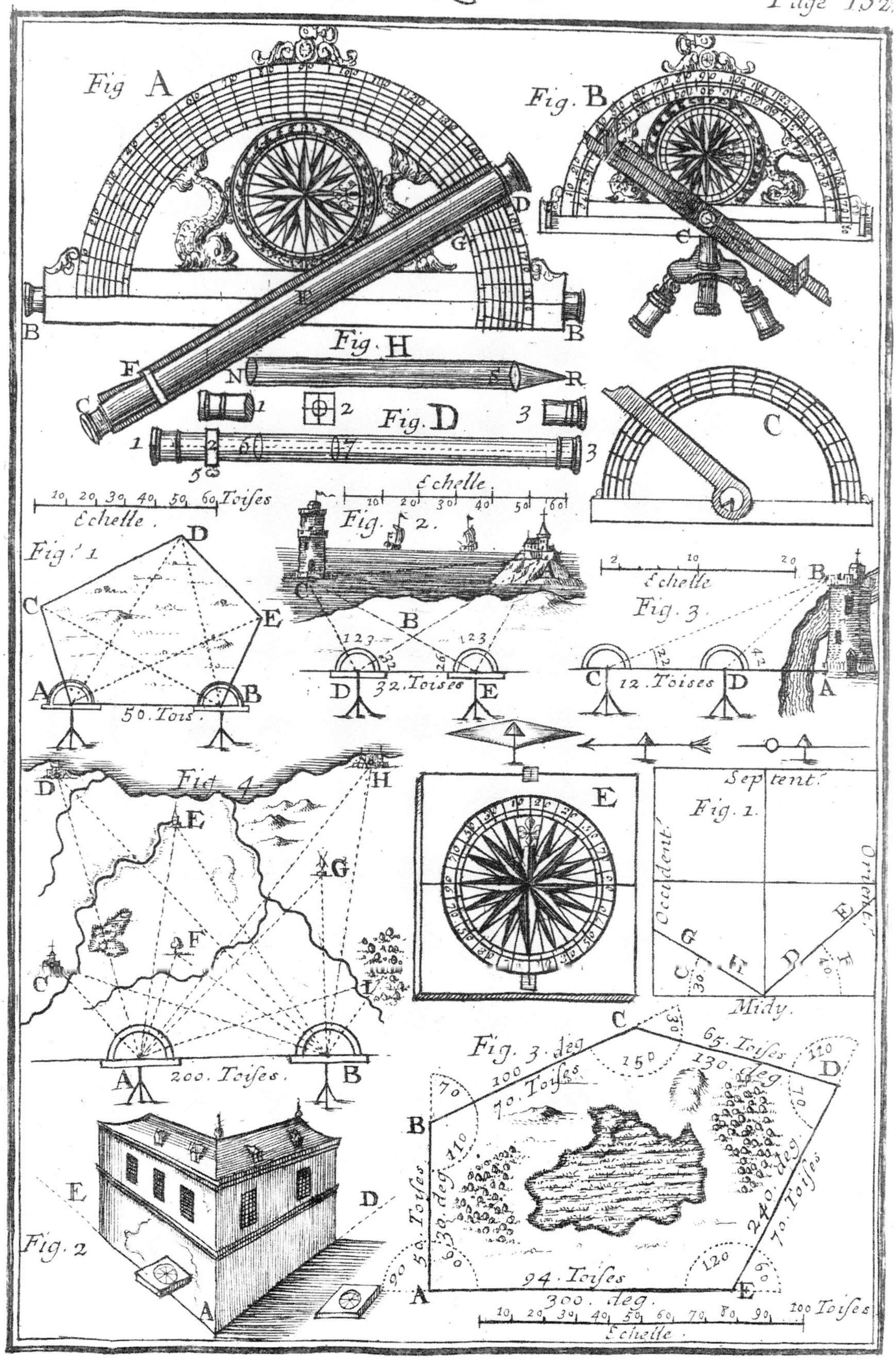
Fig A
Fig. B
Fig. H
Fig. D
C
Echelle
Toises
Fig.' 1
Fig. 2
Fig. 3
50. Tois.
32. Toises
12. Toises
Fig. 4
200. Toises
E
Septent.
Occident.
Orient.
Midy.
Fig. 1.
Fig. 3. deg
65. Toises
130. deg
70. Toises
94. Toises
300. deg.
240. deg.
Fig. 2
Toises
Echelle

features. He published about forty of them and their enormous popularity and Sanson's continuous production of maps made his name synonymous with geographic study by the time of his death in 1667.[7]

While Nicolas Sanson worked to keep his maps and tables *au courant* with geographical knowledge, his sons Adrien and Guillaume contented themselves with the re-publication of his maps without correction, in spite of recent discoveries about longitude. It was left to one of Sanson's pupils, Claude Delisle (1644-1720), historian and geographer, to realize the potential of astronomical observation and the impact it would have on cartography. Of Claude's twelve children, three would distinguish themselves in science: Guillaume, Joseph-Nicolas, and Louis Delisle de la Croyère. Guillaume Delisle (1675-1726), who was himself a student of Jean Dominique Cassini, mastered the techniques of "positive geography," extending Sanson's approach by gathering and analyzing all information — cartographic and literary — about a region. Working together with his father, Guillaume Delisle spectacularly changed the image of the world for the French public through his maps: the Mediterranean shrank by 300 leagues in breadth; the Atlantic and Indian Oceans were equally reduced and the Pacific greatly enlarged. In North America the familiar shape of California as an island now was transformed into a peninsula. Delisle's maps marked what Vaugondy and many after him called "the reform of cartography."[8] Before his premature death in 1726, Delisle published no less than onc hundred maps, correcting them constantly according to astronomical observations and information from travellers. In this he followed the practice of Sanson, who had likewise understood the impermanence of a map and had corrected his work before re-issue.

Left *From Nicolas Bion,* Traité de la Construction des Instruments Mathématiques, *Paris: Jombert & Nion fils, 1752, showing the uses of the* graphomètre, *the* rapporteur, *and the compass in triangulation surveying. (Photo courtesy of the William L. Clements Library, University of Michigan)*

Below *A map seller's shop at the end of the seventeenth century. From Sanson's* Atlas Nouveau, *1692. (Photo courtesy of the William L. Clements Library, University of Michigan)*

Just as Sanson's heirs had been content to republish his work without further corrections, so when Delisle died at the young age of fifty-one, his wife, who was his heir, continued to sell his maps without making corrections to the plates. Thus with the death of Delisle a void was left in the cartographic world of Paris. Similarly, the other names from the seventeenth century associated with maps — Jaillot and de Fer — were still represented by their heirs but with uncorrected re-issues rather than new or corrected plates. So while Paris had established itself as a cartographic capital of Europe by the middle of the seventeenth century and its pre-eminence had continued into the first two decades of the eighteenth, by 1730, French cartography was in fact languishing.

It was in this year that Gilles Robert Vaugondy had the good fortune to inherit the Sanson geographical stock from Nicolas' grandson, Pierre Moullart-Sanson. Unlike his Sanson predecessors, Gilles Robert not only re-issued the Sanson maps but by the 1740s began to produce his own corrections of them. By the time his son Didier was creating maps, in 1745, he had established himself as a geographer of some note, sought after by book publishers for illustrating historical works and for making atlases. His son Didier expanded their business with additions of globes and spheres. Together they produced two atlases — a total of 317 maps — within ten years, in addition to their other maps.

This was only the beginning of the contribution of both Vaugondys to the history of French cartography. While they embodied the philosophy of positive geography, using and analyzing as many sources as possible, from ancient texts to the latest surveys, they were not seduced by false notions of a map's permanence or its absolute accuracy. For them, accuracy was not the only criterion of a "good" map; it must also have an aesthetic appeal and be useful. These ideas offer us a framework in which to judge not only the Vaugondy cartographic achievements, but those of their contemporaries as well. Their maps underwent many changes, and their treatises often end with entreaties to the reader to inform them of mistakes. They experimented with "systematic" cartography by which a map illustrates a geographical theory rather than a reality. The desire to satisfy the mind and to please the eye underpins their work and manifests the "enlightened sense of duty" that guided all their actions.[9]

References

1. Voltaire, letter to Jacques Lacombe; another in similar vein to Gabriel Cramer, both dated March 1767. *Correspondence de Voltaire*, ed. Th. Besterman (Geneva, 1953-1965): Vol. 65, nos. 13161, 13163.
2. Josef Konvitz, *Cartography in France 1660-1848: Science, Engineering, and Statecraft* (Chicago: University of Chicago Press, 1987).
3. Le Père de Dainville, *La Géographie des Humanistes* (Paris, 1940): p. 381.
4. Mireille Pastoureau, "La France Divulguée: évolution de la Cartographie gravée du XVIe au XVIIIe siècle," *Espace Français* (Paris: Archives nationales, 1987): p. 63.
5. J. J. Rousseau, in Book Three of *Emile*, 1762.
6. D. Robert de Vaugondy, "Préface Historique," *Atlas Universel* (Paris: 1757), p. 13a-b.
7. Sanson, Nicolas, *Atlas du Monde 1665*, facsimile edition, ed. Mireille Pastoureau (Paris: Sand & Conti, 1989).
8. Didier Robert de Vaugondy, *Essai sur l'histoire de la Géographie* (Paris: 1755): pp. 158-9. It has been pointed out by both de Dainville and Broc that some of Delisle's cartographic images were preceded by the work of Nicolas de Fer whose world map of 1694 anticipates that of Delisle, published in 1700. In Italy, Father Riccioli had reduced the Mediterranean in breadth from 1160 to 882 leagues forty years before Delisle's map. Le Père de Dainville, *op cit.*: pp. 446, 456. Numa Broc, *La Géographie des Philosophes* (Paris: 1974): p. 27.
9. Didier Robert de Vaugondy's eulogy of his father, in the "Préface," *Institutions Géographiques* (Paris:Boudet & Desaint, 1766): p. iv.

CHAPTER ONE

Gilles Robert Vaugondy, *Géographe ordinaire du roi*

According to Gilles Robert Vaugondy, the attribute of a good geographer should be knowledge of mathematics, arithmetic, geometry, trigonometry, astronomy, optics, and history.[1] However, outside the military, no institution existed in the early eighteenth century specifically designed to train cartographers. Little is known about Gilles' education, but as a *professeur en mathématiques*, he was well-suited to his profession of geographer. Nor is it known how he became friends with Pierre Moullart-Sanson, the grandson of Nicolas Sanson, from whom he may have learned some of the skills of map editing, publishing, copper plate engraving and printing from the roller press.

There is no doubt that Moullart-Sanson thought well of Gilles Robert Vaugondy, choosing him as one of three friends designated to succeed to the Sanson stock in his will of 1730. Guillaume and Adrien Sanson, the sons of Nicolas Sanson, had both died without male heirs; they passed the Sanson stock on to their nephew Pierre, who added Sanson to his name in 1703. Childless himself, Moullart-Sanson wished to preserve the name and cartographic heritage of the Sanson family. His will described his choice of the three friends "whose conduct, good mores, intelligence, and capacity in [geography] qualify them to inherit all the plates, maps, prints, designs, manuscripts, collections, papers, geographical dictionaries, globes, machines, instruments, utensils and materials, and generally all that concerns geography...plus all his books, bound and unbound, his paper for printing, writing, and designing, together with his bookcases and cabinets and even the right of his privilege if it still exists at the time of his death."[2] The condition of this bounty was that the three friends would continue the work of geography and maps under the name of Sanson and no other so that the name would be perpetuated.

Who were the three friends of such intelligence and capacity? One was a priest, Jacques Simon Perrier, chaplain at the church of St. Eustache; one a lawyer, Jean Frémont, on the rue Grenelle, and the third, the professor of mathematics, Gilles Robert Vaugondy. How the four of them came to know each other is a mystery. It is possible that they had been students at the College of Louis le Grand, the Jesuit academy in Paris.[3] Soon after inheriting the equipment of the trade, Gilles Robert Vaugondy acquired the title, *géographe du roi*, and with Frémont he acquired the privilege to print and sell the Sanson maps. The three friends were now ready to embark on their new career.

The Family of Gilles Robert Vaugondy

Little in the known background of Gilles Robert Vaugondy augured a career in cartography. He was born in Paris on 24th August 1688 into a family neither scholarly nor scientific. His father, Louis Vaugondy, was a silk fabric maker. It was from his mother, Jeanne Renée Robert, that he took the name Robert with which he signed his maps. Gilles Robert Vaugondy married the daughter of a packer, Marie Magdeleine Blié, on 7th October 1719. She seems to have participated in the business for she was a signatory to most of her husband's legal transaction. Gilles' brother, Jean Vaugondy, was an enamelist who worked for many watch and clock makers. In 1723, he married Marie Magdeleine's sister, Anne Geneviève Blié.[4]

Whatever his early training and education, Gilles Robert Vaugondy signed his marriage document in 1719 as "*géographe*," and witnessed his brother's wedding in 1723 as "*professeur en mathématiques*." He was sufficiently proficient to be appointed *géographe du roi* in 1734 "in the place of the late Pierre Moullart-Sanson, in order to enjoy the quality of all its privileges, franchises, and other advantages belonging to it, just as the said Sanson enjoyed."[5]

While such a title suggests a man leading the contemplative life of a *géographe de cabinet*, a family incident preserved in the records of the prison Bicêtre paints a vivid picture of the rough and tumble environment of the artisan class from which Gilles emerged. Though concerned with his brother, it tells us something about the social circumstances of the Vaugondy family.

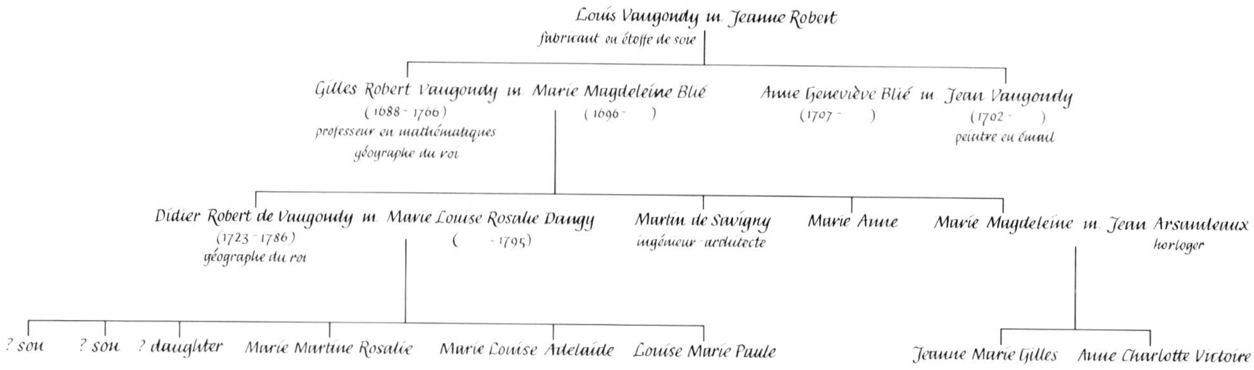

The family tree

Jean Vaugondy, the enamelist, was popular with well known clockmakers of the day. Such a craftsman was integral to clock and watchmaking, for the decorated case, painted in many colours in miniature, was the most important selling feature. The enamelling process was delicate and complicated; each colour on the watch or clock needed to be fired separately without damaging the clock's mechanism. Furthermore, enamelists were often themselves instrument makers; their guild was the principal one allowed to make barometers and thermometers, for example.

Jean Vaugondy's steady hand in this highly skilled trade seemed not to have been affected by his propensity for drink, which was marked enough to cause Marie Magdeleine Blié to warn her sister Anne Geneviève not to marry him. But Anne Geneviève was a headstrong girl and married Jean in 1723. By her account in the Bicêtre documents, things deteriorated within months after the wedding. Jean was a carouser and wife beater, who brought home "créatures à l'occasion", gave his wife venereal disease, and went off for three months to Blois with another woman. Through the intervention of the curate of St. Sulpice, a restraining order was issued but Jean Vaugondy took no notice. In desperation, at the beginning of 1731, Anne Geneviève moved out to a friend's house. By the end of March, Jean had pursued her there, demanding that she return home, and beating both her and her friend until a local officer came to arrest him. Jean Vaugondy was sent to Bicêtre, a prison on the southern outskirts of Paris.

Jean's defence told a different tale. Organized by his brother Gilles who rallied his wife and other members of Anne Geneviève's family, this version maintained that the marriage had been happy for six years, but that Anne Geneviève had never wanted to work or take care of her household. Jean had been generous, giving her five new dresses and allowing her to go to Dieppe with the wife of Fortin, a clockmaker, but when she returned with a priest in tow, her family was suspicious. They said their sister had always been stubborn and selfish and was inclined to abuse Jean for his drinking and his simple-mindedness.

As witnesses to Jean's skill and character as a steady worker, no fewer than five eminent Paris clockmakers signed a document describing Jean Vaugondy as one of the most capable enamelists in Paris and that his detention was causing their businesses to suffer. Jean was released from prison in June; the records end the story there.[6]

This colourful picture of an unhappy marriage has interesting sociological implications and makes clear the connection between the Vaugondys and a community of skilled instrument makers. Clock and watchmakers could also be scientific instrument makers, extending their skills to the manufacture of globes and spheres, a skill that became part of the Vaugondy map business.

Starting a map business

At the start, in 1731, it was enough for Gilles Robert to become acquainted with the Sanson inheritance. With Perrier and Frémont, he had inherited not only Moullart-Sanson's stock of maps, plates, globes, manuscripts, and geographical papers, but also a roller press for copper plates and a house on the corner of the rue du Bac and rue St. Dominique in a fashionable neighbourhood south of the Seine. Though the house provided collateral for many of Gilles Robert's early map ventures, it was unsuitable for the map business since it was occupied by tenants. Gilles himself was living on the rue St. Denis, north of the Seine, nowhere near the print and book trade district of the Latin Quarter on the Left Bank.

Needing a place closer to engravers and printers, who were restricted by law to a certain part of Paris, in April of 1731 the three friends rented a house for 400 *livres* a year on the quai de l'Horloge on the north side of the central Île de la Cité. For anyone entering the map business, this location had understandable attractions: it was conveniently close to the bookdealers and printing presses of the Left Bank, especially the area along the rue St. Jacques and the rue St. André des Arts. The quai des Orfèvres on the south side of the island was heavily populated by mathematical instrument makers and the quai de l'Horloge itself was the address of two globemakers: Nicolas Bion and Jacques Baradelle.

The Turgot/Bretez Plan of Paris (1739): the corner of the rue du Bac and rue St. Dominique where stood the house inherited by Gilles Robert and Jean Frémont from Pierre Moullart-Sanson. The house was to provide much needed collateral for Robert to run his business. (Photo courtesy of the William L. Clements Library, University of Michigan)

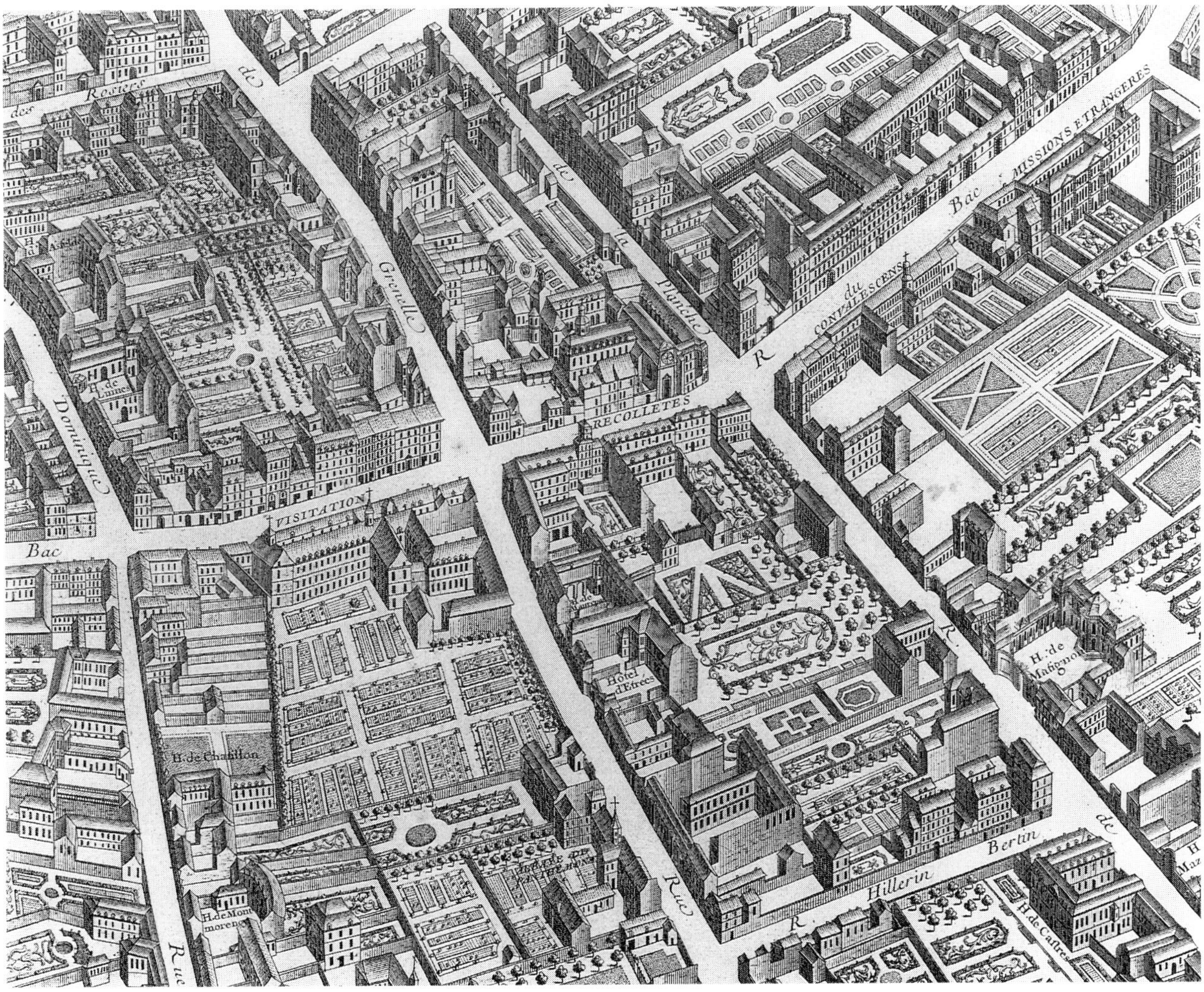

The quai de l'Horloge had many historical connections with geographers.It had been home to notable geographers and mapsellers in the seventeenth century such as Nicolas de Fer, Alexis-Hubert Jaillot, Jean-Baptiste Nolin, and Melchior Tavernier. In the early part of the eighteenth century it was the address of Guillaume Delisle, whose son-in-law and successor to the title *premier géographe du roi*, Philippe Buache, lived opposite on the quai des Megisseries. Buache moved to the quai de l'Horloge in 1745. Another eminent geographer of the period, Jean Baptiste Bourguignon d'Anville, lived within view in the Galeries du Louvre. One aspect of the dissemination of ideas which is sometimes overlooked is the proximity in which the participants lived. Five notable French geographers in the eighteenth century lived within half a mile of each other, three on one block on the quai de l'Horloge.

Almost immediately after renting the house on the quai de l'Horloge, Perrier, the priest, sold his right of succession to his co-heritors for 4000 *livres*. This sale marked the beginning of a financial partnership between Jean Frémont and Gilles Robert Vaugondy which was to last until Frémont's death in 1751.

During 1732 and 1733, the lawyer and geographer busied themselves with complicated financial arrangements. Debts incurred by Moullart-Sanson had to be cleared, the house on the rue du Bac / rue

---

The Turgot/Bretez Plan of Paris (1739): The Pont neuf and the quai de l'Horloge, also known as the quai des Morfondus. The quai de l'Horloge takes its name from the clock, the oldest public timepiece in Paris, situated at the northern corner of the Palais de Justice, which dominates the centre of the Ile de la Cité. Sometimes called the quai des Morfondus [the quai of the frozen ones], the quai was where the litigants to the courts lined up, waiting for their cases to be heard, shivering in the cold north wind, true to the other meaning of 'morfondus' — dejected. At first, the Robert Vaugondys lived at the rue de Harlay end of the quai, later near the Pont Neuf. On the southern side of the Place Dauphine was the quai des Orfèvres [the quai of the gold and silversmiths], home of many scientific instrument makers in the seventeenth and eighteenth centuries. (Courtesy of the William L. Clements Library, University of Michigan)

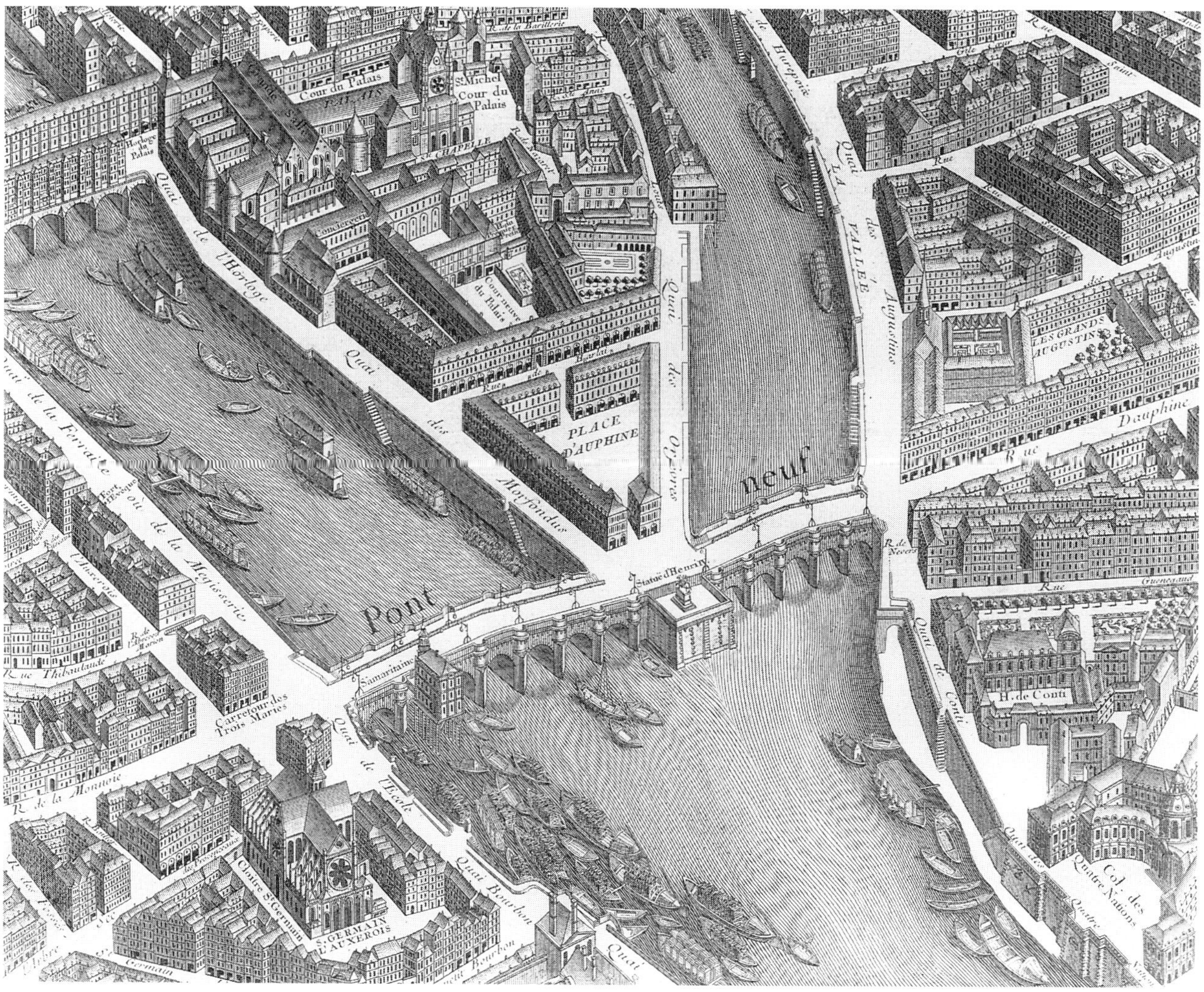

St. Dominique needed many repairs to avoid being condemned, and Gilles Robert wanted to buy a collection of Sanson copperplates which were in the hands of Jean Mariette. By using the rue du Bac house as collateral and borrowing large sums of money (a total of 14,000 *livres*), Frémont and Vaugondy were able to repair the house and purchase the stock from Mariette.

Mariette was the descendant of the Pierres (father and son) Mariette, seventeenth-century publishers of prints who had contracted with Nicolas Sanson to produce atlases. One aspect of their arrangement had been that the atlas plates should be split between the Sansons and Mariette so that neither party could profit separately since neither would have a complete set of plates.[7] This arrangement was particularly effective for two-sheet maps, with Mariette, for example, keeping the eastern half, Sanson the west.

By purchasing Mariette's plates, Gilles not only completed the Sanson collection but added to it considerably, for the sales inventory also included 152 plates of Philippe Briet's *Parallela Geographiae Veteris et Novae*, a popular geography of the mid-seventeenth century, and numerous maps by Philippe de La Rue and Pierre Duval, Sanson's nephew. Mariette sold the stock to Gilles Robert, who in turn borrowed money from Frémont, to whom he was now considerably in debt.

Though most of the Sanson material was now completely in the hands of Gilles Robert Vaugondy, there remained a portion still unobtainable: that which Guillaume Sanson had shared with Alexis-Hubert Jaillot and his heirs. By 1726, these plates, mostly for two-sheet maps, were "worn and tired" and no longer being printed by the Jaillots. Nonetheless, rancour between the Vaugondys and the Jaillots was to manifest itself in the 1750s when Jaillot's granddaughters took exception to the Vaugondys describing

Guillaume Delisle, "Plan de Paris," c. 1705, with the following places marked: 1) quai de l'horloge, 2) the Observatory, 3) the Louvre, 4) rue du Bac/rue St. Dominique, 5) rue St. Jacques, 6) Hôpital de la Salpêtrière.

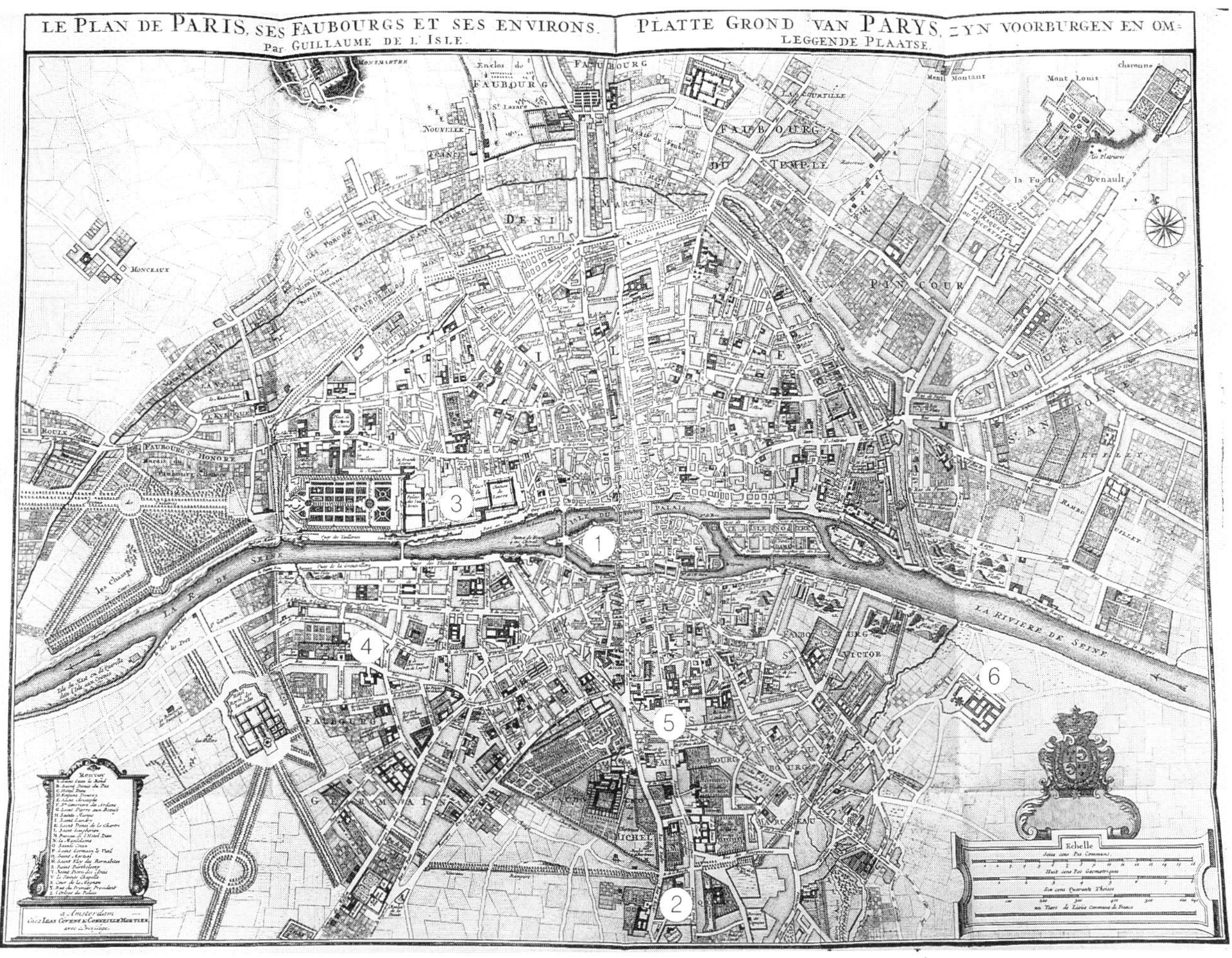

themselves as the "sole possessors" of Sanson material, since the Jaillots still possessed plates dating from the early contracts between Guillaume Sanson and Alexis-Hubert Jaillot. By the terms of the Sanson-Jaillot agreement, the maps shared were never to be enlarged or reduced from their two-sheet format and scale, but Jaillot had broken the agreement and published maps from their joint stock. This forced Guillaume Sanson to bring suit against him, creating long-standing animosity between the Sanson/Vaugondys and the Jaillots.[8]

From the evidence of documents in the notarial archives in Paris, it is clear that Jean Frémont provided the liquid capital to continue the Sanson map business and Gilles Robert Vaugondy went deep into debt to him. Gilles' summary of the map business in 1737 was not promising:

Receipts:	26,365 *livres*	
Expenses:	25,740 *livres*	
Profits:	625 *livres*	(of which half belonged to Frémont)[9]

The account is no more detailed than that. Are these figures surprising or normal? The receipts seem rather large when set against the average sale price of a map — between ten *sous* and one *livre*. Gilles Robert may well have been selling copies of the *Introduction à la Géographie*, a popular manual for schools, and issues of various Sanson maps. Given that he had the engraved plates and press in hand, where did the large expenses come from? The most likely explanation is that in 1737 Gilles Robert Vaugondy was for the first time preparing his own maps and incurring costs of engravers — as high as 750 *livres* per sheet — as well as copper plate — as much as 60 *livres* for a folio-size plate. This investment could account for the large expenses.

Until his death in December, 1751, Frémont continued to support Gilles Robert with loans, many set against the latter's share in the house on the Rue du Bac. His will made Gilles Robert Vaugondy and his sons Didier and Martin executors of his estate. He also re-organized the Vaugondy debt, making the principal over to his sisters and leaving the income to Gilles.[10]

Frémont's detailed bequests to a variety of family and friends show a lawyer of moderate means with a large interest in geography and related disciplines. He left the Vaugondys most of his library and all of his maps, plans, prints, globes, spheres, and planispheres. To other friends he left manuscripts of papers on historical subjects and a variety of scientific instruments: a barometer, two thermometers, two hydrometers, two sets of mathematical instruments (one in copper and one in silver), magnifying glasses, microscope, compass, solar quadrant. This scientific and geographical collection illuminates Frémont's friendship with Moullart-Sanson and his interest in and encouragement of the Vaugondy family in the continuation of the Sanson business.

After Frémont's death, Gilles Robert left the house at the sign of the *Tête d'Or* at the rue de Harlay end of the quai de l'Horloge and rented another house at the Pont Neuf end for 2000 *livres* a year. Though the house had no distinguishing sign like the *Tête d'Or*, their simple address of "quai de l'Horloge, près le Pont neuf" marked all their maps from 1751, with one exception, until the sale of the stock to Jean Baptiste Fortin in 1778.

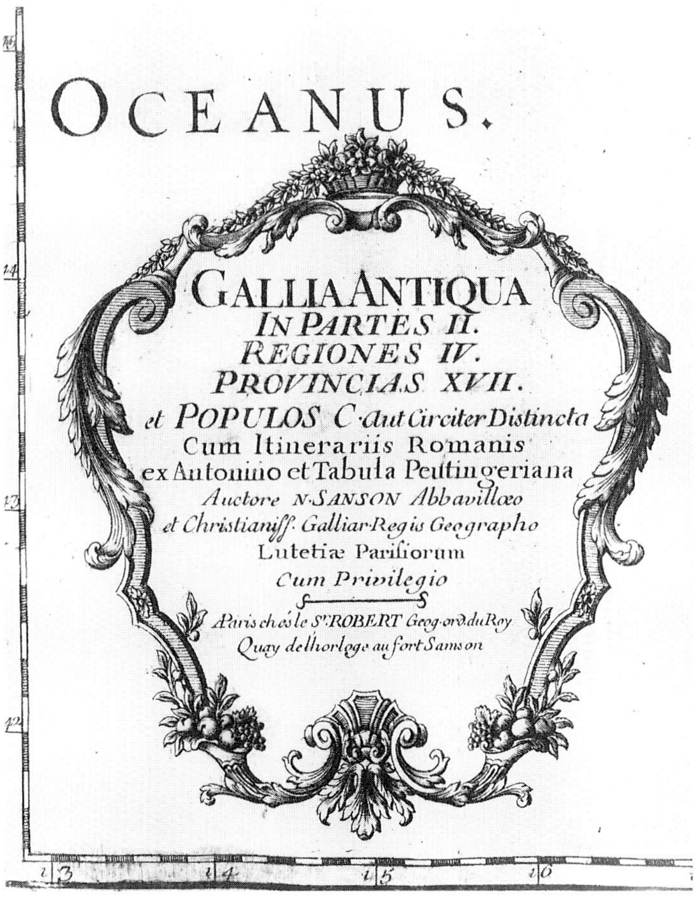

The cartouche from a Nicolas Sanson map re-issued by Gilles Robert. The address on the quai de l'Horloge is 'at the mighty Samson.' This is the only instance of the name of a sign being used by the Robert Vaugondys. The map is in the **Atlas Factice de la Géographie Ancienne**, *no. 37. (Photo courtesy of the William L. Clements Library, University of Michigan)*

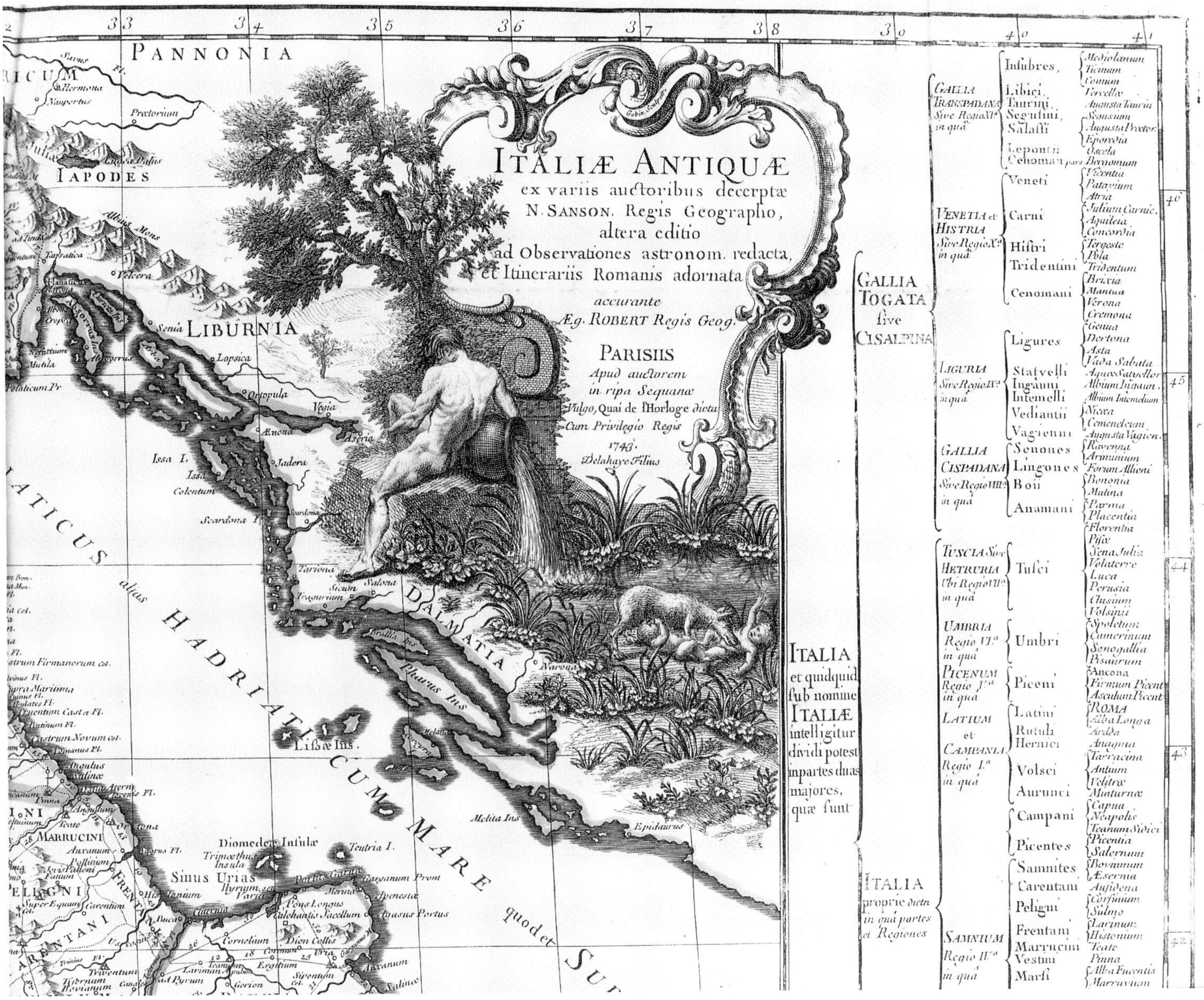

The cartouche and part of the geographical table from Gilles Robert's 'Italiae Antiquae,' 1743 [Cat. 329]. The cartouche shows the Tiber River and the young twins Romulus and Remus being cared for by the wolf. The map is engraved by 'Delahaye filius', probably Jean-Baptiste Henri Delahaye who fell out with the Vaugondys during the production of the **Atlas Universel**. *(Photo courtesy of the William L. Clements Library, University of Michigan)*

The Early Works of Gilles Robert

As Gilles Robert Vaugondy began to publish maps, he chose to sign them using his mother's maiden name: Gilles Robert. He rarely used Vaugondy on maps, though it continued to be his legal name. It was his son Didier who styled himself "de Vaugondy" on his work.

The first maps issued with Gilles Robert's imprint are Sanson maps of French dioceses, thirteen of them, issued in 1730 and 1731. Gilles Robert must have devoted a good deal of energy during the 1730s to sifting through the Sanson material, sorting and consolidating the stock. His first new issues were the four continents, published in 1738 and 1739, which were copies of earlier Sanson maps.

Gilles did not attempt an original map until 1737 when he published the "Carte pour servir la lecture de l'histoire ancienne de M. Rollin" in two sheets [Cat. 381]. Charles Rollin's *Histoire Ancienne des Egyptiens, des Cartheginois, des Assyriens, des Babyloniens, des Mèdes, et des Perses, des Macédoniens,*

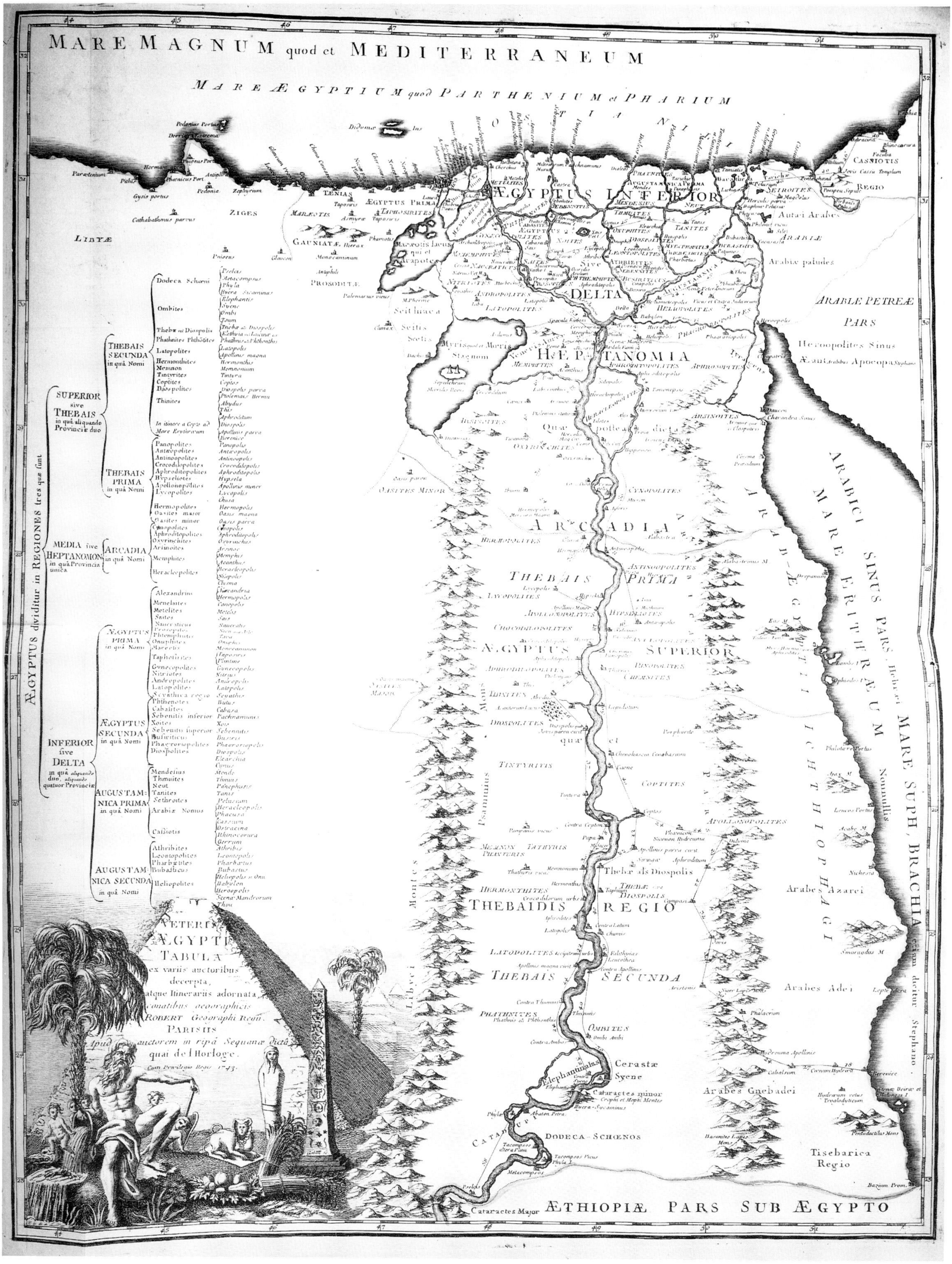
MARE MAGNUM quod et MEDITERRANEUM
MARE ÆGYPTIUM quod PARTHENIUM et PHARIUM
LIBYÆ
ÆGYPTUS INFERIOR
DELTA
HEPTANOMIA
ARCADIA
THEBAIS PRIMA
ÆGYPTUS SUPERIOR
THEBAIDIS REGIO
THEBAIS SECUNDA
ARABIÆ PETREÆ PARS
ARABICI SINUS PARS, Hebræi MARE SUPH, BRACHIA
MARE ERITHRÆUM
Cerastæ
Syene
Cataractes minor
DODECA-SCHŒNOS
Cataractes Major
ÆTHIOPIÆ PARS SUB ÆGYPTO
Tisebarica Regio
SUPERIOR sive THEBAIS in quâ aliquando Provinciæ duo
MEDIA sive HEPTANOMON in quâ Provincia unica
INFERIOR sive DELTA
ÆGYPTUS dividitur in REGIONES tres quæ sunt
VETERIS ÆGYPTI TABULA
ex variis auctoribus decerpta,
atque Itinerariis adornata,
conatibus geographicis
ROBERT Geographi Regii.
PARISIIS
Apud auctorem in ripâ Sequanæ dictâ quai de l'Horloge.
Cum Privilegio Regis 1743.

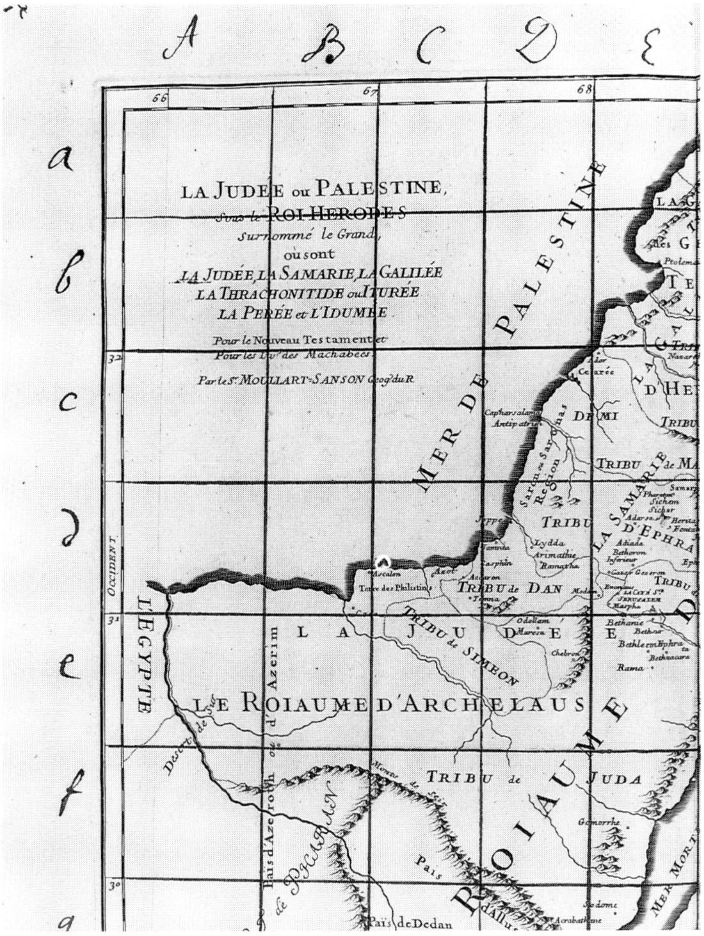

A map by Pierre Moullart-Sanson in the **Atlas Factice de la Géographie Ancienne** *which may have belonged to Gilles Robert. The manuscript addition of grid lines and letters suggests that it was one of the sources or working models for Robert for his maps of this region. (Photo courtesy of the William L. Clements Library, University of Michigan)*

des Grecs, published between 1731 and 1738, was an immensely popular work. Produced in thirteen volumes in-duodecimo, the work already had maps tipped in. Gilles Robert's large map (1015 x 656 mm. assembled) was designed to rest on a table top or be mounted on a wall to serve as a reference while one read Rollin.

In 1738, Gilles Robert published another original work to accompany a book. Derived from Sanson material, the "Description des Gaules...sur les Observations de Dom Martin Bouquet et l'Abbé LeBeuf," [Cat. 87] seems to have been commissioned by the publisher of the *Receuil des historiens des Gaules et de la France*(1738 - 1752), J.B. Delespine. The address on the two Vaugondy maps is the same as that of the LeBeuf book: "rue St. Jacques dans la maison de Mr Delespine, à l'image St. Paul." This is the only time Vaugondy's address is not on the quai de l'Horloge, and suggests that Delespine owned the plates.

Throughout the 1740s Gilles concentrated on re-issuing Sanson material: Nicolas Sanson's diocesan maps of France, the Scandinavian countries by Guillaume Sanson, and a series of maps concerning ecclesiastical history. But he did not neglect his own work. In 1742 he published his own edition of a Sanson map of France [Cat. 90]; in the Sanson tradition it has a table for religious, administrative, financial, and commercial divisions of state. The map could be coloured according to these divisions after printing. The geographical table to the right of the map soon became a feature of many Robert and Robert de Vaugondy maps thereafter.

In 1743 he published several more historical maps: "Imperium Caroli Magni" [Cat. 91] (again to accompany the works of Bouquet and LeBeuf); "Italiae Antiquae" [Cat. 329]; "Veteris Aegypti Tabula" [Cat. 430]; and he brought out a fourth edition of Guillaume Sanson's popular *Introduction à la Géographie* published by Laurent Durand in 1743.

Durand must have liked Gilles Robert's work, for he published an assemblage of Sanson material in the *Géographie Sacrée et Historique de l'ancien et du nouveau Testament* (Paris, 1747), for which Gilles prepared three new Holy Land maps: "Terre de Canaan" (1743), "Carte de la Terre des Hebreux" (1745), "Carte des Voyages de Notre Seigneur Jesus Christ" (1747). [Cat. 391, 392, 394], all based on Sanson maps and manuscripts. To complement these, he republished Moullart-Sanson's "Carte de la Géographie Sacrée pour l'ancien testament" (1740; originally, 1716) and Guillaume Sanson's "Geographia Patriarchalis." The two hemisphere mappemonde with biblical quotations of 1743 also fits into this group. [Cat. 4]

Left *'Veteris Aegypti Tabula' [Cat. 430], one of Gilles Robert's early maps (1743). The detail of the cartouche shows clearly the signature of Gobin at the base of the obelisk. A similar cartouche design was employed by Didier on his map of Egypt in the* **Atlas Universel** *[Cat. 431], p. 65. (Photo courtesy of the William L. Clements Library, University of Michigan)*

Gilles Robert's interests extended beyond the ancient world. In 1744 he published the "Theatre de la Guerre dans les Pays-Bas" [Cat. 168] which, while based on an earlier work by Moullart-Sanson and Vaultier, was clearly aimed at a market concerned with the outcome of battles in the continuing War of Austrian Succession. In 1745, he designed a map of Germany to accompany the *Histoire Générale de l'Allemagne* (1748) by R. Père Barre. [Cat. 246] This map may have inspired the "l'Empire d'Allemagne" [Cat. 240] also published in 1745, with a geographical and political table to the right. Other maps of the contemporary world followed, such as "le Cercle du Rhin" (1746) [Cat. 249] and the maps of the world and the four continents, all dated 1749. [Cat. 6, 19, 30, 373, 424].

Though based on Sanson material, the world and continent maps demonstrate that Gilles Robert incorporated new astronomical observations for latitude into his maps. He reduced the breadth of continents; Africa, for example, lost nearly fifteen degrees. He also included recent events. The world map shows the circumnavigation of the world by Admiral Anson (1740-44) and the route of the Spanish galleon from Manilla which he captured in 1743. The map of Asia is corrected according to the *Atlas Russien* of 1745, and shows the spit of land in the north Pacific seen by the Russian Tchirikow during Bering's third expedition. These maps, with their geographical tables on the right hand side, reached out to a large audience by combining topical information with geographical accuracy.

The *Atlas Portatif Universel et Militaire*

Once he had mastered cartographic techniques and established a working relationship with the publisher Durand, Gilles Robert was ready to assemble his first coherent collection, the *Atlas Portatif Universel et Militaire*. Published in conjunction with Laurent Durand and Pissot *fils*, in its final form the atlas comprised 209 maps by both Gilles Robert and his son Didier Robert de Vaugondy. A prefatory printed note in the copy in the Library of Congress tells us that the atlas came out in two editions: the first edition of 1748 consisted of general maps only. The second edition, of 1749, appearing after the Peace of Aix-la-Chapelle allowed the Vaugondys "to feel it would be agreeable to go into detail of all the kingdoms of Europe and other parts of the world."[12]

The atlas was designed to accompany the French translation of Patrick Gordon's *Geographical Grammar*, which Gilles Robert had reviewed, corrected and augmented. The "Avertissement de l'Editeur" in the French edition describes the goal of the *Atlas Portatif Universel et Militaire*: "Yet we offer to the Public precisely what the author [Gordon] demands: it is an Atlas which, although small, will furnish all the details that one could wish, because it will contain a great number of large scale maps of Provinces... Care has been taken for this Atlas to be very exact and made according to the new observations. In order to render it more useful, we have prepared an alphabetical Table of the principal names of cities with the longitude and latitude so that they may be found more easily in the Atlas,... This Table is found at the end of the [Grammaire Géographique]."[13] The organization of the atlas follows a pattern typical of atlases of the period: The World, Europe (north to south); Asia, Africa, and the Americas. As the editor's note suggested, the maps of the Low Countries (Netherlands, Belgium, and Flanders) are extremely detailed. This was only to be expected since the War of Austrian Succession (1740-1748) had been fought to a great extent on Flemish terrain. Of the 209 maps, eighty-five are signed either "Robert de Vaugondy, *fils*," i.e., Gilles' son Didier, or "Robert de Vaugondy," a style which Gilles rarely, if ever, used, and thus may also be ascribed to Didier. The atlas' military emphasis is revealed on the title page, replete with symbols of the army: horns, flags, spears, drums, a commanding officer on horseback, and a naval scene in the background. The format of the maps is oblong quarto, making it truly portable, no doubt to appeal to soldiers or officers, who were required to provide their own maps.

However, Gilles' brief Introduction tells us that his Atlas was aimed aimed at a more general public as well: "for the instruction of those who wish to learn Geography... and for the use of Merchants." He placed a map at the beginning of the atlas entitled "Introduction à la connaissance et à l'usage des Cartes." It is a practice map for those persons "who are comfortable using maps although they haven't learned Geography." It shows all the marks and symbols for every feature found on a typical map and explains how to measure the distance from one place to another, using either the scale or the latitude degrees.

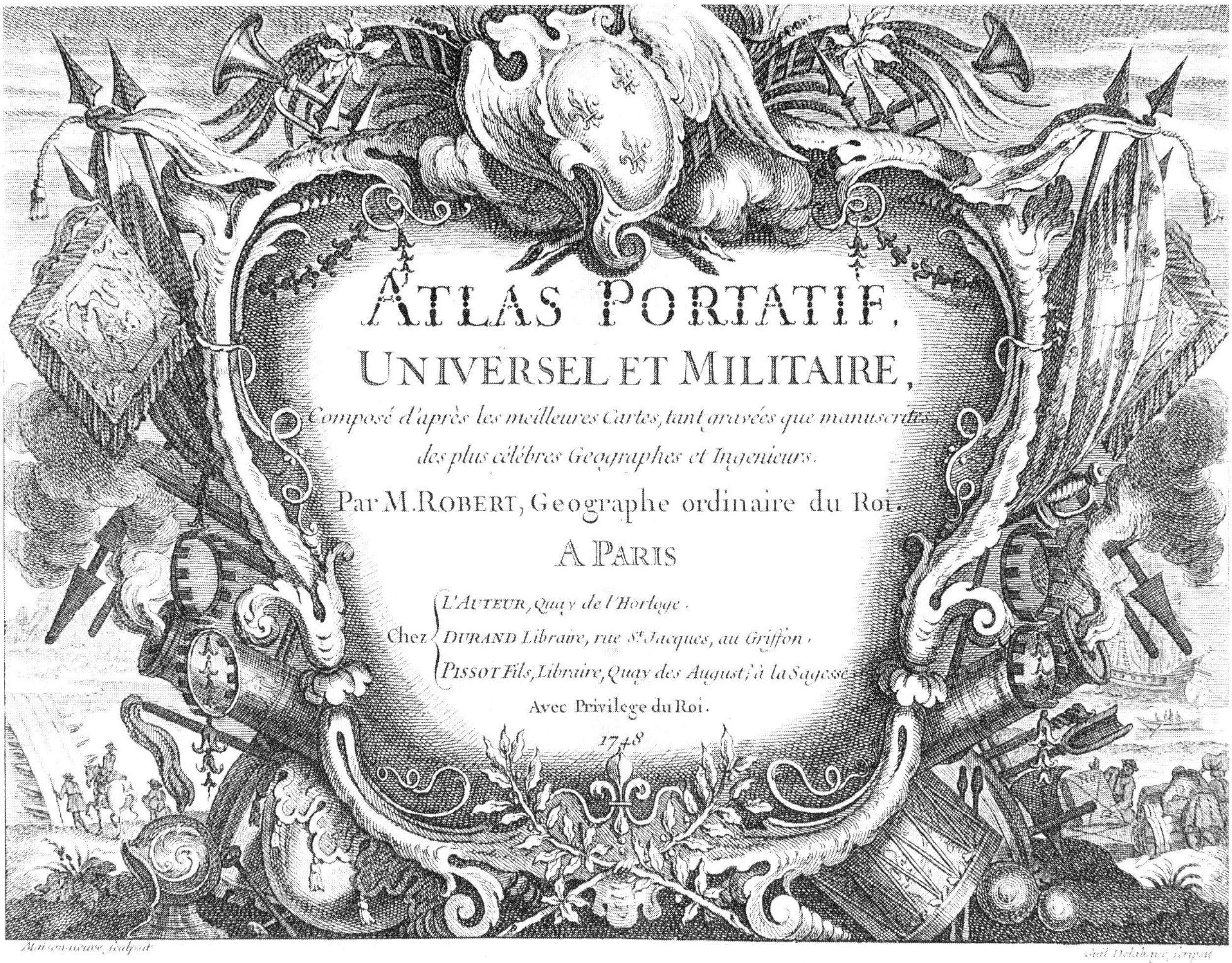

The title page from the **Atlas Portatif Universel et Militaire** *published by Gilles Robert in 1748/49, containing eighty-five maps by his son Didier. The atlas plates were later bought by Charles Panckoucke who re-issued it in 1769 with a new title page. (Photo courtesy of the Bibliothèque Nationale, Paris)*

The privilege and possibly the plates for the *Atlas Portatif Universel et Militaire* remained in the hands of the publisher Laurent Durand until Charles Panckoucke acquired the Durand stock in 1764.[14] Panckoucke brought out a later edition of the *Atlas Portatif Universel et Militaire*, a copy of which exists in the Bibliothèque Nationale (Paris) dated 1769, with the imprint of Panckoucke in Paris and Pîtel-Préfontaine in Falaise. The title is truncated to *Atlas Portatif Universel*, and the title page newly designed. The maps are for the most part unchanged, with a few exceptions: Asia, which is signed by Robert de Vaugondy and re-dated 1754, and the Mappemonde, with a change of engraver from Lattré to Dussy, and a new date of 1768. The latter also shows the discoveries by the Russians along the north west coast of North America and the Aleutian Islands. Panckoucke had hired Didier Robert de Vaugondy in the early 1770s to produce maps for his *Supplément* to the *Encyclopédie*; it is quite possible that he invited Vaugondy to re-do some of the maps in the *Atlas Portatif*. Yet by 1774, Didier described this atlas as "now rare, the plates having been used and dispersed."[15]

The *Atlas Portatif Universel et Militaire* was the first attempt of Gilles Robert Vaugondy to produce an extended series of original maps in a consistent format. The ideas of consistency and usefulness were to underpin the next Vaugondy enterprise, the folio size *Atlas Universel*. Prior to that venture, Didier Robert de Vaugondy came into his own and received his title of *géographe du roi* for another type of cartography: globes.

LA MONARCHIE DES HEBREUX SOUS SALOMON ou LE ROYAUME D'ISRAEL distingué en douze Gouvernemens
Par le Sr. ROBERT. Geog. Ord. du Roy.
Ce Royaume a duré depuis l'An du Monde 2909 jusqu'en 3029, ce qui fait 120 ans, savoir 40 ans sous SAÜL, 40 ans sous DAVID, et 40 ans sous SALOMON, après lequel il fut divisé en Rme. de Juda, et Rme. d'Israel.
Le Roy SALOMON possedoit tous les Païs qui est au deça du Fleuve [d'Euphrate] depuis Thapsa jusqu'à la ville de Gaza, et aux frontieres d'Egypte.
Echelle
AQUILON
SYRIE
GRANDE MER OCCIDENTALE
ARABIE PETREE
ARABIE DESERTE
Desert de Sur
Desert de Pharan
ROYAUME D'EGYPTE
OCCIDENT
Amalec
Madian
Mont Oreb
Mont Sinai
MER ROUGE
CARTE DE LA TERRE DES HEBREUX ou ISRAELITES partagée selon l'ordre de Dieu aux douze tribus descendantes DES DOUZE FILS DE JACOB, savoir au delà du Jourdain, deux portions qui furent accordées AUX TRIBUS DE RUBEN, ET DE GAD, et une demie AUX FILS DE MANASSE, au deça du Jourdain, une portion A LA TRIBU DE JUDA, une à celle D'EPHRAIM, et une demie AUX FILS ET AUX FILLES DE MANASSE, sept portions qui échurent au sort AUX TRIBUS DE BENJAMIN, SIMEON, ZABULON, ISSACHAR, ASER, NEPHTALI, ET DAN. les Villes que l'on désigna dans chaque tribu pour la demeure de ceux de la TRIBU DE LEVI, et les six Villes de refuge.
Dressée pour la lecture de l'Ecriture Sainte sur les cartes, et manuscrits des Srs. SANSON. Par le Sr. ROBERT, Geographe ordinaire du Roy.
A PARIS Chés l'Auteur Quai de l'horloge du palais avec Privilege du Roy 1743
TRIBU DE DAN
TRIBU DE SIMEON
TRIBU DE JUDA
Satrapies des Philistins
EDOM ou IDUMEE
ARABIE
Desert de Sin
Montagnes de Judee
MIDI
G. Delahaye sculp.

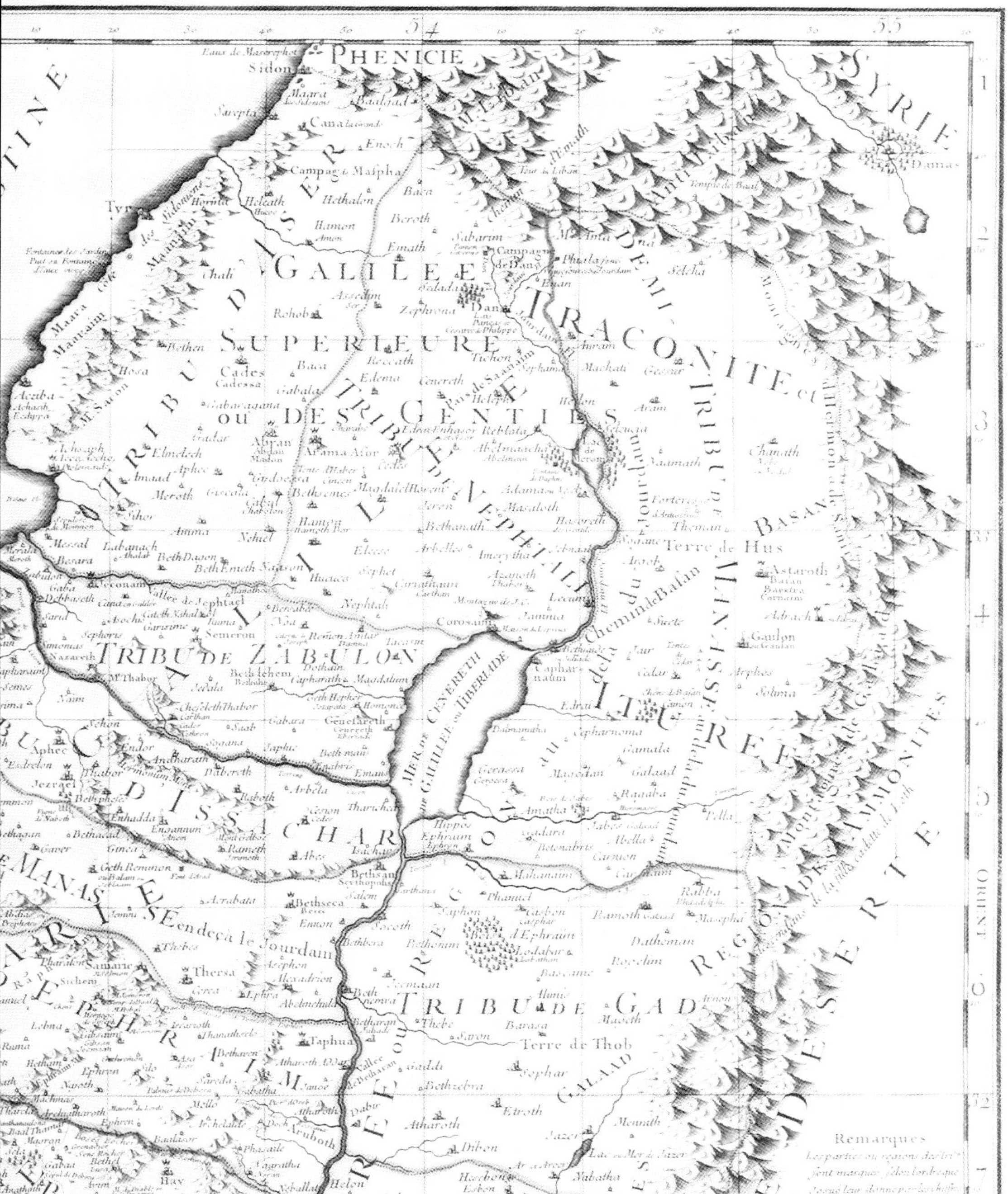

"Carte de la Terre des Hébreux ou des Israelites" by Gilles Robert, 1745 [Cat. 392]. A feature which is particularly striking about the Robert maps of this period is their rococo ornamentation, rich in religious iconography and **trompe l'oeil** *motif. (Photo courtesy of the William L. Clements Library, University of Michigan)*

Septentrion
GOLFE DU MEXIQUE
Tropique du Cancer
MER DU SUD
ou
Equateur ou
PACIFIQUE
MER
ANTILLES
S. Domingue
la Martinique
PAYS DE L'AMAZONE
BRESIL
Charcas
Chaco
Tucuman
Patagons
Avertissement
ISLE DE LA MARTINIQUE
Echelle
ISLE DE S. DOMINGUE
Echelle
MER DU NORD
Midi

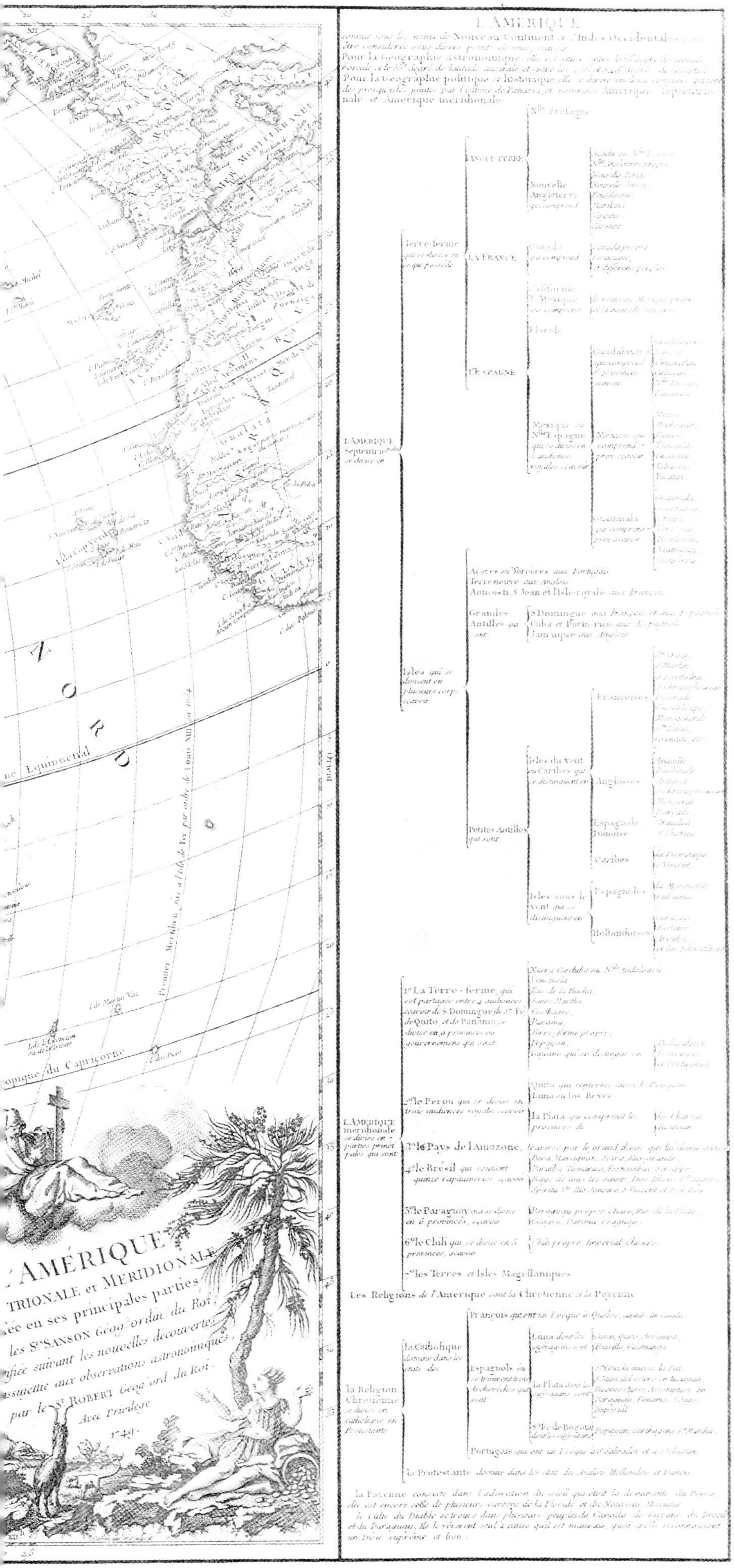

'L'Amérique Septentrionale et Méridionale,' 1749, by Gilles Robert [Cat. 19], one in the series of world and continent maps published by him in that year. Gilles' son Didier re-edited these maps and re-published them several times. This is the first state of this map. (Photo courtesy of the William L. Clements Library, University of Michigan)

MER DU NORD

OCEAN SEPTENTRIONAL
ISLES BRITANNIQUES
MEDITERRANÉE
SAHARA ou DESERT
AFRIQUE
PAYS DES NEGRES
ETHIOPIE
Equateur
Tropique du Capricorne
Ecliptique
MER DES INDES
NOUVELLE HOLLANDE
TERRES ANTARCTIQUES
Cercle polaire antarctique

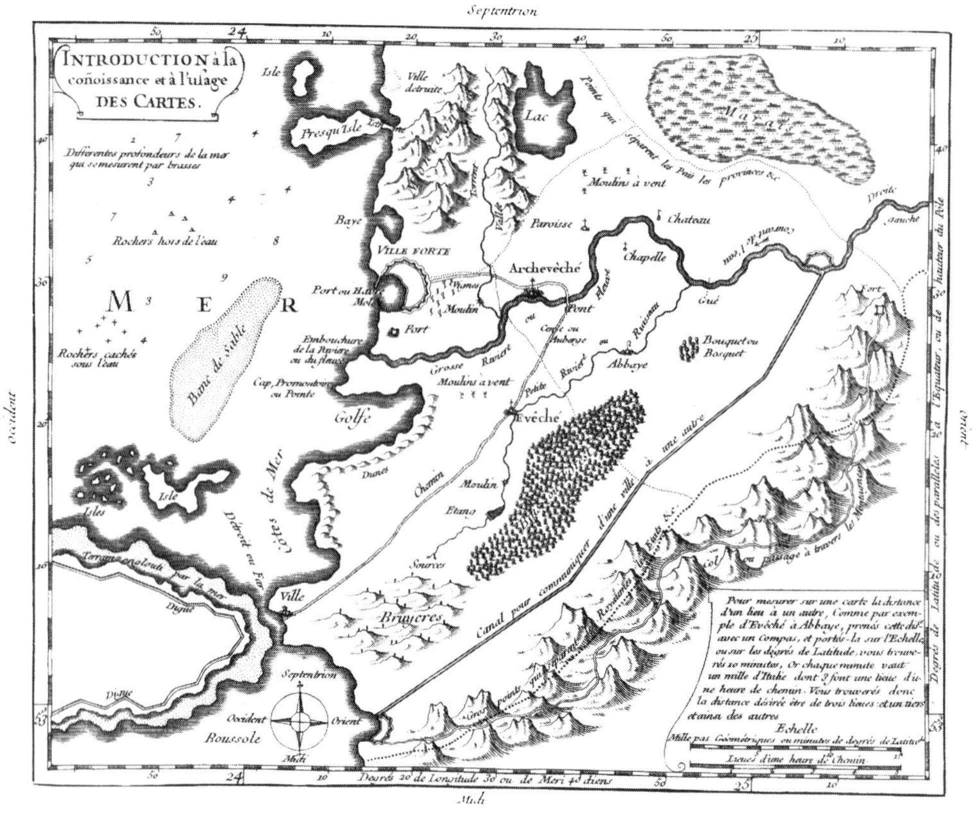

A 'practice' map from the **Atlas Portatif** *entitled 'An introduction to the understanding and use of maps' [Cat. 500]. (Photo courtesy of the Newberry Library, Chicago)*

Preceding page *The Mappemonde of 1749 [Cat. 6] on which is shown the route of the British Admiral Anson around the world (1740-44) and that of the gold-bearing Spanish galleon from Manilla which he captured. (Photo courtesy of the William L. Clements Library, University of Michigan)*

References

1. G. Robert [Vaugondy], *Introduction à la Géographie des Srs. Sanson* (Paris: Durand, 1743): p. xiii.
2. AN: Minutier Central, CXVII, 373, 22 juin 1730, will of Pierre Moullart-Sanson.
3. De Dainville cites a small group of maps inserted by a Jean Frémont, a young graduate of Louis-le-Grand, into his notes for a geography course given by le père Le Camus (1699-1700). Bibliothèque du Seminaire d'Yzeure. Cours du P. Le Camus (1699), in *Géographie des Humanistes*, (Paris, 1940) p. 457.
4. BN:Manuscrits, n.a.f. 12195 (64711, 64713)
5. AN: O^{1}78, f° 294-5, 21 décembre 1734.
6. Bibliothèque de l'Arsenal: Archives de la Bastille (Bicêtre), MS 11.166, dossier Vaugondy. Anne-Geneviève seems to have been either widowed or ultimately divorced from Jean Vaugondy; she married a second time a master mason, and in 1771 claims succession to a Pierre-Jacques Forestrier. AN:Y 11788 (28 juillet 1771).
7. Mireille Pastoureau, *Les Atlas Français XVIe – XVIIe Siècles*, (Paris: Bibliothèque nationale, 1984) pp. 387-388. The sale of the Mariette stock to Gilles Robert is in AN: Minutier Central LXXVI, 259, 19 décembre 1733.
8. Mireille Pastoureau, *op.cit.*, pp. 231-232.
9. AN:Minutier Central, XX, 561, 19 juillet 1737.
10. AN:Minutier Central, XX, 660, 12 octobre 1751.
11. AN:Minutier Central, XCI, 880, 17 décembre 1751.
12. Library of Congress: 5981, an 80-map 'Supplement' to the *Atlas Portatif Universel et Militaire* with the quoted inscription. BN: Estampes xf.23 has only 136 maps.
13. BN: G 105000. Patrick Gordon, *Grammaire Géographique ou Analyse Exacte et Courte du Corps Entier de la Géographie*, (Paris: Durand and Pissot, fils, 1748) p. x.
14. BN:Ms.Fr.21962, f°209-224, cited and published by Suzanne Tucoo-Chala, *C.J. Panckoucke et la Librarie Française*, (Paris, 1977) pp. 527-529.
15. D. Robert de Vaugondy, 'Notice Raisonnée des Ouvrages,' *Mémoire sur les pays de l'Asie et de l'Amérique*, (Paris: A. Boudet, 1774).

CHAPTER TWO

The Globes of Didier Robert de Vaugondy

The *Atlas Portatif Universel et Militaire* would have been difficult, if not impossible, for Gilles Robert Vaugondy to produce without the help of his son Didier. Born in June of 1723, Didier Robert de Vaugondy was educated as a geographer by his father.[1] His first geographical work, published in 1742, corrected and expanded Nicolas Sanson's *Tables Méthodiques pour les divisions des Gaules et de la France* (first published in 1644 in-folio; republished by Guillaume Sanson in 1696). Unlike his father, who used his mother's maiden name to sign his work, Didier reverted to the patronymic Vaugondy, but added the perhaps pretentious "de" before it.

Gilles Robert was helped in his work not only by Didier but also by his other son Martin Vaugondy de Savigny, who is mentioned both in Frémont's will of 1751 and in a house evaluation of 1767 where he is described as an "engineer-architect." While Martin signed no map that we know of, he participated in the preparation of the *Atlas Universel* (see Chapter Three) and helped Didier with the surveying for the Paris maps of the 1760s (see Chapter Six).

But it was Didier, perhaps as the eldest, who was to follow in the paternal footsteps. His first work, however, broke with the Sanson tradition of producing maps only; he turned instead to the production of globes. Though Nicolas Sanson had not been a globe maker, the Vaugondys had inherited globes from Moullart-Sanson and Frémont. The association between the Vaugondys and scientific instrument makers has been noted, too. They were neighbours on the quai de l'Horloge of the well known globe maker and mathematician, Nicolas Bion (1652-1733) whose son of the same name continued globe production in the mid-eighteenth century. In addition, the globe maker Jean Pigeon (1654-1739) had collaborated with Pierre Moullart-Sanson at the beginning of the century. According to Pigeon's daughter, the geographers Delisle and Moullart-Sanson and the globe makers Bion and Pigeon had all worked together and lived near each other: Pigeon on the Place Dauphine; Bion, Moullart-Sanson, and Delisle on the quai de l'Horloge.[2] Didier's uncle, Jean Vaugondy, the enamelist, worked for Nicolas Fortin, a well known instrument maker; Jean-Baptiste Fortin, perhaps a relation, was a globemaker who would eventually take over the Robert business.

Didier Robert de Vaugondy's first map was a set of terrestrial globe gores, published in 1745 [Cat. 12]. The gores were accompanied by his small book, *Abrégé des Differens systèmes du monde, de la sphère et des usages des globes, suivant les hypothèses de Ptolemée et de Copernic* (Paris: Durand, 1745). The work summarizes the systems of the universe according to Ptolemy, Copernicus, and Tycho Brahé, with preference given to the Copernican system. It also describes an armillary sphere, one example of which was presented by Rudolf Schmidt to the Coronelli Society Conference in Budapest, August 1989; another is in the Smith Collection of the University of Southern Maine. Neither of these spheres is dated, but the central globe of each shows none of the Russian discoveries in the North Pacific during the 1740s, made available with the publication of the *Atlas Russien* in 1745, and which became a typical feature on Vaugondy maps after 1749.

The spheres have an innovation described by Didier in the *Abrégé des systèmes du monde*: "The Zodiac on this sphere has only ten degrees: thus one might better know the right ascensions. Moreover, this size suffices for the orbit of the Moon which does not extend beyond it. The same sphere is given a crepuscular circle which is eighteen degrees wide; this circle serves as a horizon for the beginning and the end of twilight."[3] The width of a zodiac circle was usually sixteen degrees; by reducing it to ten, one could see the degrees of the equator and consequently the right ascensions and the orbit of the moon. The addition of the crepuscular or twilight circle was also something new for armillary spheres and is perhaps a symptom of concerns about light and predicting daylight in the eighteenth century.[4]

Didier Robert de Vaugondy's globe gores of 1745 also show innovations. The prime meridian (the

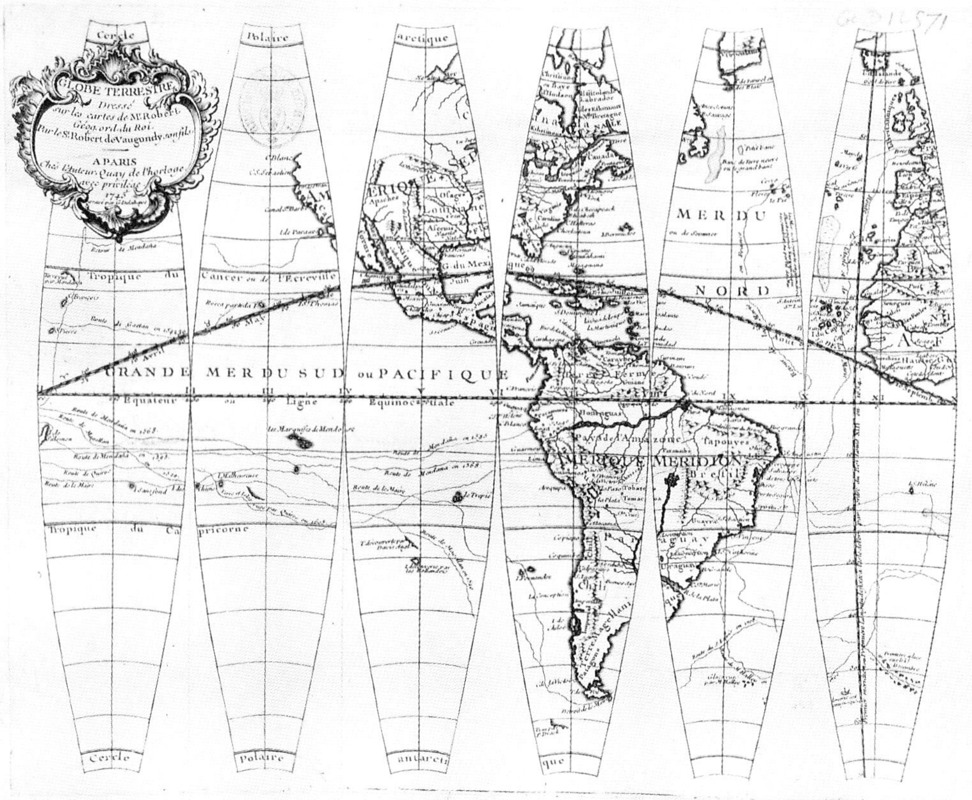
GLOBE TERRESTRE
Dressé
sur les cartes de Mr. Robert
Géog. ord. du Roi.
Par le Sr. Robert de Vaugondy son fils.
A PARIS
Chez l'Auteur, Quay de l'horloge
avec privilège
1745
Cercle Polaire arctique
Tropique du Cancer ou de l'Ecrevisse
GRANDE MER DU SUD ou PACIFIQUE
Equateur ou Ligne Equinoctiale
Tropique du Capricorne
MER DU NORD
Cercle Polaire antarctique

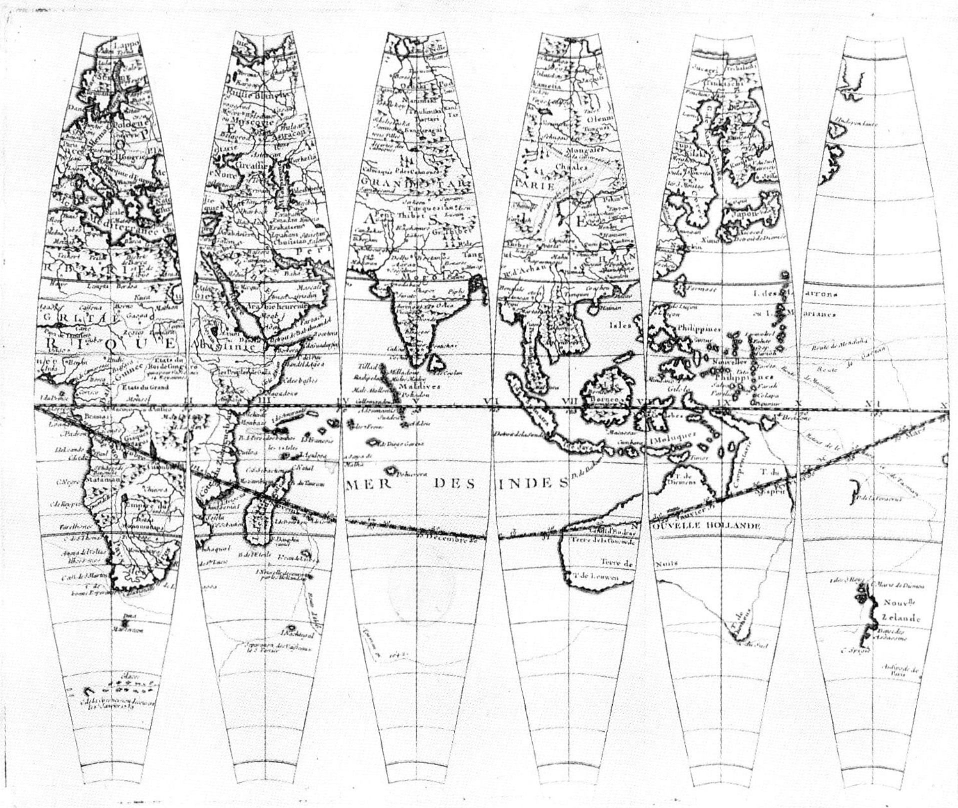
MER DES INDES
NOUVELLE HOLLANDE
Nouvelle Zelande

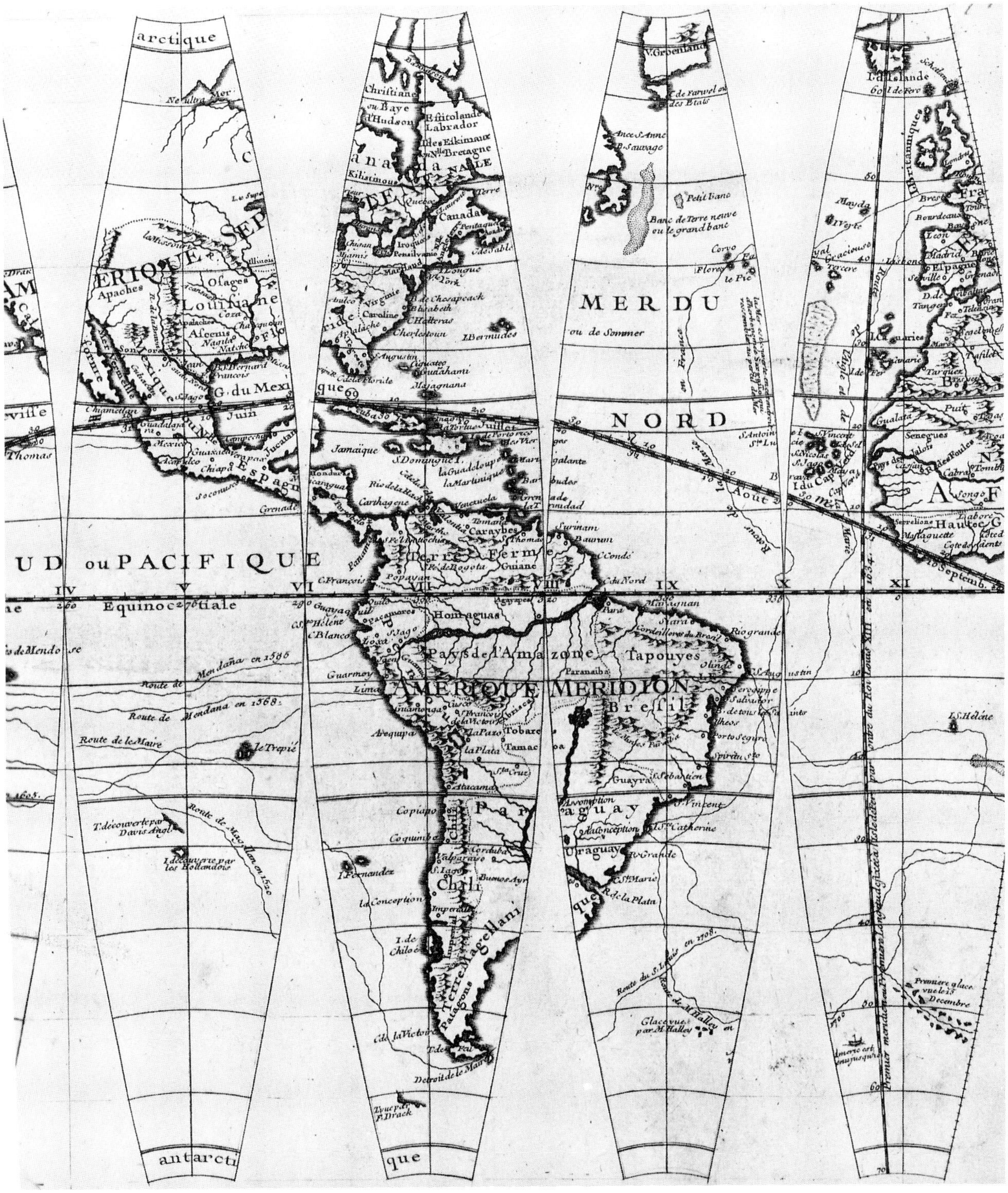

Above *A detail of Didier's 'Globe Terrestre', 1745, showing the Roman numerals used to mark longitudinal time zones. The Vaugondys used this device on many of their maps. (Photo courtesy of the Bibliothèque Nationale, Paris)*

Left *The 'Globe Terrestre' of 1745 by Didier Robert de Vaugondy. This set of globe gores is his first known work. [Cat. 12] (Photo courtesy of the Bibliothèque Nationale, Paris)*

longitude of the Île de Fero) is graduated by degrees in the normal fashion, but Didier also graduated the meridian of Paris. The ecliptic intersects the equator at the Paris meridian instead of at the prime meridian, so as not to obscure the Antilles in the West Indies, whose sugar provided twenty percent of French foreign trade. The meridian lines, at every fifteenth degree, are numbered with Roman numerals, as an aid to the problems set out in the *Abrégé* and Didier's later work, *Usage des Globes* (1751). The Roman numerals represent the time difference from Paris, a German innovation of the seventeenth century, and found on the world and continent maps of Gilles Robert Vaugondy, 1749.[5]

In 1750 Didier Robert de Vaugondy prepared a six-inch globe based on his 1745 globe gores He presented it to Louis XV as a masterpiece in petition for the title *géographe ordinaire du roi.*

Right *Portrait of King Louis XV, "le Bien Aimé" (1710-1774). He was tutored as a child by Guillaume Delisle and was a supporter of scientific projects and mapping ventures such as the Cassini "Carte de France." (Courtesy of British Museum)*

Below *The Louvre on the Bretez Plan of Paris (1739). The roofs of the east and north wings (at the top and to the left) were still unfinished in 1750. The "pavillon" where Didier wished to create his globe workshop is in the centre of the east wing. (Photo courtesy of the William L. Clements Library, University of Michigan)*

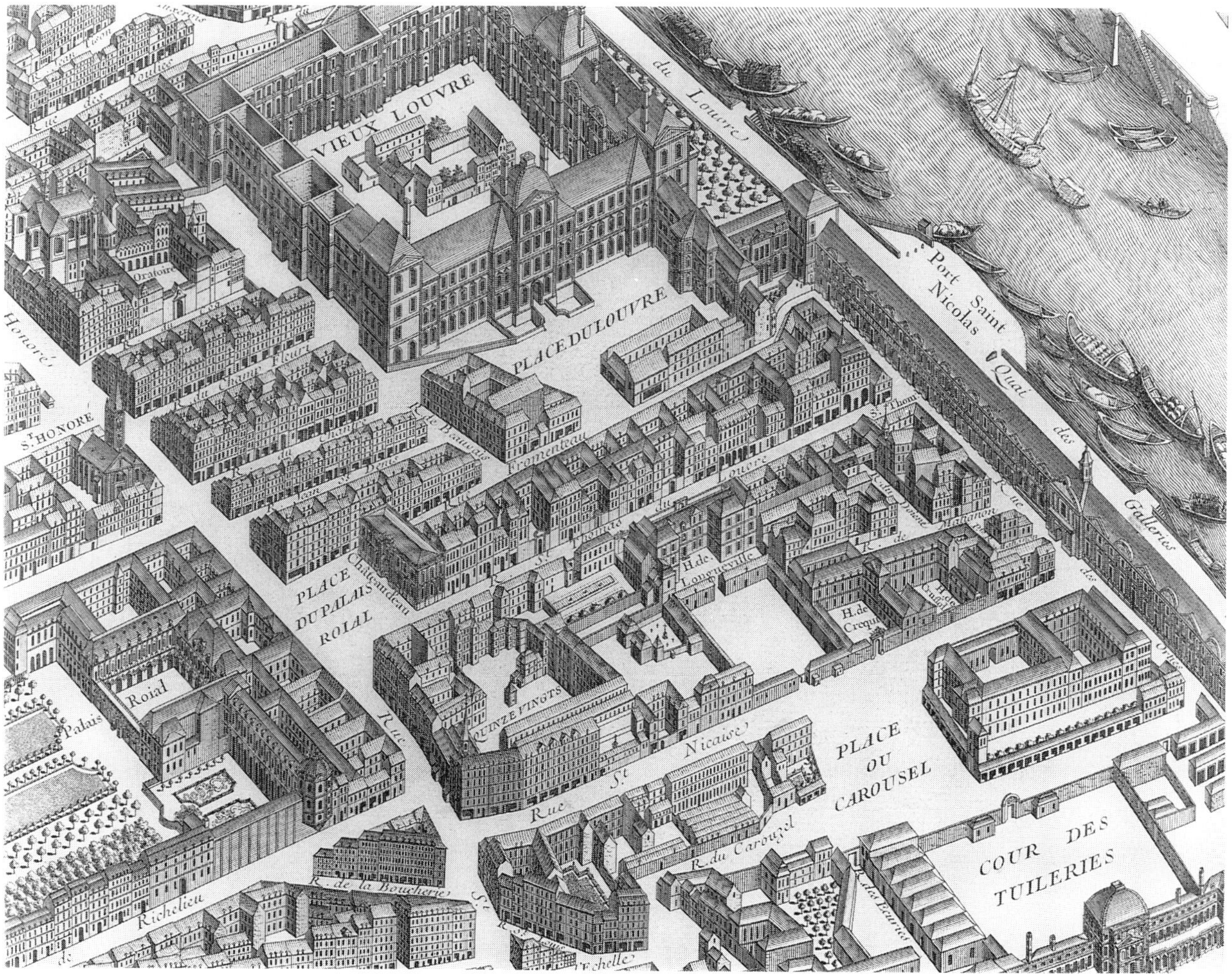

The King, who had been tutored as a child by Guillaume Delisle, was well disposed to geography and no doubt kept abreast of work done by members of the Academy of Sciences, particularly the studies of longitude carried out in Peru and Lapland which affected the understanding of the shape of the earth. At the presentation of the six-inch globe, some discussion between monarch and subject must have touched upon the inadequacies of small globes for showing the true shape of the earth as an oblate spheroid. A larger globe would be better, as Didier explained: "as this size [six inches] of globe does not permit the difference between the two axes of the globe to be felt — a difference that results from the operations which the Academicians have made at the Polar circle and at the equator in order to determine the true shape of the flattened earth — His Majesty has ordered me to construct another terrestrial globe of six feet in diameter, to be designed by hand."[6]

The King also ordered eighteen-inch globes for the Navy. At this point, the Academy of Sciences was asked to approve the globes, and in 1751 the astronomers Cassini and Le Monnier made their favourable report. Didier Robert de Vaugondy himself tells us that "the greatest number of members of the Academy came at the time to the author's house in order to see the globes mounted and they witnessed to him their satisfaction with them."[7] The interest of the Academy was no doubt roused since Vaugondy's eighteen-inch globes were the largest to be built in Paris since the three-foot globes of Vincenzo Coronelli, sponsored by the Academy in 1693.

Didier presented the eighteen-inch globes and a special copy of the *Usage des Globes* to the King in November or December of 1751. He assured him that "the six foot terrestrial globe is at the moment my only occupation."[8] The Academicians Cassini and Le Monnier had reservations, however, about the six-foot globe, noting that the the flattening of the earth on a globe of this size would only be five *lignes* (c. 11 mm.) or two and one-half *lignes* (c. 6 mm.) at each pole. They suggested that the meridian of the globe be exactly circular, with the same dimension as the equator; thus, the flattening would be visible by means of the gap between the globe and the meridian.[9]

The odd shape of the globe was not the only problem; its size was a hindrance too. The small Robert Vaugondy residence on the quai de l'Horloge could hardly accommodate the equipment and space required by this project. Close by — a five minute walk across the Pont Neuf — was the Louvre, not just a royal residence but home to courtiers, administrators, artisans and royal institutions.

Didier Robert de Vaugondy would have been familiar with the Galeries du Louvre along the bank of the Seine which served as apartments, workshops, and boutiques for a wide variety of craftsmen. Guillaume and Adrien Sanson had lived there at the turn of the century, and Jean Baptiste Bourguignon d' Anville was living there in 1750. The Louvre accommodated the Academy of Sciences as well as a gallery with the famous military relief plans. The vast twelve-foot Coronelli globes were there until 1722 and a School of Naval Construction was housed there from 1750. So it was no surprise for Didier to propose building a workshop in the centre of the east wing of the Louvre, known as the Colonnade.

Didier's proposed workshop ran immediately into a bureaucratic wall. The Director of the Louvre foresaw many difficulties with current residents of the east wing, not least of whom was the duc de Nevers whose fifteen horses, six carriages, two apartments for his grooms and storerooms for forage were all housed there. The

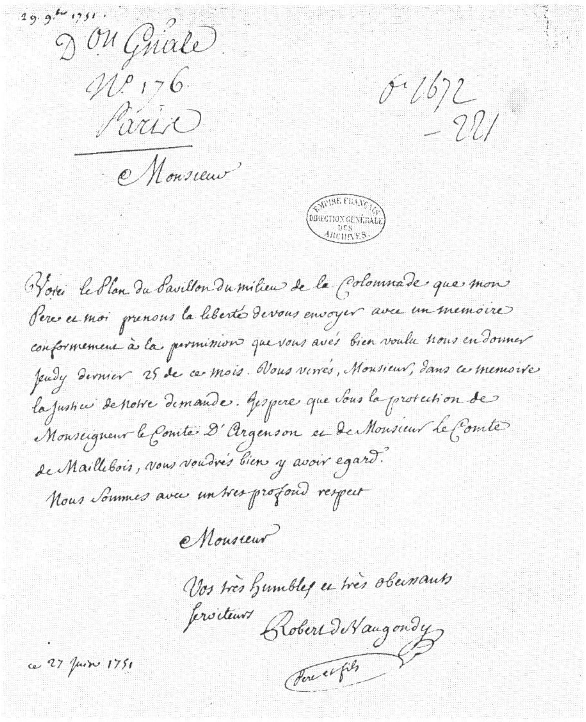
29. 9.bre 1751.

D On Gnale

No 176

Paris

O1 1672 - 221

Monsieur

EMPIRE FRANÇAIS DIRECTION GÉNÉRALE DES ARCHIVES

Voici le Plan du Pavillon du milieu de la Colonnade que mon Pere et moi prenons la liberté de vous envoyer avec un memoire conformement à la permission que vous avés bien voulu nous en donner Jeudy dernier 25 de ce mois. Vous verrés, Monsieur, dans ce memoire la justice de notre demande. J'espere que sous la protection de Monseigneur le Comte D'Argenson et de Monsieur Le Comte de Maillebois, vous voudrés bien y avoir egard.

Nous sommes avec un tres profond respect

Monsieur

Vos très humbles et très obeissants serviteurs

Robert de Vaugondy

Pere et fils

ce 27 9bre 1751

Didier Robert de Vaugondy's cover letter to the King's Director of Buildings regarding his proposed workshop for building globes in the Louvre (AN: O[1]1672 (221)). (Courtesy of the Archives Nationales, Paris)

A detail of one of the manuscript celestial globe gores showing Didier Robert de Vaugondy's instructions to the designer and engraver Gobin. [Cat. 3] (BN: Rés.Ge.DD.2967 f.4) (Photo courtesy of the Bibliothèque Nationale, Paris)

duc de Nevers would not be pleased at any disruption. Furthermore, the Director estimated the total cost of workshop conversion at nearly 26,000 *livres*, more than the King was willing to pay for the luxury of seeing a large oblate spheroid.[10] The six foot globe project was stopped.

However, Vaugondy globe production on a smaller scale continued. Gilles and Didier decided to sell their globes by subscription. They advertised in the *Journal des Scavans* of April, 1753 (pp. 254-255): "These globes have been constructed by order of the King. Mr. Robert [Vaugondy] has fixed their diameter at eighteen inches, because larger than this, they would be too cumbersome on board ship. This size has been agreed to by the King who has ordered that they be constructed for the use of the Navy. The Artist has brought to their construction every precision possible: they conform to the latest observations and he has employed the most capable master craftsmen for their design and engraving."

The terrestrial globe was engraved by Guillaume Delahaye, who had also engraved the gores for Vaugondy's presentation globe and several of the maps in the *Atlas Portatif Universel et Militaire*. The celestial globe was designed and engraved by Gobin, "the delicacy of [whose] burin equals the elegance of his pencil."[11] Both engravers had worked for Gilles Robert on the maps produced during the 1740s.

The manuscript gores for the celestial globe are in the Bibliothèque Nationale today and justify Didier's praise of Gobin (see endpapers of this volume). On the manuscript sheets there are many instructions written in Robert de Vaugondy's hand, showing his attention to detail: "Here the cartouche," "only a single line without being doubled," "dotted line [crossed out], a connected line."

Subscribers to a pair of globes also received a copy of Didier Robert de Vaugondy's *Usage des Globes* (Paris: Boudet, 1751) and for an additional 50 *livres* they could have an unusual instrument called a *Spherico-gnonomètre* or *rapporteur sphérique*, invented by Louis-Jacques Goussier (1722-1799). This instrument, a spherical protractor, when applied directly to the celestial globe, could measure parallactic angles and solve spherical trigonometrical problems — important for navigators.

The globes also came with a *chape du vertical*, a half-circle of copper that rolled freely around the globe's meridian and carried a small gilded ball which represented the sun in the Copernican system, similar to the quadrant of altitude sometimes found on globes of this period: a thin flexible slip of metal attached to the meridian, divided into degrees. Finally, the globes boasted a unique mounting system which allowed them to be turned in three ways — on the polar axis, by the meridian, and by the horizon. Yet no turn in one direction disturbed the orientation achieved by the other two.

Right *The Powderham Globe by Thomas Heath, London, made in 1750 for the Earl of Devon. This is the only known parallel to the glass celestial globe of Didier Robert de Vaugondy. (Photo courtesy of Trevor Philip and Sons)*

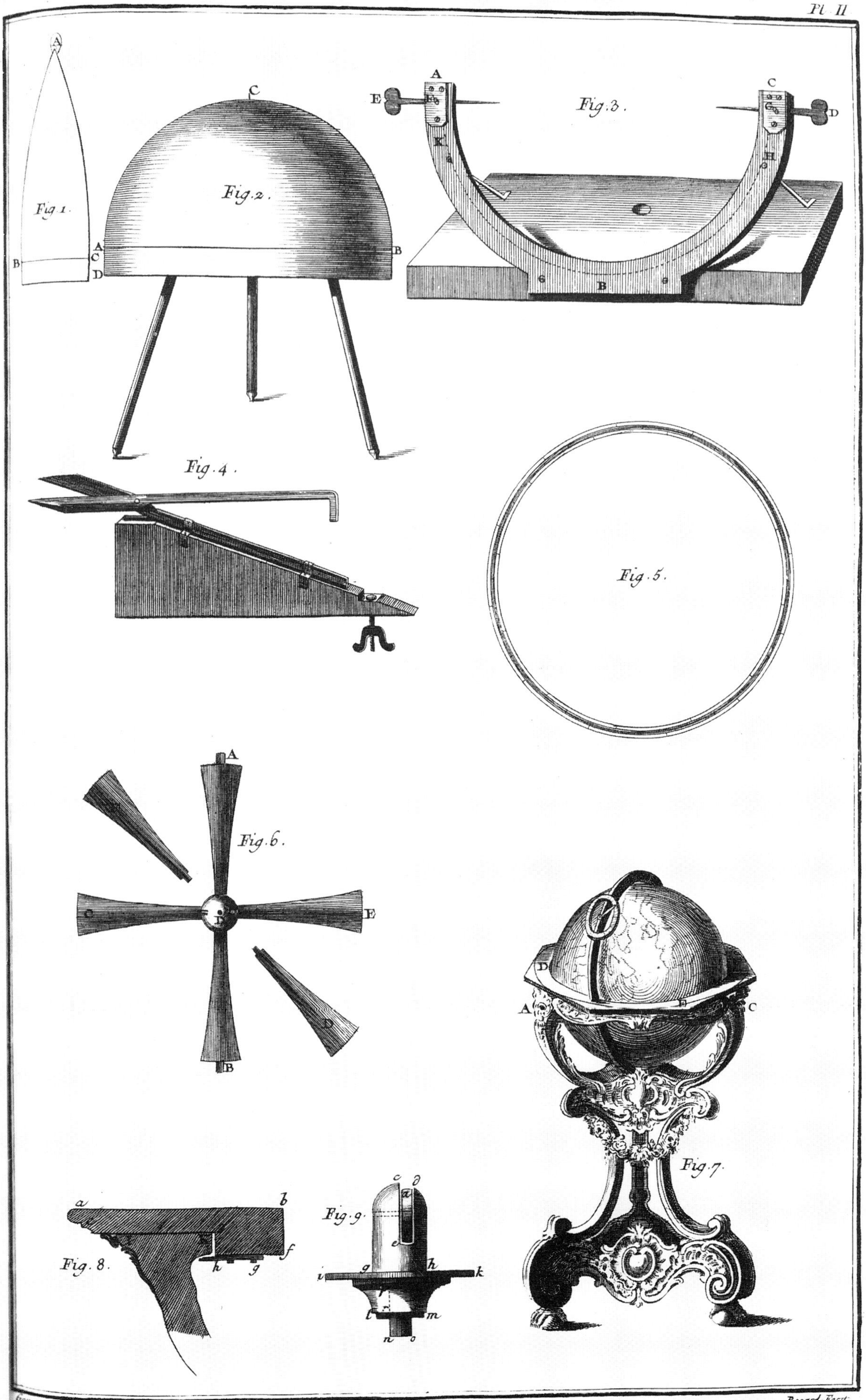

Goussier Del. Benard Fecit

Géographie, Construction Méchanique des Globes.

A small six-inch celestial glass globe by Didier Robert de Vaugondy, c. 1750 [Globe Cat. 4]. It may have had a small terrestrial sphere suspended inside it. (Photo courtesy of the Mariner's Museum, Newport News, Virginia.)

Thus Didier sought to establish himself as a "compleat" globemaker. The visitation of the members of the Academy of Sciences to his studio and presentations made by him to that body in 1752 and 1753 on other subjects must have drawn the attention of Diderot and d'Alembert, editors of the *Encyclopédie*. Didier Robert de Vaugondy wrote a lengthy article on "Globes" for Volume VII of this work which describes in detail the process of globe making from the shaping of the rounded root of the knotty elm (particularly water resistant) to the ultimate preparation of a good paste for glueing on the gores. The plate which illustrates the article shows various stages in the construction of a globe.

Another Robert de Vaugondy innovation with globes may be found in a pair of six-inch glass globes. The constellations are painted or inked

Left *A plate from Volume V of the plates accompanying the* **Encyclopédie** *of Diderot and d'Alembert for which Didier Robert de Vaugondy prepared the articles on "Géographie" and "Globes". The latter was a "how-to" treatise on constructing globes. (Photo courtesy of the Department of Rare Books and Special Collections, The University of Michigan Library)*

The eighteen-inch globe now in the office of the Director of Archives and Documentation, Ministère des Affaires Etrangères, Paris. Didier Robert de Vaugondy himself said that this globe and its celestial mate were transported to Versailles to the Ministry of Foreign Affairs. They are dated August, 1751, "dressé par ordre du Roi." [Globe Cat. 6, 8] Their stands are replicas of the globe pictured in the **Encyclopédie**. *The globes were the largest to be made in France since the three-foot globes of Coronelli, commissioned by the Académie des Sciences in 1693. (Photo by permission of the Ministère des Affaires Etrangères, Paris.)*

The elegant stands described in the **Encyclopédie** *and seen today on the pair of globes in the Ministry of Foreign Affairs were not cheap. The prices quoted for subscribers in the* **Journal des Scavans** *of April, 1753 (pp. 254-255) were as follows:*

simply mounted .... 460 livres
mounted on plain feet but in good taste, varnished à la capucine .... 600 livres
on more elaborate feet and set off in bronze .... 800 livres
on feet richly ornamented, engraved as shown on the vignette, and with the arms of the subscriber engraved .... 1000 livres

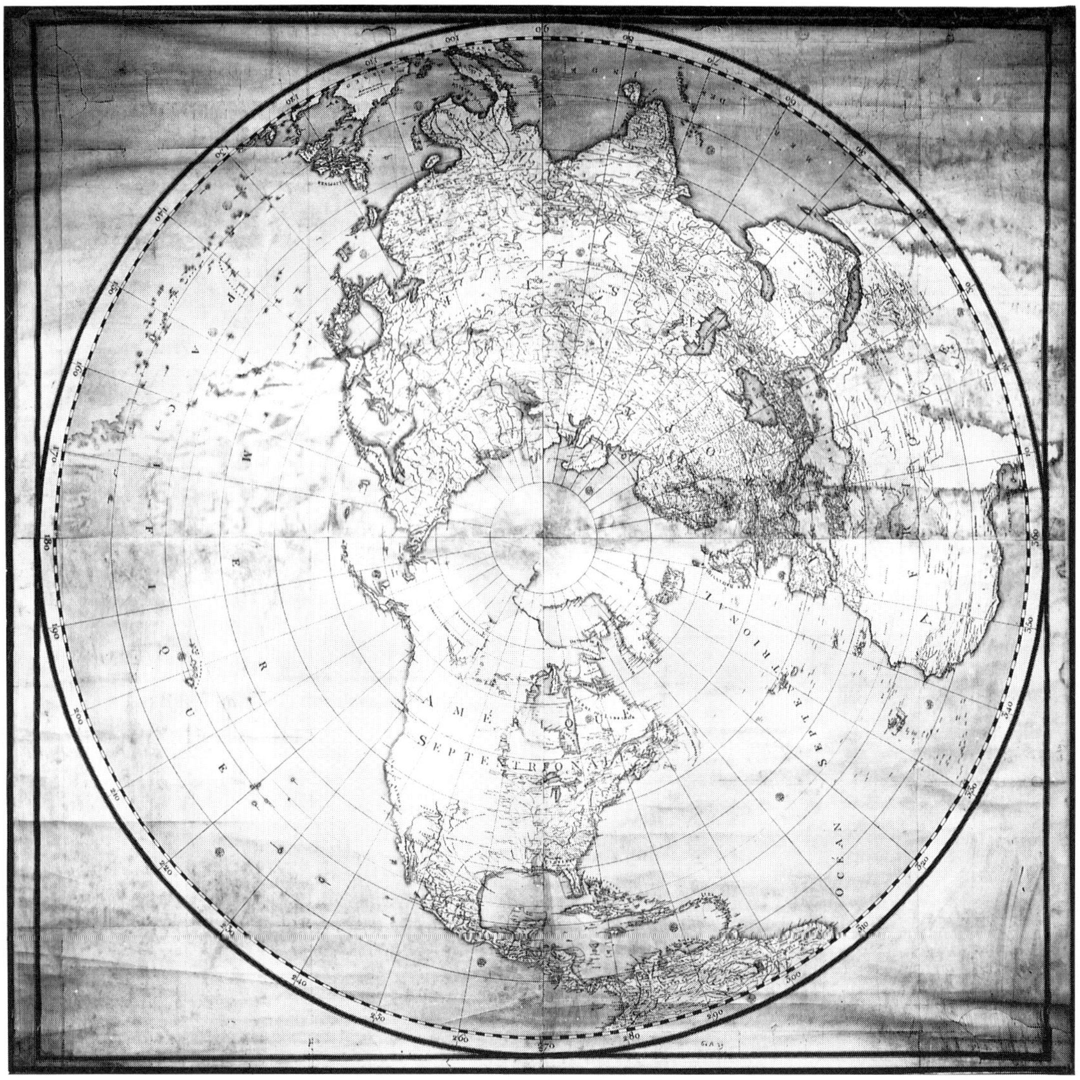

Above *The eight-foot hemispheres prepared under the supervision of Didier Robert de Vaugondy and N.G. Le Clerc for the 'Globe Bergevin.' (BN: Ge.AA 29/30) (Photo courtesy of the Bibliothèque Nationale, Paris)*

Left *The Globe Bergevin, eight feet in diameter. Didier Robert de Vaugondy collaborated on the geographical content with Nicolas-Gabriel Le Clerc. Didier died before its completion in 1795. (Photo by permission of the Musée nationale du Château de Versailles)*

by hand on the celestial globe. Now in the Mariners Museum in Virginia (USA) and undated, they are signed by Robert de Vaugondy, with the quai de l'Horloge address. The celestial globe has openings at the top and bottom, perhaps for a light source to be placed in the interior. This glass globe was not advertised in any of the Vaugondy catalogues nor has any parallel example been seen.[12] The closest may be the Powderham sphere, created by Thomas

Heath in London in 1750 for the Earl of Devon. In this apparatus, a transparent celestial sphere revolves around a small terrestrial globe.

A similar device, though not of glass, is described in the Vaugondy catalogue of 1777: a Copernican sphere of twenty-nine inches made for an *amateur* in 1773. It carried a terrestrial globe of six and a-half inches, mounted on its own copper meridian and horizon and on its own gear, showing "with more clarity and precision the principles of astronomy than armillary spheres do."[13] The sphere and a pair of eighteen-inch globes were priced together at 1200 *livres*.

Shortly before his death, Didier Robert de Vaugondy embarked on a final globe project. Around 1784, Nicolas-Gabriel Le Clerc (1725-1798) presented a paper to the comte de Vergennes, then minister of Foreign Affairs, which proposed the construction of an eight-foot globe which would show the latest geographical discoveries of La Pérouse and Captain Cook. Didier collaborated with Le Clerc on the geographical content of the globe, but the supervision of the construction was entrusted to Dom Claude Bergevin (1743-1789), after whom the globe was finally named. A model of eighteen inches was built and Didier Robert de Vaugondy supervised the creation of two eight-foot hemispheres showing the relevant geography. Didier's death in 1786 prevented him from witnessing the globe's completion in 1795. Today the globe is at Versailles.[14]

The young Didier Robert de Vaugondy had shown his inclination for useful geography early, his talent for innovation, a strict attention to detail, and an instinct for attractive presentation. These attributes were characteristic of all his later works, from the *Atlas Universel* to the great controversies over the de la Fuente discoveries to the publication of tourist maps of Paris. The Vaugondy philosophy, well expressed in the preface of the *Atlas Universel*, was that a map should be "beautiful and useful" (*bel et utile*), but this lofty credo did not always to lead to critical success.

References

1. Didier Robert de Vaugondy, "Préface," *Institutions Géographiques* (Paris: Boudet and Desaint, 1766): p. iv.
2. Mme. de Prémontval, *Le Mécaniste Philosophe* (The Hague: Pierre van Cleef, 1750): p. 43.
3. Quoted from the *Mémoires de Trévoux*, April, 1746, pp. 907-911.
4. Rudolf Schmidt, "Eine Armillarsphäre von Didier (?) Robert de Vaugondy," *Internationale Coronelli-Gesellschaft für Globen-und instrumentenkunde: Information No. 16*, December 1989, pp. 4-8.
5. See Rodney Shirley, *The Mapping of the World* (London: The Holland Press, 1983): nos. 330, 331, and 432.
6. Didier Robert de Vaugondy, "Préface," *Usage des Globes* (Paris: Antoine Boudet, 1751): p. ii.
7. Didier Robert de Vaugondy, "Notice raisonnée des ouvrages," *Mémoire sur les pays de l'Asie et de l'Amérique* (Paris: Robert and Boudet, 1774).
8. Didier Robert de Vaugondy, *Usage des Globes*, BN: Rés. V. 2108. This is the only copy seen with the "Epître." The binding has the royal coat of arms and is stamped Bibliothèque du Roi; it was probably the presentation copy, especially as it mentions the birth of the duc de Bourgogne which took place in November of 1751.
9. Académie des Sciences, *Procès-verbaux*, 19 mai 1751, and also Pochette des Séances for 19 mai 1751.
10. All the documents concerning Vaugondy's proposal and the Director of the Louvre's reaction in Paris: AN: Série O[1] (Maison du Roi), 1671 (120-128), 1672 (221, 265, 269), 1907B (1751: 46). Also see Mary Pedley, "Globes for a King: the six-foot globe of Robert de Vaugondy," *Globusfreund* (Vienna, 1987): no. 35-37, pp. 145-156.
11. 'Préface,' *Usage des Globes*, p. viii.
12. A "globe transparent 4° of 1746" is listed in the 'Etat de planches cuivre,' from the sale of the *fonds* of Buache de la Neuville to Dezauche, AN: MC: XXVIII, 418, 5 juin 1780.
13. BN:Ge. 7808, Didier Robert de Vaugondy, Catalogue, 1777.
14. Marcel Destombes, "De la chronique à l'Histoire: Le Globe Terrestre Monumental de Bergevin 1784-1795," *Archives Internationales d'Histoire des Sciences* (Juin, 1977): Vol. 27, No. 100, pp. 113-134. The original designs for the terrestrial globe gores are in the Observatoire de Paris: Ms. 1043.

CHAPTER THREE

Bel et Utile: The Atlas Universel

The work done together by Gilles and Didier on the *Atlas Portatif Universel et Militaire* naturally developed their ideas about the nature and function of maps and atlases. Didier's work on globes also refined his sense of innovation. Their next joint work, the *Atlas Universel*, embodied their philosophy of cartographic design and purpose, that an atlas should be "commode, complet, uniforme, et suivi" (easy to handle, complete, uniform, and consistent). Didier explained this in more detail in his preface to the atlas, in the section titled "l'Entreprise de nouvel atlas."[1]

First and foremost, an atlas should be easy to handle or *commode*. The very large sizes of some contemporary maps seemed bizarre to the Vaugondys, who were thinking particularly of the maps by d'Anville and the Jaillot atlases. In their opinion, maps should be no higher than sixteen or eighteen inches, the size required to fit onto shelves designed for folio-sized volumes in any library.

An atlas that is *complet* has a sufficient number of maps of both small and large scale to allow all aspects of ancient and modern history to be studied. No sovereignty, however small, should be omitted. Atlases should be *uniforme et suivi* — uniform and consistent — not only in style, but in scale, an innovation introduced by Nicolas Sanson. Every effort should be made to keep the same scale for neighbouring areas. Atlases should be simple and easy to comprehend, using a minimum number of symbols and the same ones throughout. "To what end does one encumber the positions of big towns with so many hieroglyphic marks which only offer a confused mass to the eyes? [Such marks] demand a special and discouraging study of nomenclature which is usually placed in the corner of a map for explanation."[2] By using the same symbols on all their maps, the Vaugondys avoided explanatory tables which could make map reading so irksome.

The Vaugondys also wanted their maps to be "distinct and well engraved." Their stringent demands for consistent size, slant, and spacing of lettering on their maps led to a lawsuit between their publisher, Boudet, and their engravers, the family Delahaye. This lawsuit was only one of many difficulties encountered in the publication of this atlas. Its first years of production were dogged by charges of plagiarism, criticism in the journals, and attacks on the Vaugondys for betraying their fellow geographers.

The Publication of the *Atlas Universel*

It is not clear who first conceived the idea of producing the *Atlas Universel*. After publishing the *Atlas Portatif Universel et Militaire*, the Vaugondys doubtless felt capable of a larger venture, in terms of the size of map. However, the notion of a folio-size atlas probably came not from the Vaugondys but from an enterprising but somewhat unscrupulous printer and publisher named Antoine Boudet, whose previous experience with maps had been unfortunate.

Boudet was a considerable figure in the publishing world of mid-century Paris. Stepson of Jean Baptiste Coignard, *imprimeur du roi* and head of the Imprimerie Royale, Boudet bought half Coignard's business in 1751 when the latter purchased the office of *secrétaire du roi* and turned his attention to business other than printing. Helped by family connections, Boudet made a fortune in his own right with the

---

Portrait of the Marquise de Pompadour, one of the subscribers to the **Atlas Universel**. *Note the globe and maps on her work table. By Maurice Quentin de La Tour (1704-1788) (Courtesy of the Musée du Louvre, Paris)*

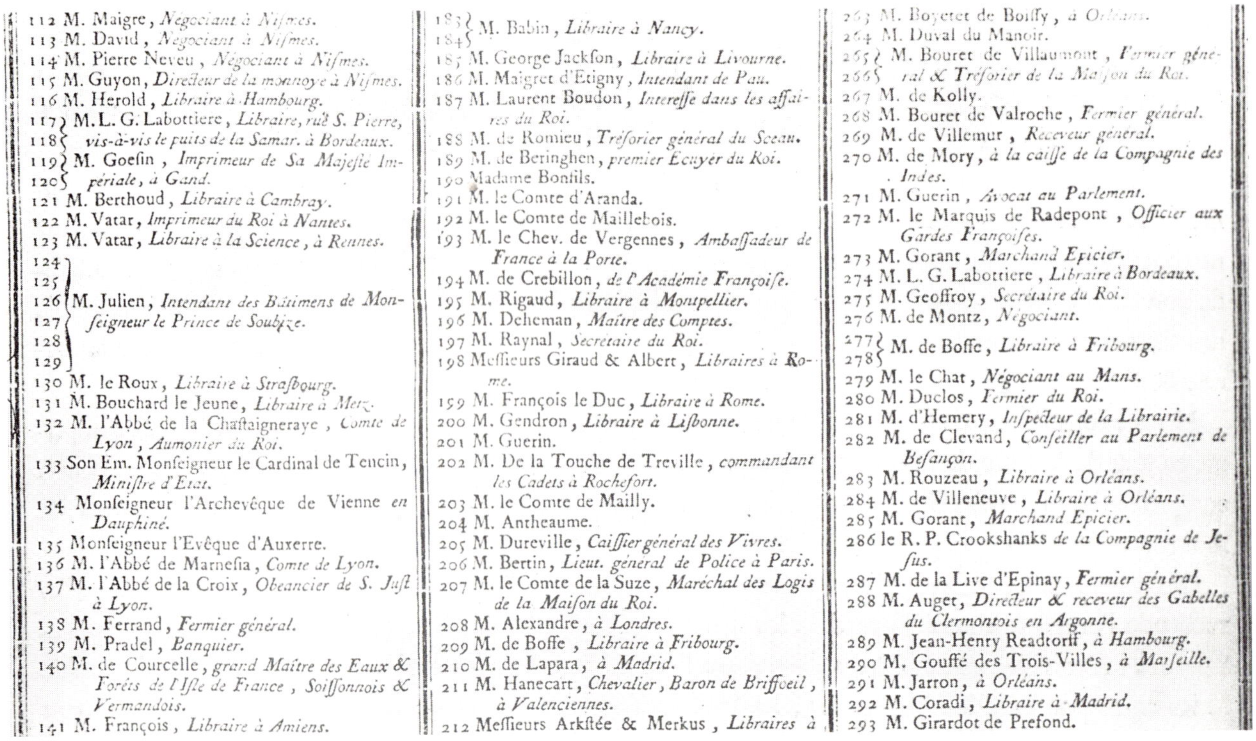

112 M. Maigre, *Negociant à Nismes.*
113 M. David, *Negociant à Nismes.*
114 M. Pierre Neveu, *Négociant à Nismes.*
115 M. Guyon, *Directeur de la monnoye à Nismes.*
116 M. Herold, *Libraire à Hambourg.*
117, 118 M. L. G. Labottiere, *Libraire, ruë S. Pierre, vis-à-vis le puits de la Samar. à Bordeaux.*
119, 120 M. Goesin, *Imprimeur de Sa Majesté Impériale, à Gand.*
121 M. Berthoud, *Libraire à Cambray.*
122 M. Vatar, *Imprimeur du Roi à Nantes.*
123 M. Vatar, *Libraire à la Science, à Rennes.*
124, 125, 126, 127, 128, 129 M. Julien, *Intendant des Bâtimens de Monseigneur le Prince de Soubize.*
130 M. le Roux, *Libraire à Strasbourg.*
131 M. Bouchard le Jeune, *Libraire à Metz.*
132 M. l'Abbé de la Chastaigneraye, *Comte de Lyon, Aumonier du Roi.*
133 Son Em. Monseigneur le Cardinal de Tencin, *Ministre d'Etat.*
134 Monseigneur l'Archevêque de Vienne *en Dauphiné.*
135 Monseigneur l'Evêque d'Auxerre.
136 M. l'Abbé de Marnesia, *Comte de Lyon.*
137 M. l'Abbé de la Croix, *Obeancier de S. Just à Lyon.*
138 M. Ferrand, *Fermier général.*
139 M. Pradel, *Banquier.*
140 M. de Courcelle, *grand Maître des Eaux & Forêts de l'Isle de France, Soissonnois & Vermandois.*
141 M. François, *Libraire à Amiens.*
183, 184 M. Babin, *Libraire à Nancy.*
185 M. George Jackson, *Libraire à Livourne.*
186 M. Maigret d'Etigny, *Intendant de Pau.*
187 M. Laurent Boudon, *Interessé dans les affaires du Roi.*
188 M. de Romieu, *Trésorier général du Sceau.*
189 M. de Beringhen, *premier Ecuyer du Roi.*
190 Madame Bonfils.
191 M. le Comte d'Aranda.
192 M. le Comte de Maillebois.
193 M. le Chev. de Vergennes, *Ambassadeur de France à la Porte.*
194 M. de Crebillon, *de l'Académie Françoise.*
195 M. Rigaud, *Libraire à Montpellier.*
196 M. Deheman, *Maître des Comptes.*
197 M. Raynal, *Secretaire du Roi.*
198 Messieurs Giraud & Albert, *Libraires à Rome.*
159 M. François le Duc, *Libraire à Rome.*
200 M. Gendron, *Libraire à Lisbonne.*
201 M. Guerin.
202 M. De la Touche de Treville, *commandant les Cadets à Rochefort.*
203 M. le Comte de Mailly.
204 M. Antheaume.
205 M. Dureville, *Caissier général des Vivres.*
206 M. Bertin, *Lieut. géneral de Police à Paris.*
207 M. le Comte de la Suze, *Maréchal des Logis de la Maison du Roi.*
208 M. Alexandre, *à Londres.*
209 M. de Bosse, *Libraire à Fribourg.*
210 M. de Lapara, *à Madrid.*
211 M. Hanecart, *Chevalier, Baron de Brissoeil, à Valenciennes.*
212 Messieurs Arkstée & Merkus, *Libraires à*
263 M. Boyetet de Boissy, *à Orléans.*
264 M. Duval du Manoir.
265, 266 M. Bouret de Villaumont, *Fermier général & Trésorier de la Maison du Roi.*
267 M. de Kolly.
268 M. Bouret de Valroche, *Fermier général.*
269 M. de Villemur, *Receveur général.*
270 M. de Mory, *à la caisse de la Compagnie des Indes.*
271 M. Guerin, *Avocat au Parlement.*
272 M. le Marquis de Radepont, *Officier aux Gardes Françoises.*
273 M. Gorant, *Marchand Epicier.*
274 M. L. G. Labottiere, *Libraire à Bordeaux.*
275 M. Geoffroy, *Secrétaire du Roi.*
276 M. de Montz, *Négociant.*
277, 278 M. de Bosse, *Libraire à Fribourg.*
279 M. le Chat, *Négociant au Mans.*
280 M. Duclos, *Fermier du Roi.*
281 M. d'Hemery, *Inspecteur de la Librairie.*
282 M. de Clevand, *Conseiller au Parlement de Besançon.*
283 M. Rouzeau, *Libraire à Orléans.*
284 M. de Villeneuve, *Libraire à Orléans.*
285 M. Gorant, *Marchand Epicier.*
286 le R. P. Crookshanks *de la Compagnie de Jesus.*
287 M. de la Live d'Epinay, *Fermier général.*
288 M. Auget, *Directeur & receveur des Gabelles du Clermontois en Argonne.*
289 M. Jean-Henry Readtorff, *à Hambourg.*
290 M. Gouffé des Trois-Villes, *à Marseille.*
291 M. Jarron, *à Orléans.*
292 M. Coradi, *Libraire à Madrid.*
293 M. Girardot de Prefond.

A part of a page of the list of subscribers as published in the front matter of the Atlas Universel *(1757). The apparently random order of names is in fact chronological. Subscribers were enrolled as their orders came in; the first subscribers received the first pulls of the maps from the copperplates. Thus low numbers received darker, sharper impressions. Subscribers to the* grand papier *edition (21 by 30 inches) received maps before subscribers to the* petit papier *(19 by 26 inches). [A. Boudet, "Prospectus du Atlas Universel...proposé par Souscription," Paris 1752, 4 pp. BN: G.1538, in-fol.] (Photo courtesy of the William L. Clements Library, University of Michigan)*

innovative periodical sheet, *Petites Affiches*, which earned him 100,000 *livres* in the six years during which he published it. Boudet himself credited the periodical with providing him the means to undertake the "enterprise of the grand atlas of M. Robert which England alone supported to the tune of more than 500 copies."[3] Boudet's shrewd business sense marked him for some as too clever by half. The Inspector of the Book Trade, Joseph d'Hemery, described him as a "Petit finot" [sic] — a crafty little fellow — and more than a "little suspicious."[4]

Always looking for new enterprises, Boudet had flirted with the idea of an atlas before his involvement with the Vaugondys. In 1746-47 he had hired the geographic engravers Dheulland and Vallet to engrave several large maps. Some of these maps were copies of the work of Guillaume Delisle, who had died in 1726. The engraver Dheulland worked for J.N. Bellin, engraving maps for the Dépôt de la Marine, and Bellin himself was employed by Boudet to correct the copies made by Dheulland. In spite of the lack of specific copyright laws, Delisle's son-in-law, the *premier géographe du roi* Philippe Buache, claimed the Delisle maps as his property and brought suit against Boudet. Gilles Robert, along with several other geographers, signed an affidavit on Buache's behalf in this case against plagiarism. The case was settled out of court and Boudet's engravers returned seven maps and plates to Buache.[5]

These legal entanglements did nothing to diminish Boudet's interest in geographical works. In 1750, as one of the "Libraires Associés," he embarked on publishing a new edition of Bruzen de la Martinière's *Le Grand Dictionnaire Géographique*, a massive ten-volume work in-folio first brought out in the Hague between 1726 and 1739. The Paris edition was to be six volumes and Boudet may have thought that a folio atlas would make an excellent accompaniment to the reference work. Among his fellow Libraires Associés was Durand, one of the joint publishers of Gilles Robert's *Atlas Portatif*, itself an accompaniment to the reference *Grammaire Géographique*.

Boudet's introduction to the Vaugondys may have come when the "Libraires Associés" published a new edition of the *Méthode abrégée et facile pour apprendre la Géographie*, a small manual containing seventeen maps, all dated 1750, by Didier Robert de Vaugondy (with one perhaps by his father). Similar in scale to the maps of the *Atlas Portatif Universel et Militaire*, these small maps were designed and engraved to fit the text they illustrate. Boudet was certainly familiar with Didier by 1751, when he published the latter's *Usage des Globes*.

The Subscription List of the *Atlas Universel*

With a sharp eye on the economics of producing an atlas, Boudet decided that rather than risk capital on such a large enterprise, he would offer it by subscription. He published a large prospectus in 1752 and placed notices in the *Journal des Sçavans*. The atlas was offered in two sizes, *petit papier* at 96 *livres* and *grand papier* at 120, a considerable savings on the post-subscription prices of 120 and 150 *livres* respectively. The sizes referred to the paper, not the map size: *grand papier* simply had larger margins and, when bound, a more imposing aspect on the shelf.

Over 1100 copies of the atlas sold by subscription, 601 in *grand papier* and 517 in *petit papier*. Booksellers made up nearly half the list, ordering copies in large numbers. Many of them lived outside Paris; quite a few abroad. There is a significant absence of geographers from the list. Those in Paris lived close enough to the quai de l'Horloge to buy a copy directly from the Vaugondys, with the exception of Roch-Joseph Julien, who subscribed to fourteen, no doubt to stock his map store in the Hôtel de Soubise. Two English geographers subscribe: J[ean] Rocque (sixteen copies) and T[homas] Jefferys (four copies), but no orders are placed by Dutch or German geographers, only booksellers.

The list of subscribers to single volumes included several illustrious names: the marquise de Pompadour (who ordered two), the comte de Maillebois (who introduced Robert de Vaugondy's globe project to the Academy of Sciences), the Cardinal de Tencin, the Maréchal duc de Belle-Isle. The *secrétaires du roi*, lawyers, tax-farmers, and government functionaries who made up twelve percent of the subscription pool were all part of that socially mobile group for whom building a library was a mark of arrival in the upper classes.[6] Such subscribers in these numbers presaged another commercial success for Boudet.

Buache's Criticism of the *Atlas Universel*

However, Philippe Buache had not forgotten the incident of the plagiarized Delisle maps and Boudet's infringement of privilege in 1746. When the Director of the Book Trade asked Buache and Jean Baptiste Bourguignon d'Anville to assess Boudet's prospectus and the first seventeen maps of the atlas for official approval, Buache did not hesitate to criticize the atlas project severely.

Buache's first concern was whether a bookdealer like Antoine Boudet had the legal right to publish and sell an atlas, an act which "stripped business from geographers into the hands of bookdealers." He accused the Vaugondys of having "betrayed their profession by depriving their colleagues of their business."[7]

Turning to the maps in the atlas, Buache pointed out many and various problems. There were inaccuracies on the map of France along the meridian of Paris: cities west of the Paris meridian which should have been to the east; the distance between Bayonne and Perpignan was short by nine minutes. According to Buache, the maps of the Holy Land and Gaul were mere copies of earlier maps by Gilles Robert (Holy Land, 1745; Gaul, 1738) which were faulty upon publication and had not been corrected since. Finally, the Vaugondys' map of Russia had a note stating that a scale was impossible on such a northerly conic projection; Buache claimed it could bear a scale. On the other hand, he complained that the projection for South America was defective and NOT able to bear a scale, though in fact it did.

Buache recommended that the production of the *Atlas Universel* should be stopped. It was not. In light of the controversy about the atlas, M. de Malesherbes, the magistrate who oversaw the legal affairs of the book trade, engaged Cassini to examine a number of the engraved maps before they were delivered to the subscribers. According to Didier Robert de Vaugondy, "The favourable report of this judge — a man of integrity and a *savant* astronomer — closed the mouth of envy and allowed the execution of this grand enterprise."[8]

Didier answered Buache's objections by altering some maps, leaving others unchanged, and by explaining the construction of all of them in the preface to the atlas. In spite of these conciliatory efforts, more criticisms appeared in the *Journal des Sçavans* of May, 1752 (pp. 554-555) after the first instalment of thirty maps was published. Similar in tone to the remarks of Buache, they focus on problems of accuracy in the maps themselves. They pointed out erroneous political features: the regions of Scanie and Babus in south Sweden and Norway, for example, were shown belonging to Denmark which had not owned them for eighty years; Arcadia (i.e. Acadia) was shown as part of Canada when it had been ceded to the English by the Treaty of Utrecht (1713). These were problems of colouring and easily corrected. Other problems would require re-engraving: Mt. Vesuvius was shown as a city; the Scilly Isles were not on the map of the British Isles nor was the name of the Orcades (i.e. Orkneys). These problems were corrected on the plates, though early states exist of the British Isles [Cat. 46] with no Scilly Isles shown.

The Lawsuit with the Delahayes

Engraving alterations were neither cheap nor easily executed. Such corrections led to further problems and disputes, this time between the engravers and Antoine Boudet. In 1754 Jean-Baptiste Henri Delahaye and his wife brought suit against Boudet, claiming that Boudet owed them money for the engraving of two maps from the *Atlas Universel*: "Imperium Caroli Magni" (Empire of Charlemagne) and "Le Tyrol."[Cat. 93, 280] Boudet had refused to pay, maintaining that the maps needed many corrections and possible re-engraving before they could be published.

In early February 1754, the case was brought before the Consul-Judges of Paris who appointed three arbiters: Philippe Buache (hardly a dispassionate observer), Jean-Baptiste Bourguignon d'Anville,

---

"La Chaussée du Pont Neuf et la Pompe de la Samaritaine," 1777, by J.-B. N. Raguenet looks directly at the quai de l'Horloge, where the Vaugondys lived. (Courtesy Musée Carnavalet, Paris).

'Britannicae Insulae' [Cat. 46] from the **Atlas Universel**. *State 1, without the Scilly Isles. (Photo courtesy of the Bibliothèque Nationale, Paris)*

geographer and member of the Académie des Inscriptions et Belles-Lettres, and Jean-Thomas Hérissant, a printer and bookdealer with geographical interests. The arbiters interviewed Boudet, the Vaugondys, the Delahayes, and several other map engravers in order to assess the claims of the two parties.[9]

Because the lawsuit contended that Antoine Boudet owed money to the Delahayes for completed and corrected maps, much of the evidence revolved around the processes of map engraving, checking proofs, and making corrections. The question which lay at the heart of the suit was who was responsible for overseeing the work of engravers: the geographers (authors concerned with the accuracy of the plates), the publisher (owner of the plates), or the engravers (craftsmen assuring the work's quality)? As so often

'Imperium Caroli Magni' [Cat. 93] from the **Atlas Universel**. *The lettering on the map caused a lawsuit between the engravers, the Delahayes, and the publisher of the atlas, Antoine Boudet. (Photo courtesy of the William L. Clements Library, University of Michigan)*

---

happens, no one was willing to take complete responsibility. The testimony in the suit demonstrates how easily errors were made on maps, because of the number of hands involved in production, and how they were left uncorrected, because of delays in payments for work.

The spokesman for the Delahayes was Jean-Baptiste Henri's wife. She represented not only her husband, but her brother-in-law, Guillaume-Nicolas Delahaye, her father-in-law, Jean-Baptiste Delahaye, and the entire Delahaye workshop of engravers unrelated to the family. The Delahayes had engraved maps for Gilles Robert's *Atlas Portatif Universel et Militaire* so were no strangers to Vaugondy work.[10] Mme Delahaye's testimony described the agreement made on 15th February 1753 between her husband and Boudet for the engraving of the remainder of the Vaugondy maps for the *Atlas Universel*. She noted the family's reluctance to take on more work, for they were already occupied with Vaugondy

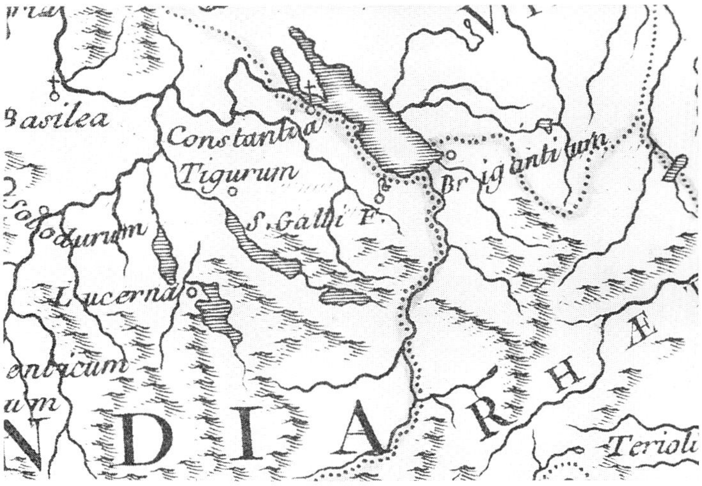

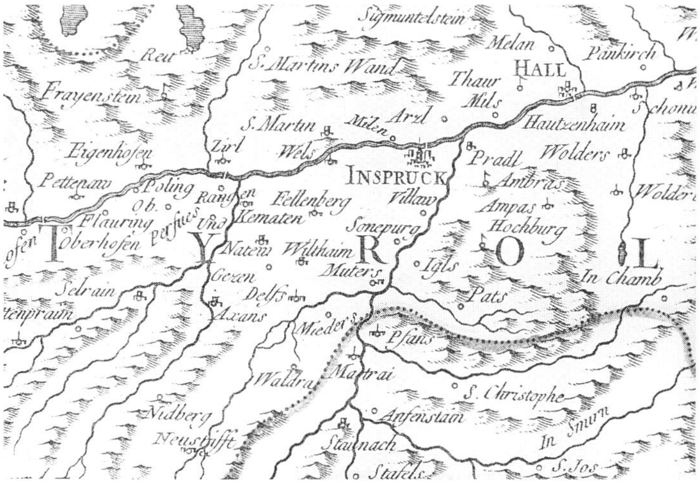

Problems with italic lettering on maps in the **Atlas Universel.** Left *The italic lettering on a later state of the 'Imperium Caroli Magni,' showing signs of correction. [Cat. 93]* Right *Italic lettering on 'Le Tyrol' [Cat. 280] showing the 'squeezed' style the Vaugondys objected to. (Photos courtesy of the William L. Clements Library, University of Michigan)*

maps, "but we accepted the agreement all the same, inasmuch as it was my dear father who had agreed to it and it was a point of honour for us to finish it."

By the terms of the agreement, Boudet was responsible for buying the copper plates; the Vaugondys were to deliver the manuscript maps to the Delahaye workshop near the Louvre on the rue Fromenteau

An engraver's signature did not mean he or she was the sole engraver. Engravers ran considerable workshops in which the work was divided between a number of specialists — one person might engrave the mountains, another the rivers, another the lettering. Such a division of labour led to the difficulties in the suit brought by the Delahaye family against Antoine Boudet, the publisher of the **Atlas Universel.** *This illustration of engravers and etchers at work from Volume V of the plates accompanying the* **Encyclopédie** *might well illustrate what the Delahaye workshop looked like. (Photo courtesy of the Department of Rare Books and Special Collections, The University of Michigan Library)*

for engraving either by the Delahayes themselves or their assistants. The plates were then brought back to the Vaugondys' house on the quai de l'Horloge for proofs to be pulled. (The Vaugondys had inherited a rolling press from Moullart-Sanson.) If the map itself was correct, the Vaugondys were to give a *bon* or *vu* written on a note (*billet*) to be presented to Boudet. An example of such a *billet* was presented as evidence in the lawsuit; it ran as follows: "I am persuaded, Monsieur, that you will be content with this new map...which M. Delahaye brings you. We have found it to be well executed. Robert de Vaugondy."

On receipt of such a *billet*, Boudet would pay 190 *livres* to the Delahayes. If corrections were necessary, the plates went back to the Delahayes, then were returned to the Vaugondys for further proofs. When the plate needed no further corrections, the Vaugondys would give a final *billet* to the Delahayes to authorize payment of the final thirty *livres* from Boudet for correction work. The total cost of engraving was thus 220 *livres* per map. (This might be compared with the price of the completed atlas in its most expensive form, 150 *livres*. For 108 maps, Boudet faced a total outlay of 23,760 *livres* just for engraving; but since he stood to make 121,752 *livres* on the subscription alone, he was not likely to lose money.)

In the case of the two maps in dispute, the Delahayes contended that the Vaugondys were pleased with both. True, the Vaugondys had found the italic on the Charlemagne map to be a "little squeezed," but they could not have been so dissatisfied, for after proofs had been pulled, they had given Mme Delahaye the designs for the titles of both maps and the scale for the Tyrol map. But, said Mme Delahaye, when the titles had been placed on the maps, the Vaugondys had complained that the author's name was not quite right in the title of the Charlemagne map: the "Vaugondy" was not in leaning capitals like the word "ROBERT." The fault was repaired and the plate brought back to the Vaugondys, followed three or four days later by the map of the Tyrol. This was on 30th December 1753, two months prior to the lawsuit. The Vaugondys promised to have proofs of these maps ready by 2nd January, when the proofs were returned to the Delahayes along with some corrections. Yet they also gave Mme Delahaye a *billet* for payment by Boudet.

On 4th January, M. Delahaye went to Boudet's house with the note, but Boudet did not pay him, telling him to come back on the following Monday, for he wanted to talk to Vaugondy first. On the Monday, Boudet announced to Mme Delahaye that he would pay for the Tyrol map only after some corrections had been made and would not pay at all for the Charlemagne map. Mme Delahaye expressed some dismay at this since the Vaugondys had, in her presence, approved the map. She said they had even overseen the engraving of both maps, to the extent of choosing the particular engravers, Laurent and Baisiez, to work on them. Therefore she claimed that the maps were the Vaugondys' responsibility. The Delahayes had fulfilled their part of the agreement and should be paid, especially in light of what she saw as the unfair conditions of the agreement: "So it's all right for him that for this work — plates with six thousand words and filled with detail, costing 220 *livres* — he [Boudet] holds back 30 *livres* as an 'indemity.' There it is, you can see the justice of it: He wants to indemnify himself and yet he has the plates, the designs, and still has our money. You might say he is really making dupes of us."

In his turn, Gilles Robert denied most of Mme Delahaye's description of events. He said he had never specified Laurent to engrave his maps since he found his work mediocre, nor had he asked for Baisiez, nor did he ever oversee the work of engravers. It was true, he allowed, that his younger son, [Martin] Robert de Savigny, had been at the Delahayes during the engraving of the Charlemagne and Tyrol maps but had not been happy with what he had seen. When proofs were pulled from these plates, Gilles Robert himself found them unacceptable. However, the Delahayes had remonstrated with the Vaugondys about their complaints, and unbeknown to Gilles, his elder son Didier had given Delahaye a note for Boudet, written in equivocal terms: "I hope, Monsieur, that you will be content with the two proofs which M. Delahaye brings you. Robert de Vaugondy." It is clear from this ambiguous language that Didier did not want to be the reason for Boudet not accepting the plates. (The Vaugondys may not have wanted to antagonize the Delahayes, for the engravers, or at least Guillaume, were working on other Vaugondy maps not related to the *Atlas Universel*.)

At this point the judges decided to inspect the two maps closely, particularly the "squeezed" italic. Buache asked that the maps be compared to earlier maps made for the Academy of Sciences by Berey, Lièbaux, and La Pointe, engravers who had worked for his father-in-law, Guillaume Delisle. After these comparisons, the arbiters agreed with the Vaugondys that the italics on the Delahaye maps were indeed too thin and too tall, but they did not agree with the other Vaugondy objections. The Vaugondys claimed that on the Charlemagne map, the title "was worthless, with no equality in the initial letters, no body, too enlarged: it was not necessary to give it as a trial to an apprentice." In addition, many words were badly placed and needed erasing and better arrangement; all the longitude and latitude numbers needed to be rubbed out. They complained that certain italic words were "détestable" and required re-engraving, and capital letters reinforcing.[11]

A detail from the "Royaume d'Irlande" [Cat. 56] from the **Atlas Universel**, *a map which the Delahayes thought was to be their model of italic lettering. In fact, the Vaugondys did not like this kind of lettering at all, thinking it too tall, too thin, too squeezed. (Photo courtesy of the William L. Clements Library, University of Michigan)*

The arbiter-judges carefully cross-examined the Vaugondys as to their precise involvement in the engraving of their maps, putting several questions to them which had arisen from Mme Delahaye's testimony. Had they not specifically requested the young engravers Laurent and Baisiez, urging the Delahayes to hire the latter on the basis of his work for the publisher Daumont? Were they not so impressed with Laurent's work for the publisher Durand that they encouraged him to sign his name to maps and promised him many years' work in the future? As for italics, had they not directed the Delahayes to use as a model the italics from the map of Ireland, signed by Guillaume Delahaye? Had not Didier Robert de Vaugondy's younger brother Martin Robert de Savigny been sent to oversee and even to place certain designs on the copper and to give advice about the engraving? And finally, had not the Vaugondys, with their ambiguous note, prejudiced Boudet against the Delahayes?

The Vaugondys denied these suggestions: "We have never proposed to le Sr. Delahaye that he use lettering by Baisiez and Laurent as a model; we only know the lettering in Roman and capital of the former and have never seen the work of the latter." As for their recommending Baisiez to the Delahayes, they said that Mme Delahaye had mentioned the need for some young people to be taken on in the workshop. "We thought we were doing her a favour in pointing out Baisiez to her. [We had heard] some recommendations...of his sense and good will, but nothing at all about his talent at engraving."

As for the map of Ireland, the Vaugondys replied that they had meant this as an example of how **not** to make italics; indeed, they had complained many times that Guillaume Delahaye's italics were too tall. With regard to sending his son to supervise the engraving, Gilles answered that he had only sent him to chivvy the Delahayes along in order to keep the work moving. Didier admitted visiting the Delahaye workshop where he had been surprised to see Baisiez working on ninety-five words of italic when Gilles had expressly wished him not to. He also had seen Laurent working on the Charlemagne map and had had words with him: "The italic and Roman lettering appeared to me thin and squeezed, without liaisons, or at least the liaisons were so fine that one could not see them. On this point Laurent said to me that these liaisons were very deep and that they would be visible; as for the squeezed characters, they had been obliged by the narrow space in which they were contained; which caused me to say that when one was constrained to squeeze the lettering, it was necessary to diminish the height, in order that all be proportional."

Martin Robert de Savigny denied ever setting the designs on copper except in one instance when a copper plate was much larger than necessary and his advice was sought by the Delahayes on whether to place the map in the middle or on the side of the plate. Lastly, Gilles denied saying anything prejudicial to Boudet about the Delahayes, "not being of a character to denounce workers, and only passing definitive judgement when it is required of me."

On 1st March 1754, after three weeks of testimony, evidence, and consultation, the arbiters presented the disputants with a "Projet d'accommodement." To begin with, they pointed out that the Vaugondys, in their capacity as authors of the maps in question, had both a natural right to oversee the engraving of their maps and a legal right by the terms of the agreement passed between Boudet and Delahaye. Then the arbiters listed seven specific things to be done. They asked that the Delahayes make all the corrections asked for by the Vaugondys, that the proofs would be reviewed both by the Vaugondys and one of the geographer-arbiters (Buache or d'Anville) and that a new payment schedule be followed for the plates which included an extra forty *livres* to compensate Delahaye for the extra corrections; they agreed to split the cost of any cracked plates. The arbiters' report to the consul-judges emphasized the need for geographers to supervise the engraving of maps. The blame for any faults in engraving therefore belonged not so much to the engravers as to the Vaugondys.

Yet the struggle did not end with the "Projet d'accommodement," passed on 8th March 1754. On the 29th of the same month, the Consul-Judges sent Delahaye and Boudet back to the arbiters with new difficulties. The Delahayes had duly erased all the italic from the map of the Empire of Charlemagne and had given proofs of the now nearly blank map to the Vaugondys to redistribute the place names. In the meantime, they awaited a list of corrections for the Tyrol map which had not appeared. Delahaye was so frustrated that he offered to return the extra forty *livres*, freeing Boudet to find some other engraver. He further requested that the agreement between the two of them be cancelled as soon as possible.

The arbiters agreed with Delahaye and urged Boudet to ask the Vaugondys to submit their list of corrections for the Tyrol map to the arbiters. When he received it, Buache expressed some surprise and displeasure, for the requested corrections "are infinite and without consideration for the other maps of the Atlas, where one could find just as many corrections if they were examined as scrupulously." Buache suggested that all that was necessary on the Tyrol map was to re-do the 'squeezed' italics, and suggested the Charlemagne map as a model. He advised Boudet that "if these Messieurs [Vaugondy] have the same confidence as you do in the works done by the hand of the Sr. Delahaye, and were they engaged in writing to approve what he does, they would not be obliged for this exactitude."

This of course is the same Buache who criticized the early maps of the atlas for their inaccuracy and who had signed the recommendation that geographers be responsible for their engravers' errors. Now he was unhappy with the demands made on engravers. The lawsuit points out the difficult balance between geographical accuracy, aesthetic appeal, and financial viability. For Buache, it further confirmed his misgivings about the entry of bookdealers into the map business and his disdain for the work of the Vaugondys. It is less clear how the result of the suit affected the Vaugondys. It did not seem to sour relations between them and the Delahayes, for Guillaume Delahaye continued to be a regular engraver for them; moreover, Didier joined d'Anville as a witness at Guillaume Delahaye's wedding in 1758.[12]

Another wedding in the Delahaye family also suggests an interesting subplot. In 1755 Jean Baptiste Henri's wife, now a widow, married Pierre Laurent, engraver, perhaps the same Laurent whose italics had been considered worthless? It may account for Mme Delahaye's defence of the the young man's work.[13] Despite the friendly relations between Vaugondys and Delahayes, no map later than 1753 is signed Delahaye in the *Atlas Universel*, allowing one to conclude that Boudet's agreements with the family were at an end.

The *Essai sur l'Histoire de la Géographie*

Although the engraving lawsuit was over, it did not finish the troubles over the *Atlas* or criticism of the Vaugondy's work. Didier Robert de Vaugondy had written an *Essai sur l'Histoire de la Géographie*, published by Boudet in 1755. He presented the manuscript to the Academy of Sciences in 1754 and it received approbation and privilege from the Syndics of the Book Trade in August and October of 1754. It was a comprehensive history of geography and cartography from its ancient origins to the eighteenth century present and included an extensive discussion of the sources used for the maps in the *Atlas Universel*.

However, on its publication, Buache sent a critical review of it to Malesherbes, the Director of the Book Trade. His harsh criticisms were never published, but their sting further soured relations between the Vaugondys and Buache for some time.[14] Buache accused Didier of taking most of his material from a work of the same or similar name by de la Martinière: "Essai sur l'Origine et les progrès de la Géographie," published in the *Mémoires Historiques et Critiques*, October and December, 1722, and brought out in fascicules by J.T. Hérissant in Paris in the 1750s. (Hérissant, one recalls, was one of the arbiters in the engraving lawsuit.)

Similar accusations made their way into the review of the *Essai* in the *Journal des Sçavans* of November, 1755 (pp. 66-79). The author of the review was Joseph de Guignes (1721-1800), a well known orientalist, and while his chief criticism concerned the orthography of Arabic names and the location of the Ptolemaic capital of China, de Guignes implied that Didier had relied too much on de la Martinière without giving enough credit.

Further volleys were fired by the Jaillot family. The granddaughters of Nicolas Sanson's collaborator, Alexis-Hubert Jaillot, and sisters of Bernard-Antoine (who had died in 1749), the Demoiselles Jaillot implied that some of the maps in the *Atlas Universel* were simply copies of theirs. They further complained that in advertising the *Atlas*, the Vaugondys were described as the "sole inheritors" of the Sanson geographic business. They, the Jaillots, possessed part of the Sanson stock.[15]

The Vaugondys replied, pointing out that the Jaillots' work, like that of the Vaugondys, was *based* on Sanson and did not necessarily *copy* Sanson. The connection between the Jaillots and Sanson dated to 1673 when Alexis-Hubert Jaillot contracted with the Sanson brothers, Guillaume and Adrien, to produce a series of maps in two sheets. Jaillot was enjoined by this contract not to reduce these Sanson maps to one sheet. The Jaillot family continued to publish these same maps. As to who were the "sole inheritors" of Sanson material, Didier argued that the Jaillots were only the owners of a certain series of two sheet maps, while the Vaugondys had inherited all the other Sanson maps, books, and manuscripts.

The Sources for the Maps of the *Atlas Universel*

These questions of ownership, authorship, plagiarism, and protection of rights had led Didier Robert de Vaugondy to write a final section into his *Essai*, describing his sources, in order to deflect the charges and accusations rendered against himself and his father. It follows the tradition of the sixteenth-century geographers Ortelius and Mercator, and for modern map historians, this is perhaps the most helpful and interesting part of the *Atlas Universel*. French mapmakers of the seventeenth century had not indulged in this sort of 'cartobibliography,' but by the eighteenth century a new format emerged: the lengthy *analyse*. As its name suggests, it analyzed the sources for a map in great detail and minutely described its construction. It was a form employed particularly (if not created) by Jean Baptiste Bourguignon d'Anville and Jacques-Nicolas Bellin, whose new maps and charts for the Dépôt de la Marine were often accompanied by analytical mémoires. However, Didier Robert de Vaugondy realized certain dangers inherent in the lengthy *analyse* form: "I will sacrifice to brevity the erudition that the *analyse* of each of the maps could support. These sorts of literary works find very few readers. It suffices to give the public the means of adding to the confidence which it has already given to our enterprise...."[16]

A careful reading of the source descriptions gives some insight into the way in which the Vaugondys worked and how they chose their material. Their most common sources, not surprisingly, were Sanson and Delisle, but they did not blindly follow Sanson because of their easy access to the material. On the contrary, Didier frequently criticizes Sanson's maps. For example, the Sanson map of Germany was rejected as undetailed, inaccurate, and out of date. "If our design were only to make maps, without regard to the utility one ought to draw from them, we would content ourselves with following the detail Sanson gave to Germany; but we preferred to use the originals which this savant geographer employed, and to use other more recent and more detailed pieces not used before."[17]

Their frequent use of eighteenth-century sources, often from the 1740s, provided their atlas with up-to-date information. While their preference was for maps that had been surveyed in the field and maps published in the region itself, they did not hesitate to turn to older sources when more recent maps were found to be lacking. "The desire for perfection has always led us to original sources....It is why the conformity which can be found in some places between our map and that of Sanson comes from our using the same sources: we have been obliged to have recourse to them for lack of more recent materials....These parts, however, are more detailed in our map than in Sanson's, which renders our work superior and makes one see the utility that one draws from sources when one consults them carefully."[18]

Didier's citation of no less than fifty eighteenth-century geographers as sources for the maps of the *Atlas Universel* suggests not only the thoroughness with which the Vaugondys searched for geographical information, but also the accessibility of such works (by no means all French) in Paris at mid-century, and an efficient network of communication between geographers of different countries. The Preface also reveals that Vaugondy had an entrée to the Dépôt de la Marine where he was able to use maps in their collection for the coasts of South America [Cat. 487] and for Canada [Cat. 459]. Sometimes, however, information was surprisingly inaccessible. Didier Robert de Vaugondy bemoans that even a publisher in Avignon cannot procure for him a recent map of the west part of Provence which shows in detail the mouth of the Rhône.[19]

How the Vaugondys used their sources reveals an abiding conservatism. They did not adopt new information before determining its accuracy and assessing agreement with all other maps, ancient and modern. Their first concern was to find a map accurate enough to serve as a base map, and which conformed best to all existing maps of the area. This was their *carte citée*, which was then adjusted to more recent astronomical coordinates. They were more interested in perfecting existing geography rather than in starting *de novo*.

In order to refine their *carte citée* to achieve greater accuracy they needed recent observations of longitude and latitude. Their chief source was the *Connaissances des Temps*, a manual of bearings used chiefly by sailors. They also used the astronomical observations made by Cassini de Thury in France and other astronomers working abroad for the Academy of Sciences, especially in Africa and North America.

New observations sometimes meant new locations of towns, cities, political borders, and topographical features. In the case of Austria, the mountainous regions on earlier maps had extended too far north and south to coincide with more recent data; the Vaugondys corrected this on their map of the Tyrol[20]. In North America, recent observations made at Fort Rupert moved the Fort and James Bay seven degrees west of Quebec City; three degrees further west than on Bellin's map in Charlevoix's *Histoire de la Nouvelle France* (1744).[21] Though new co-ordinates rarely moved a town more than one degree, more often just minutes or seconds, this of course could equal many miles on the ground.

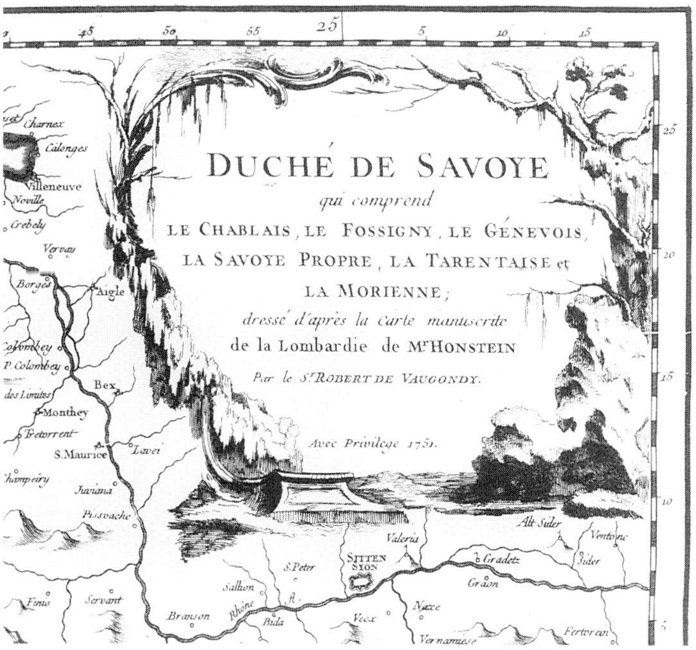

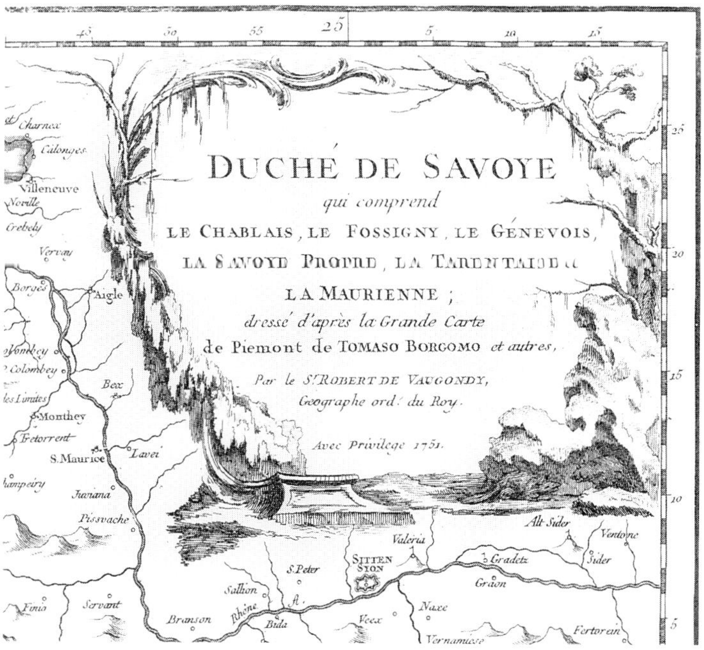

Title cartouche from the 'Duché de Savoye' [Cat. 333]. The alteration of the source shows the care with which both Vaugondys took in citing their sources. Left State 1: "dressé d'après la Carte manuscrite de la Lombardie de Mr. Honstein" (BL: c.38.f.4) (Photo courtesy of the British Library) Right State 3: The change of source for the map and the addition of **géographe ordinaire du roi** *to the title are new. On the map itself, note the change of 'Bex' from Roman to italic letters. (Photo courtesy of the William L. Clements Library, University of Michigan)*

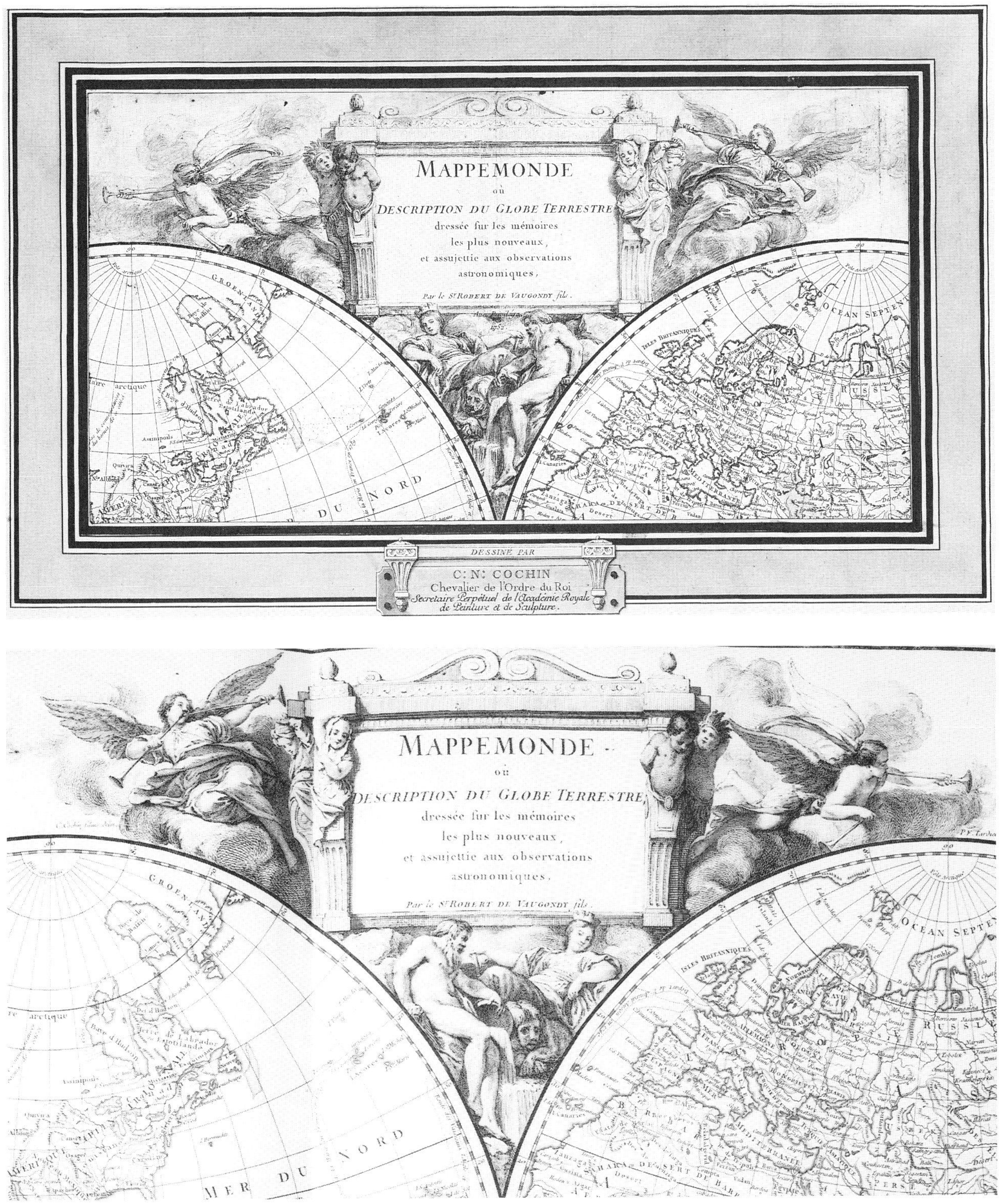

A rare example of the design for a cartouche on the map itself. The drawing is by the designer Charles Nicolas Cochin on a proof state of the Mappemonde of 1752 [Cat. 8]. Top *The drawing by Cochin. (Courtesy of the Department of Prints and Drawings, British Museum).* Above *The Mappemonde with the cartouche etched, in reverse, by P. F. Tardieu. (Courtesy of the William L. Clements Library, University of Michigan)*

More problematic were the conflicting new data. To resolve discrepancies, the Vaugondys employed a Cartesian rationality by taking a mean: "When there are many observations made at the same place and differences between them, one takes a proportional mean, in adding up all the results and dividing the sum by the number of results which made it up."[22] While mathematically sound, the result did not necessarily correspond with reality. The Vaugondys seemed to suppose all maps to be equal in their inaccuracies without regard to their sources.

Such rationality led them to reject new observations on occasions if they were too gravely out of step with all previous and contemporary work. The map of Savoy [Cat. 333] uses longitudes for the major towns of Nice, Genoa, Turin, and Milan based on observations made for earlier works: G. Delisle's Corsica (1721) and Chafrion's Genoa (1685); they rejected Bellin's bearings in the *Histoire de la Révolution de Genes*, 1750, and those in the *Connaissances des temps*, but their decision was confirmed by Tobie Mayer, a German geographer whose map of the same area in 1749 used similar longitudes. On the other hand, the Vaugondys used Bellin's same work, the *Histoire de la Révolution de Genes* without cavil for their depiction of Corsica.

Their critical approach caused them to disregard d'Anville's placement of the source of the Nile at 5° North and to follow instead Delisle's configuration which was based on the explorations of Father Paez in 1618.[23] "It is better to follow a received sentiment than to forge a chimera in order to bring something new to geography."[24]

In spite of this rational conservatism, Didier fully appreciated that a map is always in need of improvement and can never be definitive. As he said of his map of Saxony: "Our map is superior to those which French geographers have produced until now; it is left to the course of time to provide us with more certain clarifications."[25] Their trust in "more certain clarification" extended even to the criticisms they had endured: "Nevertheless, we know that works of this sort [our atlas maps] are susceptible to improvement; that is why we will gratefully profit from the clarifications so willingly given us; they will augment those which we already received after we made this map, even before we had published it."[26]

Didier's desire to improve closes the 'Préface Historique': "If our work seems susceptible to criticism and remarks...we will profit from them with pleasure. The form in which our faults are presented to us will not affect us; only the substance will draw our attention. May this work only occasion another superior to it! The benefit of the public will always be our greatest satisfaction and the sweetest recompense of our works."[27] Herein lies the core of the Vaugondy philosophy about their atlas. It was designed for the public and not for the scrutiny of geographers. "We only seek to give in our atlas what is most interesting for common use."[28] It was to be an illustrated guide for "studious people, those who read history as well as those who are only aware of current events," for whom the maps furnish a means by which "one might be transported to the scene and find oneself on the threshold of judging events and even of reflecting upon the future."[29]

Cartouches

If the Vaugondys determined to make their atlas useful, they did not neglect its beauty. A feature of the maps of the *Atlas Universel* which attracted unanimous praise from critics was the cartouches. The initial prospectus issued by Antoine Boudet, the publisher, said the cartouches would be designed and engraved by Babel, "célèbre par ses ouvrages d'ornemens."[30] In fact, none of the cartouches are signed by Babel, though he may have designed some. The greatest number seem to have been designed by one of Babel's students, Pierre Philippe Choffard (1730-1809), a prolific ornamentalist, who at age twenty probably was still working in Babel's atelier. In six existing notebooks of designs by Choffard are many cartouches that show marked similarities to those used in the *Atlas Universel*; yet although more than fifty cartouches in the atlas can be credited to Choffard, none were signed by him.[31] Choffard had trained early in his career with Dheulland, the map engraver who worked for Boudet in the latter's first foiled attempt at an atlas in 1746. Choffard later had lessons from Babel, and was responsible for both designing and engraving a great many map cartouches.

While the designs may have been his, most of the cartouches are engraved and signed by the Haussard sisters, twenty by Elizabeth, seven by Marie Catherine, ten signed simply "Haussard."[32] Of the seventy-one remaining cartouches, two are signed as designed and engraved by Moreau.[33] The Mappemonde [Cat. 8] is designed by C. Cochin filius, an eminent designer and engraver, and engraved by P. F. Tardieu.[34]

a

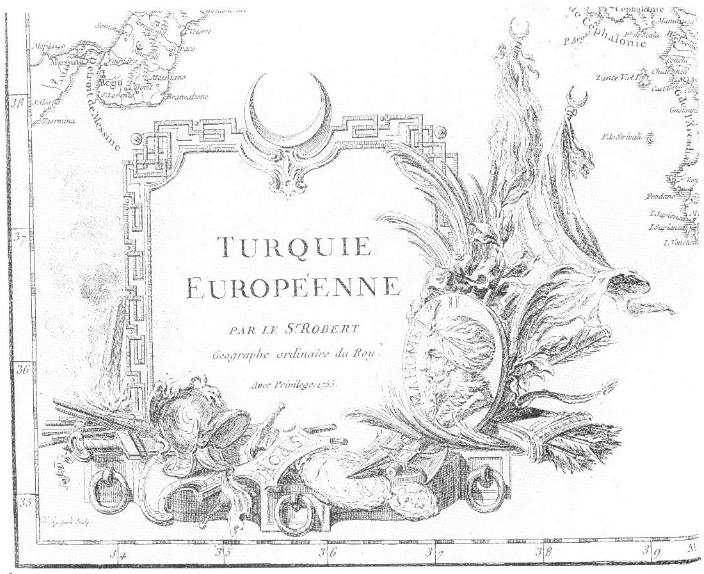

b

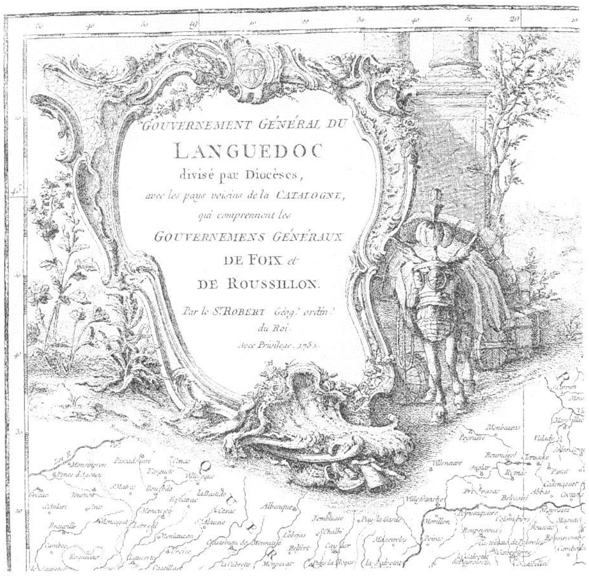

c

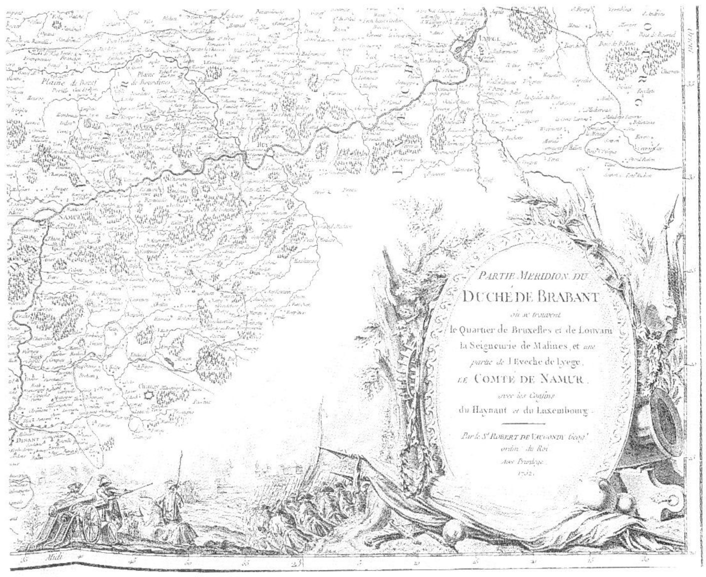

d

Various cartouches from the maps of the **Atlas Universel.** *Despite the several hands, the cartouches of the* **Atlas Universel** *are remarkably consistent in tone and restraint. They are not the lavish rococo designs found on the Sanson maps, nor do they splash great historical and allegorical scenes across the topography as on contemporary Homann maps. Rather, the title of a Vaugondy map is usually framed by flowers, swags, and symbolic elements which evoke some aspect of the place mapped. (Photos courtesy of the William L. Clements Library, University of Michigan)*

a) 'Carte de l'Egypte Ancienne et Moderne' [Cat.431]

b) 'Turquie Européene' i.e. Ottoman Greece and the Aegean Sea [Cat.367]

c) 'Gouvernement Général du Languedoc' [Cat.155]

d) 'Partie Méridionale du Duché de Brabant' [Cat.192]

e) 'La Judée ou Terre Sainte' [Cat.393]

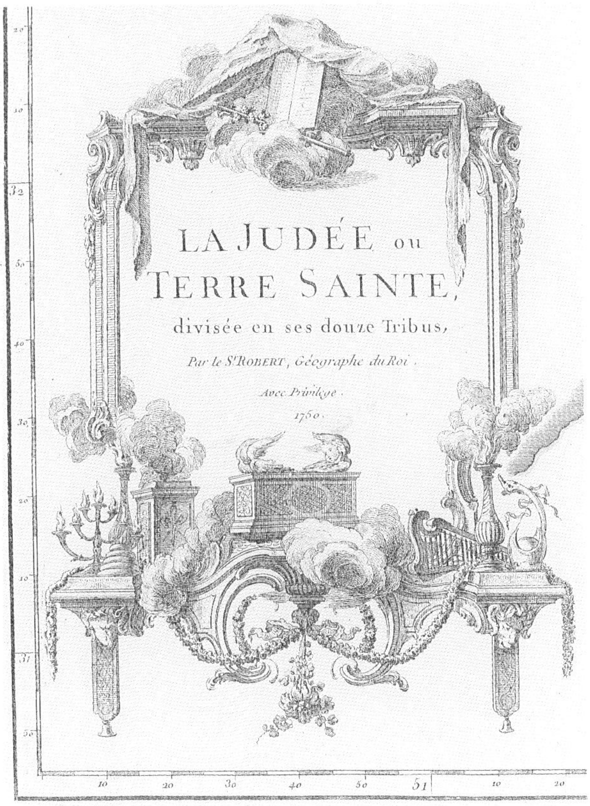

e

a

b

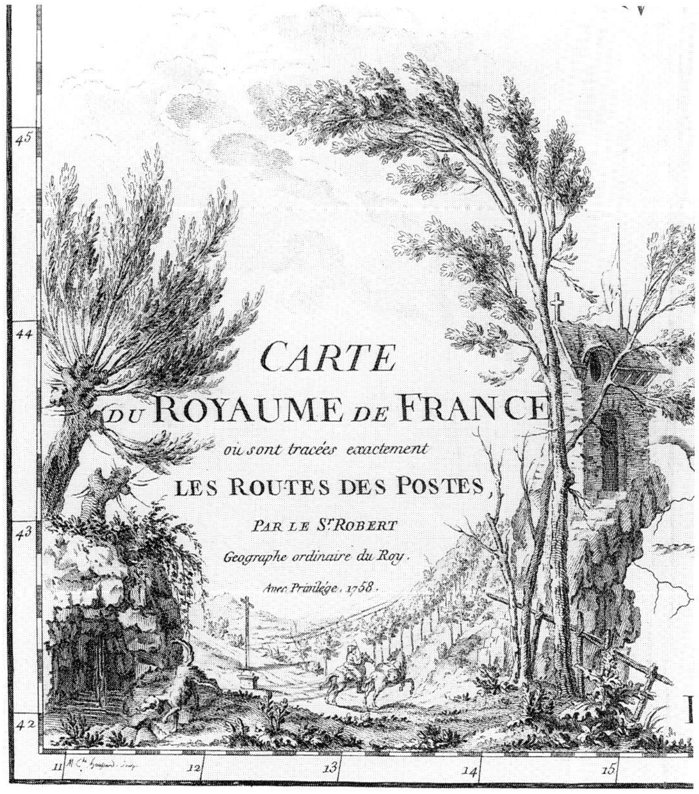

c

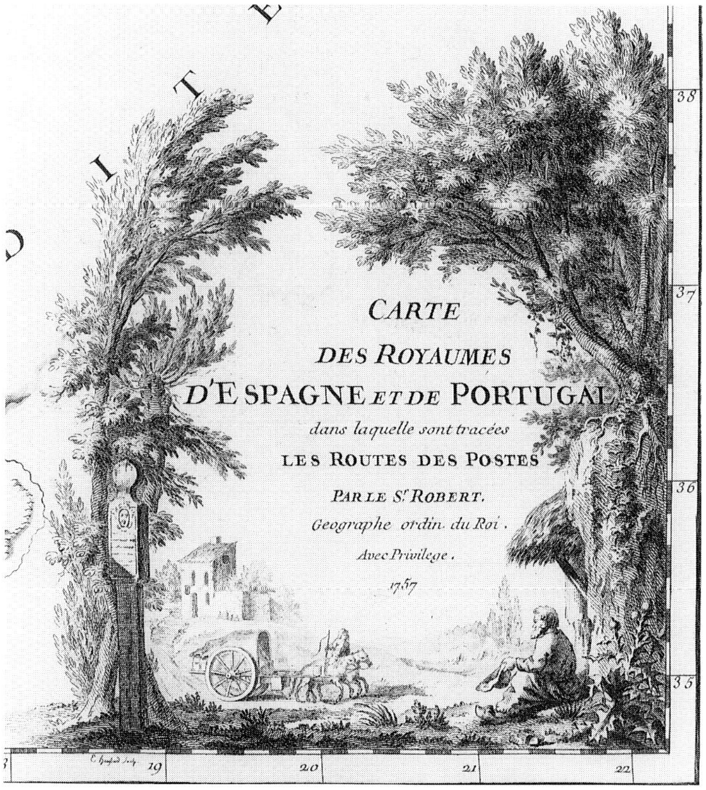

d

Cartouches from the road maps of the **Atlas Universel**. *They form a narrative of loading of the mail, dispatching it across country, the difficulties of terrain. Very often these maps, prepared as a unified post script to the atlas, are bound in the atlas in their correct geographical location, causing the reader to miss the cumulative effect of the cartouches. (Photos courtesy of the William L. Clements Library, University of Michigan)*

a) England, Scotland and Ireland [Cat.45] b) Germany [Cat.245] c) France [Cat.86] d) Spain and Portugal [Cat.303]

The cartouches are etched rather than engraved, and in later pulls have become much fainter than the engraved map.

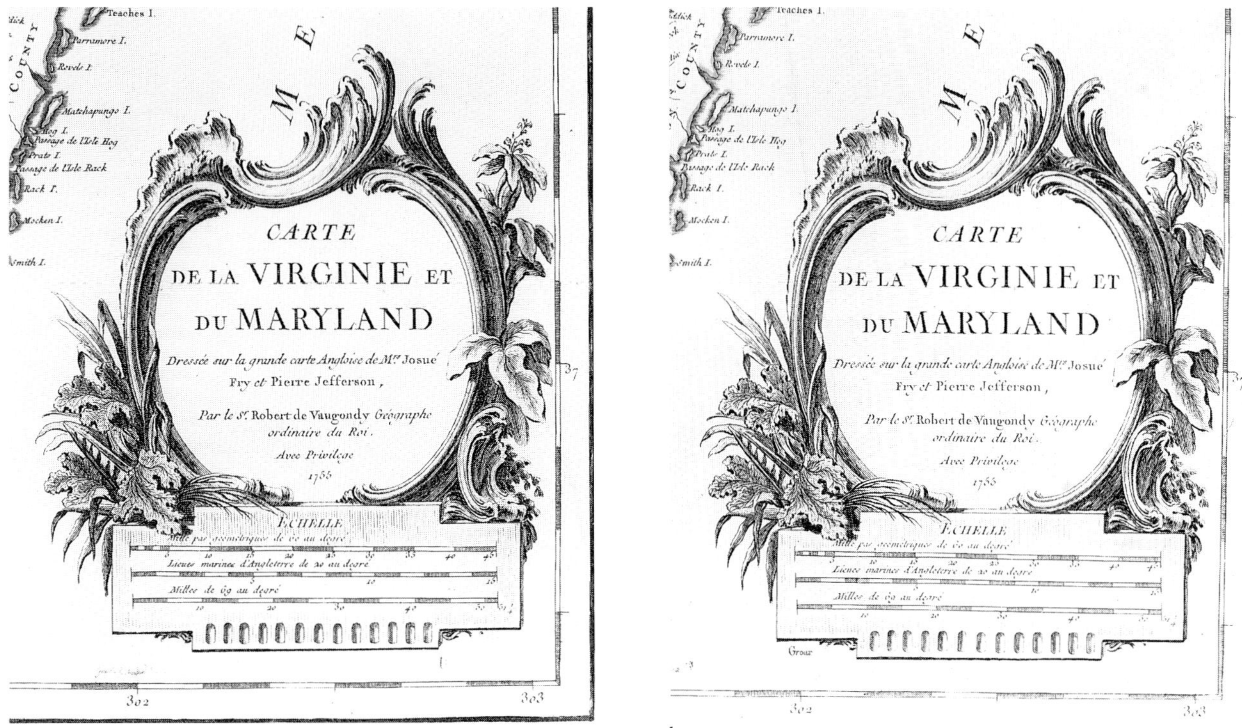

a b

Two states of the 'Carte de la Virginie et de Maryland' [Cat.470]: a) signature by Haussard b) signature by Groux. (Photos courtesy of the William L. Clements Library, University of Michigan)

Four cartouches are engraved by Gobin (Burgundy, Russia, Africa, South America, [Cat. 146, 405, 426, 487]. On the Burgundy map, Gobin's signature, "Gobin fecit" is replaced by "E. haussard sculp." on the second state. On the map of South America, the first state shows a cartouche with Gobin's signature, but a later state obliterates the signature with a greater elaboration of the cartouche. No biographical record of Gobin has been found, but he may have died around 1753, for there is no Gobin signature later than this date (the celestial globe gores of 1753) on Robert maps; his earlier work is on the Holy Land maps of 1745 and the continent maps of 1749.

A final point about the cartouches is the change of signature when the atlas was re-published by Delamarche. The previous signatures of Haussard, Moreau, *et al.* were erased and the cartouches signed by Groux. Both the identity of Groux and why he signed some of the cartouches, which were otherwise unaltered, remain a mystery. Little documentary evidence exists for Groux; Thieme and Becker list a Charles-Jacques Groux, son-in-law of the painter J.B. Noël, but no detailed reference is made to his work.[35] One can speculate that this otherwise unknown engraver seized the opportunity of claiming credit himself for the cartouches of both the *Atlas Universel* and the *Nouvel Atlas Portatif* whose title cartouches also display Groux's signature. What the relationship was between Groux and Fortin or Delamarche is not yet known. His signature, however, clearly dates any Robert map as later than 1778 when Fortin took over the Robert stock.

The *Atlas Universel*, "commode, complet, uniforme et suivi," survived an uncertain birth and came to be the choice of the public, as the Vaugondys had hoped. It belonged, however, to Boudet, so it is unclear how much the Vaugondy family profited from its sales.[36] It continued to be published by Vaugondy's successor, Charles Delamarche, well into the nineteenth century. As late as 1863 it was described as "though very much behind in geographical information, it has been for a long time very well thought of."[37] But the rancour engendered by the disputes between the Vaugondys and Buache was to sour the young Didier Robert de Vaugondy 's next geographical efforts.

References

1. The full title of the preface is the "Préface Historique dans laquelle on traite de l'origine, des progrès, et de l'état actuel de la Géographie." It re-prints Didier Robert de Vaugondy's 1755 work, *Essai sur l'histoire de la géographie, ou sur son origine, ses progrès et son état actuel* (Paris: Boudet) including the section about the publication of the atlas and the sources for all the maps.
2. 'Préface,'*Atlas Universel*, p. 19b.
3. Boudet to M. de Bombarde, 10 février 1763, quoted in H.J. Martin et al., *Histoire de l'Edition Française: Le Livre Triomphant 1660-1830* (Paris:1984): Vol. II. p. 276.
4. H.J.Martin, *loc. cit.*.
5. Details of this lawsuit are among the papers of Philippe Buache in the Bibliothèque Nationale, Paris: Ge.FF.13375. Buache's claim resulted from a lawsuit only recently won (1745) against Delisle's family (Guillaume's sister Angélique and brother Joseph-Nicolas) which granted him Delisle's proofs, plates, and privilege.
6. A more complete analysis of this subscription list in M. Pedley, "The Subscription List of the 1757 Atlas Universel: A study in cartographic dissemination," *Imago Mundi* 31 (1979), pp. 66-77.
7. The full text of these criticisms in BN: Ge.FF.13374; most are reproduced in M. Pedley, "New Light on an Old Atlas: documents concerning the Publication of the *Atlas Universel* (1757)", *Imago Mundi* 36 (1984), pp. 48-63.
8. "Notice Raisonnée des ouvrages de l'Auteur," *Mémoire sur les pays de l'Asie et de l'Amérique* (Paris: Boudet, 1774).
9. Copies of the affidavits and of Buache's own notes about the case comprise a manuscript now found in the Bibliothèque Nationale (Paris), BN: Ge.FF.13374. See note 7.
10. The father, Jean-Baptiste Delahaye, was an engraver as early as 1708, working for Guillaume and Adrien Sanson and for Pierre Moullart-Sanson. The son of a "compagnon de rivière," he became an engraver, taking his brother Mathieu as an apprentice. (AN: Minutier Central VII, 206, 20 août 1714) He had at least three children by his wife, Marie Denise Gervais de la Rivière, who is mentioned in the lawsuit as an engraver: Guillaume-Nicolas, Jean Baptiste Henri, and Marie Denise.
11. The full text in Pedley, "New Light on an Old Atlas," n. 7.
12. BN: Manuscrits, n.a.f. 12082 (18095), 8 may 1758, Guillaume Nicolas Delahaye married Anne Françoise Petremont.
13. BN: Manuscrits, n.a.f. 12131 (38381), 9 april 1755: "Pierre Laurent, engraver in geography, aged 20 years...and Marie Anne Mallet, aged 26 1/2, widow of Jean Baptiste Henri de la Haye, engraver in geography, both of the rue de la Petite Sonnerie, were married."
14. BN: Manuscrits, ms. fr. 22136 (20), 7 June 1755.
15. *Mémoires pour l'Histoire des Sciences et des Beaux Arts* [Mémoires de Trévoux], Juillet 1752, pp. 1520-1525.
16. *Atlas Universel*, "Préface,"p. 19b.
17. *ibid.*, p.27a.
18. *ibid*, p. 27b, referring to the map of the Circle of the Four Electors, Cat. 267.
19. *ibid.*, p. 26b.
20. *ibid*, p. 28b.
21. *ibid*, p. 33a.
22. *ibid*, p. 27a, of the map of Germany.
23. d'Anville, Dissertations, Academie des Inscriptions et Belles Lettres, 1752-54, pp. 46-63; cited by Broc, *Géographie des Philosophes*, pp. 103-105.
24. 'Préface,' *Atlas Universel*, p. 31b.
25. *ibid*, p. 27b.
26. *ibid*, p. 25a.
27. *ibid*, p. 33b
28. *ibid*, p. 29a.
29. *Atlas Universel*, preface to the list of subscribers.
30. Baron Roger Portalis and Henri Beraldi, *Les Graveurs du XVIIIe siècle* (Paris, 1880): Volume I., p. 68. P.-E. Babel (1720-1761), designer and etcher but also a jeweller, all in the rococo style.
31. *ibid*, pp. 412ff, pp. 459-477 for the catalogue of Choffard's cartouches.
32. The Haussard sisters were the daughters of Jean Haussard (b. 1696); they worked not only for the Roberts but also for Buffon, engraving the bird plates for the *Histoire Naturelle* (a work for which Didier Robert de Vaugondy also prepared the maps).
33. Languedoc, [Cat. 155], Orleanois, Beauce, Gatinois, [Cat. 112]. P. Moreau was an architect whose signature in rounded characters is very distinctive.
34. Pierre François Tardieu (1711-71) was a member of a well-known engraving family many of whose later scions specialized in geographic engraving. Portalis and Beraldi, *op. cit.*, Volume III, pp. 581-586.
35. Ulrich Thieme and Felix Becker, *Allgemeines Lexicon der bildenden Künstler von der antike bis zur gegenwart* (Leipzig: E. A. Seemann, 1922): Vol.15, p. 114.
36. Didier Robert de Vaugondy replied to an order for the *Atlas Universel* from the London firm of Jefferys and Faden in 1773. He told the English mapsellers that Boudet was the proprietor of the atlas and all inquiries must be directed to him. William L. Clements Library: Letters to Jefferys and Faden, no. 1, 26th August 1773.
37. Brunet, Jacques-Charles, *Manuel de Librairie* (Paris, 1863): Vol. 4, p. 1327.

CHAPTER FOUR

Patrons, Politics, and Polar Exploration

A treacherous reef one must carefully avoid and against which many authors are not sufficiently en garde is that particular prejudice, whether on behalf of one's nation or of some seductive ideas, to which one sacrifices all other knowledge and which one defends so sharply. Geography is one and by consequence true. Geography is degraded when it is used to support false conjectures or unjust pretensions... Certainly love of Country is no more an excuse for Geography than Geography's way of representing things creates a law for neighbouring nations.

—- Jacques Nicolas Bellin, *Rémarques sur la Carte de l'Amérique Septentrionale comprise entre le 28 et le 72 dégré de latitude, avec une description géographique de ces parties* (Paris: Didot, 1755): p. 5.

The struggles with the *Atlas Universel* were only a prelude to larger clashes in matters geographic. Both the lawsuits and the analytical publication of the sources of the atlas highlighted the issues with which eighteenth century geographers were wrestling. What do maps portray? Do they illustrate the world as it is known or as it is perceived by a particular interest? How does one assess the measurements of others? How does one determine the truth in a traveller's hyperbole? The questions of 'what is true' and 'what is knowable' were at the core of the eighteenth-century geographer's concern, and they were expressed in heated debate.

Didier Robert de Vaugondy's tenacity of mind and somewhat dogged character guaranteed continued conflict with his fellow geographers in these debates. In 1753, he was thirty years old and had already participated in three major geographical productions: the *Atlas Portatif* (1748), the *Atlas Universel* (1750-1757), and the globes for Louis XV. He had written the *Essai sur l'histoire de la Géographie* and articles for Diderot and d'Alembert's *Encyclopédie*: "Géographie," "Fuseau" and "Globes." He had attracted the attention of major Enlightenment authors — Buffon, Montesquieu and de Brosses — who employed him to illustrate their books. (See Chapter Five) Though young, he was ready to establish himself in the milieu of the Academy of Sciences and to seek patronage by dedicating maps to influential ministers.

The "Carte de Pays...de Canada" (1753)

While preparing maps for the *Atlas Universel*, Didier designed a map of Canada, not included in the atlas, which he dedicated to the comte d'Argenson, Louis XV's Minister of War. Among d'Argenson's other administrative responsiblities was looking after certain aspects of the city of Paris, including the Book Trade, the Royal Printing Works, the King's Library and the Royal Academy of Sciences. As a protector of that group of intellectuals in the French Enlightenment known as the *philosophes*, d'Argenson firmly supported the publication of the *Encyclopédie*, the first seven volumes of which were dedicated to him. D'Argenson certainly knew of Didier Robert de Vaugondy, having ordered the report on the Vaugondy globes in 1751. The globes illustrating Robert de Vaugondy's article in the *Encyclopédie* were presented to the comte d'Argenson's brother, the marquis d'Argenson, Minister of Foreign Affairs. In 1754, the comte d'Argenson became the arbiter of the lawsuits over the *Atlas Universel*. He was, for Vaugondy, a patron worth cultivating.

Didier Robert de Vaugondy's map of 1753, the "Carte des Pays / Connus sous le nom de / Canada / dans laquelle sont distinguées / les possessions françoises et angloises..." [Cat. 458], shows a part of North America which would have been very much on d'Argenson's mind: the area south of the St. Lawrence River known under the two names of Acadia and Nova Scotia. Upon publication, the map nearly provoked an international incident. The English objected strongly to the way in which Vaugondy

CARTE DES PAYS
connus sous le nom de
CANADA,
dans laquelle sont distinguées
les possessions françoises, & angl.s
DÉDIÉE ET PRÉSENTÉE
A MONSEIGNEUR LE COMTE D'ARGENSON
Ministre Secrétaire d'Etat
pour le département de la guerre,
par le Sr ROBERT DE VAUGONDY fils
Géographe ordinaire du Roi.
JAMES-BAY
ASSINIBOILS
Lac des Assiniboils
Lac des Cristinaux
LAC SUPERIEUR
Isle Royale
LAC HURON
Manitoualin Isle
LAC MICHIGAN
LAC ERIE
LAC ONTARIO
OUTAOUAIS
Algonquins
IROQUOIS DU NORD
PAYS DES IROQUOIS
RENARDS
MASCOUTEN
ILLINOIS
CAROLINE
Philadelphie
ECHELLE.
Mille Pas géometriques de 60 au degré
Lieues communes de France de 25 au degré
Lieues marines de France et d'Angleterre
A Paris chés l'Auteur, Quai de l'horloge du Palais
Avec Privilege 1755.

The 'Carte des Pays Connus sous le nom de Canada' (1753) [Cat.458]. Dedicated to the comte d'Argenson, Minister of War, this map was misinterpreted by the British delegation to the boundary commission on Acadia as a government-inspired attempt by the French to push back the boundaries of British Acadia towards the Atlantic peninsula. (Photo courtesy of the William L. Clements Library, University of Michigan)

A detail from Didier Robert de Vaugondy's "Carte des Pays Connus sous le nom de Canada" 1753 [Cat. 458] showing the problematic Acadian peninsula (modern Nova Scotia). Vaugondy shows "ACADIE" as only a coastal strip along the Atlantic. The English claimed Acadia all the way to the St. Lawrence River. The French were ready to concede the whole of the peninsula to them, but not the area between the peninsula and the river. (Photo courtesy of the William L. Clements Library, University of Michigan)

---

had shown French claims to Acadia/Nova Scotia. Only the eastern shoreline of the small peninsula had been left to the English.

At the time the map was published in August, 1753, the possession of Acadia was in hot dispute. A French and English boundary commission had been established in 1750 to determine more exactly the borders that the Treaty of Utrecht (1713) and the Treaty of Aix-la-Chapelle (1748) had left purposefully vague. The question was whether Acadia and Nova Scotia formed two distinct regions. Was Acadia the peninsula, and Nova Scotia the mainland between the peninsula and the St. Lawrence River? Or were they different names for the same large area? The Latin in the Treaty of Utrecht had been intentionally ambiguous when it told the French to cede to the English "all of Nova Scotia or Acadia."[1] The French viewed Acadia as a peninsula; the English saw Nova Scotia and Acadia as interchangeable names for the entire region.

When Didier Robert de Vaugondy's map appeared, the English boundary commissioners in Paris sniffed French propaganda: Acadia was shown as the peninsula only. Because of the dedication to d'Argenson, the English took the map to be an official publication. Their concern was aggravated by the advertisement for the map in the weekly *Mercure de France* which declared it to be "A New map of Canada dedicated to Mons. le comte d'Argenson, Minister and Secretary of State etc. The author has distinguished the French and English possessions on this map with great detail and exactitude, especially with regard to Acadia. It has been prepared based on papers that the Ministry had very much wanted to communicate in order to correct the article on this peninsula in the Dictionary of Trévoux."[2] Published

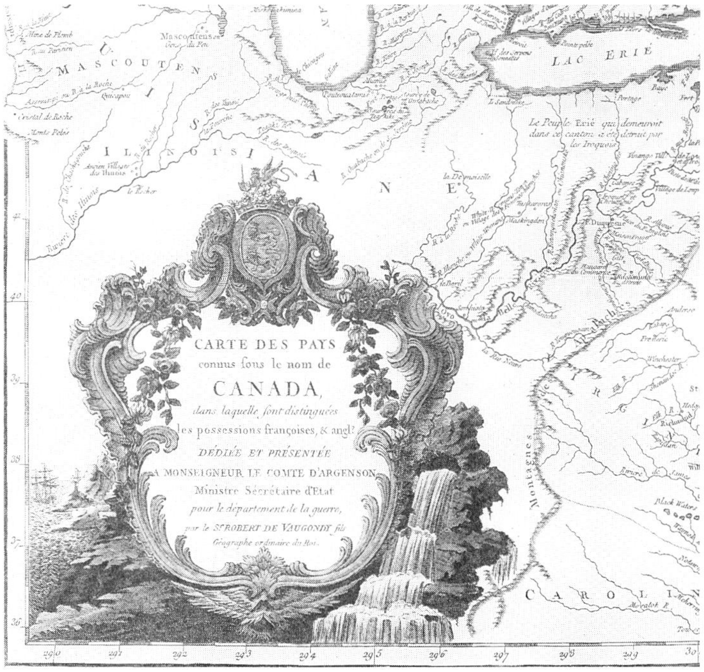

The Ohio River on the 'Carte des Pays...de Canada' (1753/5). Left *The title cartouche from the "Carte des Pays Connus sous le nom de Canada" (1753) [Cat. 458] from State 2 which shows the 'Belle Rivière' (the Ohio River) in a manner similar to that on a manuscript from the Dépôt de la Marine. Fort DuQuesne (modern Pittsburgh) was not on State 1 (1753); it was begun late in that year and completed in 1754. Its presence kindled the first great fire of the French and Indian War. To retaliate, the English built Fort Necessity, which was destroyed by the French in 1755; it is shown on State 2: "Ft. de Necessité détruite."*

Below *The "Cours de l'Ohio" from the Dépôt de la Marine. (Karpinski Collection 707; photostat of Service Hydrographique de la Marine 4044B 87) Note the positions of Ft. Du Quesne and Ft. Necessité along the river, built by French and British as disputes along the Appalachian frontier increased, leading finally to the French and Indian War. Fort Necessity was destroyed in 1755, giving the earliest date for the alterations made to State 2 of Vaugondy's map. (Photos courtesy of the William L. Clements Library, University of Michigan)*

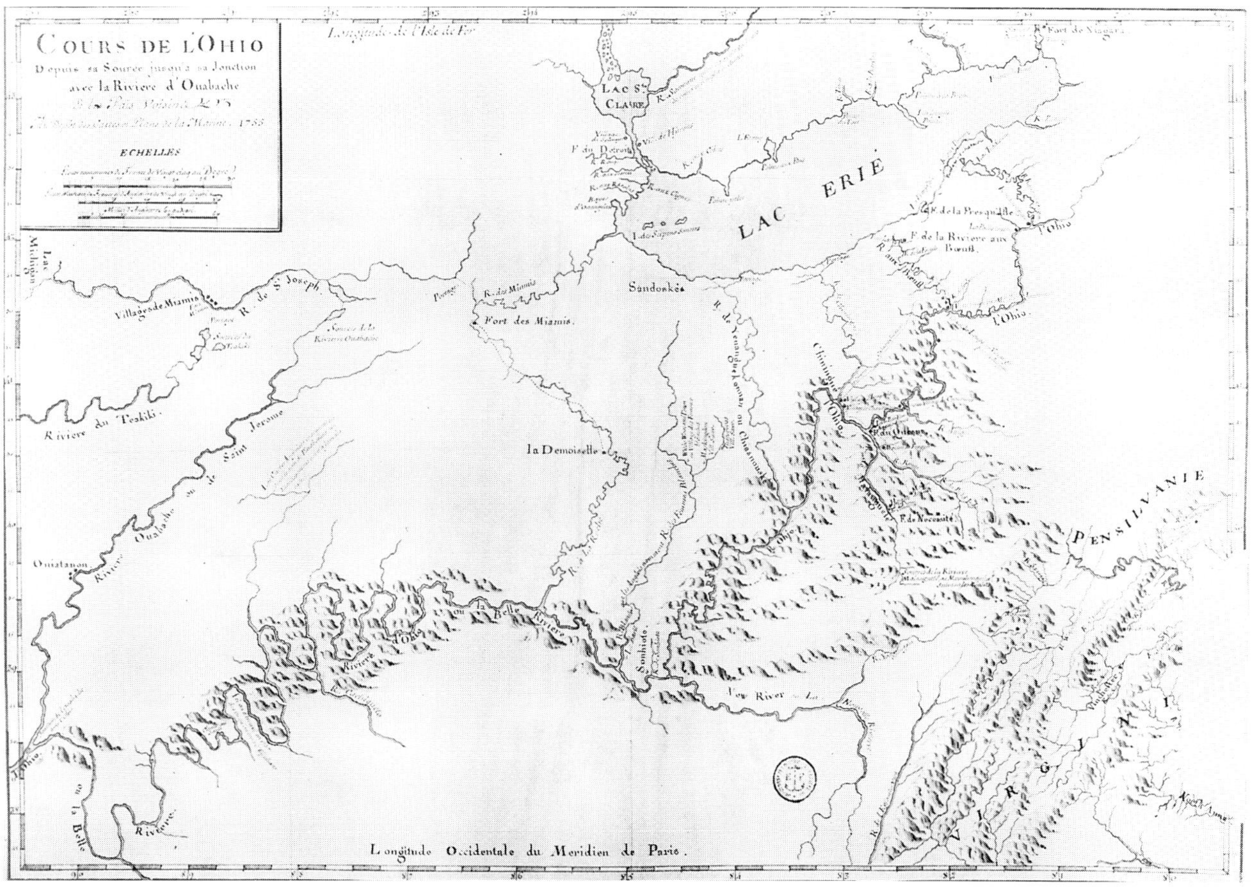

in August, the map drew such a strong reaction that by September Vaugondy had to insert a retraction in the *Mercure de France*, in which he denied receiving any official papers from the Ministry. In fact, the source for much of his information was from observations made by the marquis de Chabert at Louisbourg in 1750 and 1751, reported to the Academy of Sciences in 1752 and published in 1753; hardly state secrets.

Geographers on both sides of the Channel entered the fray. Thomas Jefferys, Royal Geographer in

London, chided Vaugondy for making Canada too large, objecting to the inclusion of the Great Lakes, the St. Lawrence, and parts of New England. For the English, "Canada" only applied to French possessions north of the St. Lawrence.[3] By contrast the French *premier géographe du roi*, Philippe Buache, now a familiar nemesis, scolded Vaugondy for not making Canada large enough by including the area west of Lake Superior. Vaugondy answered neither of his critics but did re-issue the *Carte du Canada* in 1755 with significant changes, not in Canada but along the Ohio River.

The second state (1755) of the "Carte du Canada" suggests Robert de Vaugondy's access to maps in the Dépôt de la Marine. The delineation of the Ohio River on both Didier's 1755 state and Bellin's 1755 edition of his Great Lakes map for Charlevoix' *Histoire* (first published, 1744) is based on a manuscript map of the Ohio River in the Dépôt de la Marine. A third state of the "Carte des Pays...de Canada" was published in 1778 by Jean Baptiste Fortin, who had bought the Vaugondy stock shortly before. He changed the title to "Carte du Canada et des Etats-Unis de l'Amérique Septentrionale," thereby making it one of the first maps (if not **the** first) to recognize the existence of the United States. Its publication must have followed the signing of the alliance between France and the fledgling United States on 6th February 1778.

Whether the "Carte des Pays... de Canada" garnered the ministerial patronage and protection Didier wanted is unclear, but the ripples it produced were nothing compared to the deep waters Vaugondy was to enter with his first presentation to the Academy of Sciences: the discoveries of Admiral de la Fuente.

The North American Discoveries of Admiral de la Fuente

In May 1753, Vaugondy read a paper to the Academy of Sciences entitled *Observations critiques sur les nouvelles Découvertes de l'Amiral de la Fuente.* This caused deep offence by disputing the putative discoveries of Admiral de la Fuente which had been presented to the Academy in 1752 by Joseph-Nicolas Delisle and Philippe Buache. The voyage of the Spanish admiral Bartholomew de la Fuente and the ensuing debate about its veracity have been described and documented many times.[4] As an apocryphal voyage, de la Fuente's travels in northwest America had a significant impact on both the English and the French in their continuing search for a Northwest Passage. The Spanish, however, had taken little note of their compatriot's adventures and this fact only added to the dubious nature of the account.

Admiral de la Fuente supposedly made a voyage in 1640 along the western coast of North America at the command of the Spanish Viceroy of Peru. No account of this voyage was ever published in Spain, but in 1708 a short report appeared in English in a popular London magazine, *The Monthly Miscellany, or, Memoirs for the Curious.* According to this account, de la Fuente was sent from Lima up the west coast of North America to search for a passage from the Pacific Ocean to the northern Atlantic. He left Peru in April, and by the 14th of June, 1660, he had reached 52° north latitude where, the story recounts, he met two Jesuit missionaries and named a river Rio de Los Reyes. There he split up his party, sending Captain Pedro de Bernarda northwards to explore while he himself continued in a north easterly direction along the Rio de Los Reyes. He came to Lac Belle, which supposedly had an Indian village on its shores and a peculiar cataract in its centre that caused the two rivers which had their source in this lake to run in opposite directions.

De la Fuente continued to the north east, up the Rio de Parmentiers to Lac de la Fuente (both named by the Admiral), a large lake well supplied with fish and islands, then on through more difficult countryside to Lac Ronquillo. There, at an Indian town, he encountered a certain Captain Shapley and his crew from Boston. After exchanging pieces of eight for Shapley's maps and charts, de la Fuente returned to Lac Belle. Here Captain Bernarda reported his discoveries to the north of Lac Velasco and Lac Bernarda, the latter nearly reaching the eightieth parallel. All that could be seen beyond this lake was frozen waste with icy mountains in the distance. On 5th September, de la Fuente and his crew returned to Peru, not to be heard of again until over sixty years later, when the *Monthly Miscellany* published the story, the only known account.[5]

The narrative attracted no attention whatsoever until 1744 when a certain Irishman named Arthur Dobbs used it in his *Account of Countries Adjacent to Hudson's Bay.* Dobbs' book encouraged private investors to search once again for a northwest passage to the Pacific. Dobbs had already mounted an

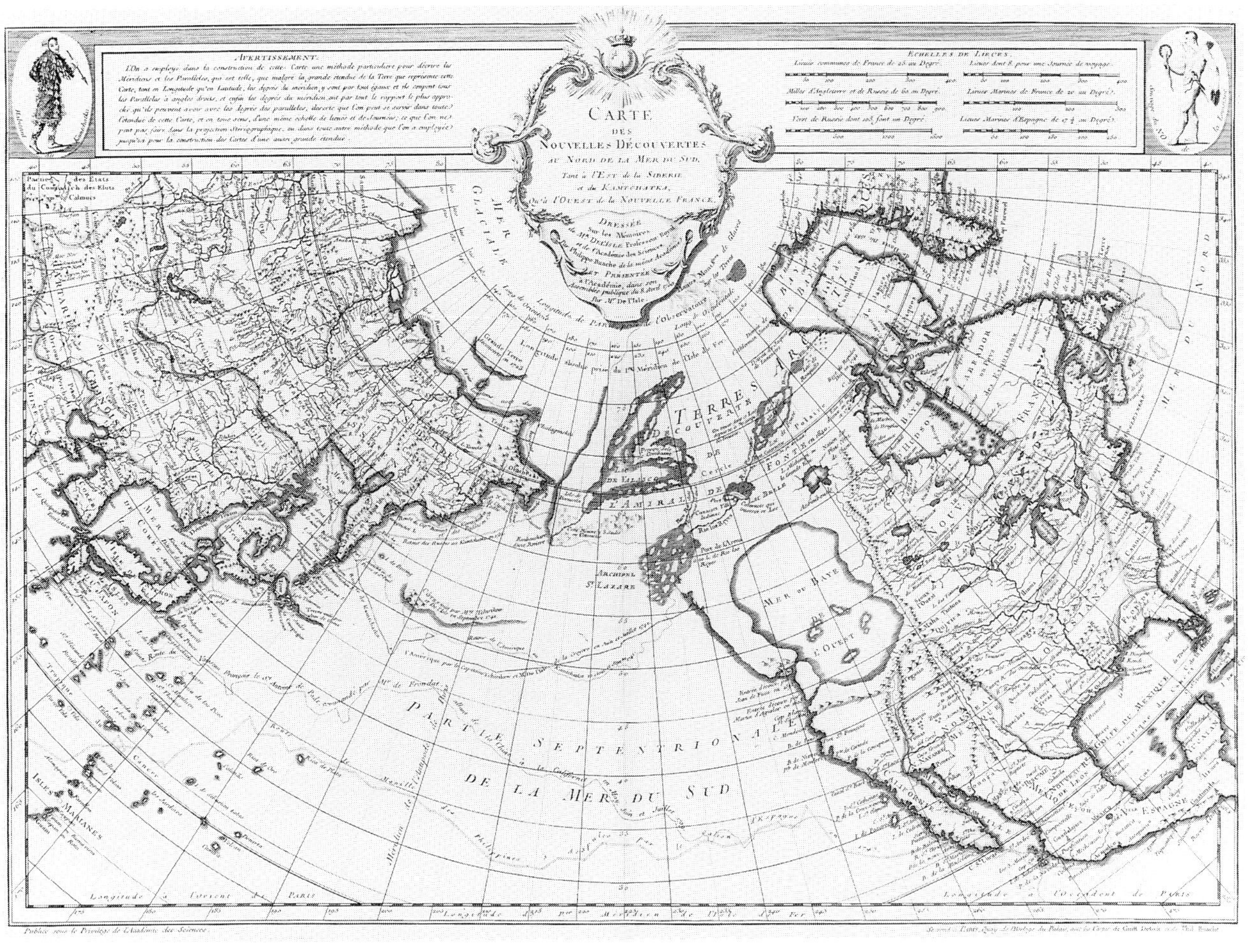

"Carte des Nouvelles Découvertes" by J.N. Delisle (drawn by Philippe Buache) in 1750. It shows the controversial 'Mer de l'Ouest,' a creation of Claude and Guillaume Delisle and north of it the putative discoveries of Admiral de la Fuente. The purpose of the map was to show the coherence of Russian discoveries with those of de la Fuente. (Photo courtesy of the William L. Clements Library, University of Michigan)

---

expedition led by Captain Christopher Middleton in 1741 which had failed to find the passage, and in 1746-1747 Dobbs tried again, sending Captain Henry Ellis to Hudson's Bay to search for the river leading to the Lac de la Fuente. Ellis' futile efforts were published in 1748 and both this account and Dobbs' work of 1744 were sent to Joseph-Nicolas Delisle in Russia.

Joseph-Nicolas Delisle (1688-1768), elder brother of Guillaume Delisle and uncle (by marriage) of Philippe Buache, had gone to Russia as an astronomer at the invitation of Peter the Great in 1726 to establish the observatory at St. Petersburg and an Academy modelled on the Academy of Sciences in Paris. During his tenure there, Delisle was largely responsible not only for the observatory but also the preparation of the *Atlas Russien* (1745), the first compilation of consistent maps of the kingdom. In addition, he looked after the organization of the final expedition led by the Dane Vitus Bering to north east Siberia (1734-1742). Upon his return to France in 1747, Delisle began to organize and digest the great mass of geographical and astronomical information he had acquired in Russia. On 8th April 1750 he presented a paper and map to the Academy of Sciences in Paris which caused an immediate sensation. The map, drawn by Philippe Buache, demonstrated the coherence of the Russian discoveries along the Aleutian islands and the Alaskan coastline with the 'discoveries' of Admiral de la Fuente.

While Delisle's paper was a success, his map was criticized for not representing the description of de la Fuente's voyage accurately. Delisle hired Jacques Nicolas Bellin of the Dépôt de la Marine to re-work the map, which was presented anew to the Academy in 1752. Buache drew his own map and presented it to the Academy in August of 1752 as part of a series of papers and maps which were ultimately published under the title *Considérations Physiques et Géographiques sur les Nouvelles Découvertes au*

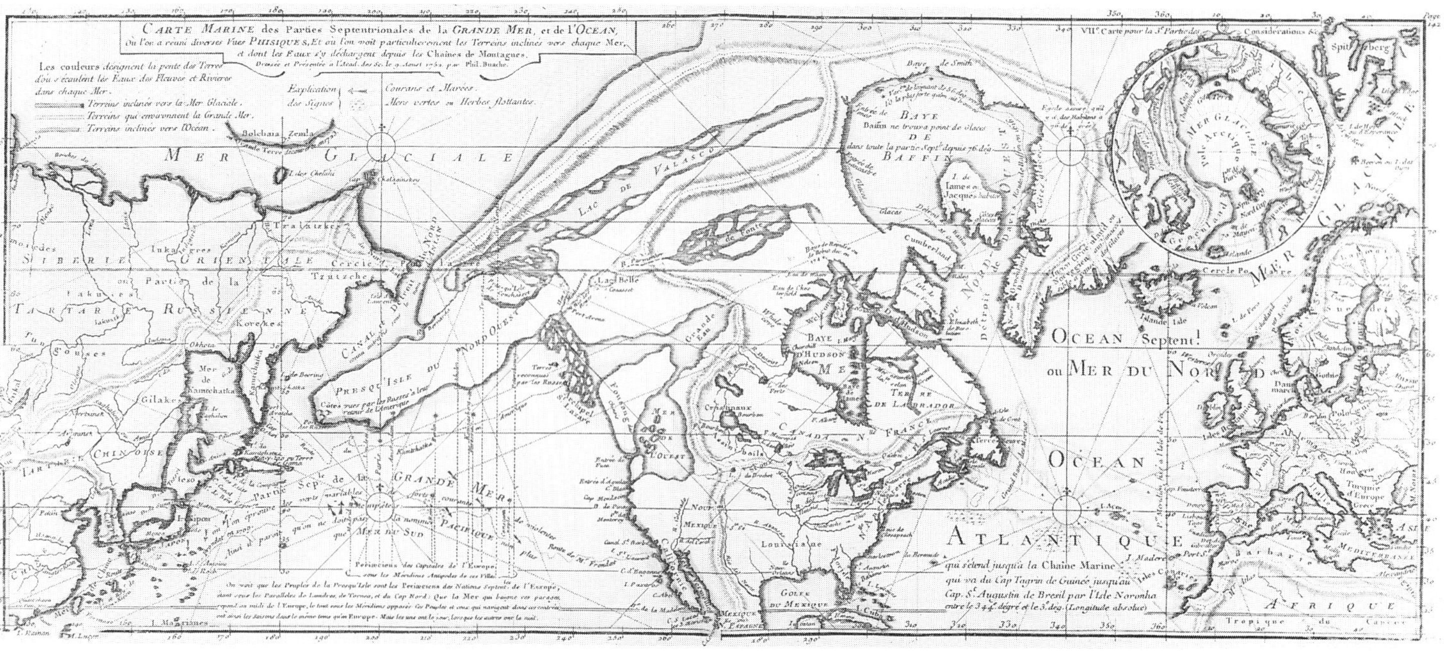

Philippe Buache's "Carte Marine" from the **Considérations Physiques et Géographiques** *(1753) showing his idea of a continuous chain of mountains, a concept supported by the de la Fuente geography of northwest America. (Photo courtesy of the William L. Clements Library, University of Michigan)*

Nord-Est de l'Asie et au Nord-Ouest de l'Amérique, avec les Mémoires relatifs. Buache included the de la Fuente discoveries in this major work of speculative geography, because he felt they supported several ideas about north American geography, particularly his theory of a continuous chain of mountains which ran both over land and under the sea around the earth.[6]

When Didier Robert de Vaugondy came to the Academy in the spring of 1753, he expressed his doubts about the truth of the de la Fuente explorations and his criticisms of the maps illustrating it. Didier distrusted the narrative on a number of counts: the long silence of the Spanish on the subject, the apparent absence of any Spanish original manuscript, the bizarre nature of the geography described by de la Fuente, and the speed with which the Spanish admiral had covered such a large amount of territory.

Robert de Vaugondy even wrote to an Italian friend resident in Madrid who replied that the Spanish thought the whole story to be a figment of British imagination and that no de la Fuente was mentioned in the archives of Signor Leyra, the Viceroy of Peru in 1640. However, the Jesuits had profited by using the story to ask for and receive men, money and help from the king of Spain to explore the California coast.[7] It was especially to present this reply from Madrid that Didier Robert de Vaugondy approached the Academy with his own paper, which received their approbation and was published shortly thereafter by Antoine Boudet as *Observations Critiques sur les Nouvelles Découvertes de l'Amiral de la Fuente...* (Paris: A. Boudet, 1753).

Buache responded sharply and in detail to Didier's paper the same day, which implies that he had seen it or at least heard the arguments beforehand, a likely possibility given the proximity of both geographers on the quai de l'Horloge. Buache maintained that the falsity of the de la Fuente document could not be shown by simple arguments; only "pièces authentiques" could show indisputably that it was false. If its falsity could not be proved, it was then gratuitous of Vaugondy to declare that the tale was fabricated in England. Finally, contrary to Didier's Madrid informant, Buache reminded the Academy that Antonio d'Ulloa, the Spanish officer who accompanied the La Condamine expedition to Peru in the 1740s, had told the Academy that he had heard the account of de la Fuente's journeys when in South America and that it had been found in the hands of pilots in Peru.

Buache's willingness to be convinced by the de la Fuente story, and his insistence that its falsehood, not its truth, be proved were reinforced by the editors of the *Journal des Sçavans* for September 1753. They were not persuaded by Didier Robert de Vaugondy's arguments, but expressed gratitude to Delisle and Buache for making the discoveries known, and gratitude to anyone who examined them with a

a

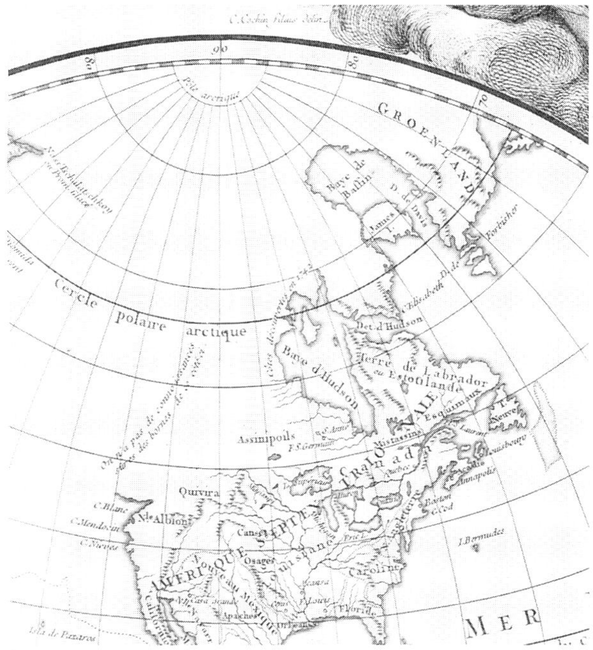

b

Two versions of the de la Fuente discoveries on Robert and Robert de Vaugondy maps. (Photos courtesy of the William L. Clements Library, University of Michigan)

a) Gilles Robert's Mappemonde of 1749, State 2, which has the 'Avertissement' concerning the de la Fuente discoveries added. (The engraving was apparently the work of Baisiez, one of the engravers involved in the **Atlas Universel** *lawsuit.) [Cat. 6]*

b) But Didier Robert de Vaugondy's Mappemonde of 1752 from the **Atlas Universel** *does not acknowledge the de la Fuente geography at all. "One has no certain knowledge of these coasts here." [Cat. 8] In 1782 on his map of North America [Cat. 449] in the* **Nouvel Atlas Portatif**, *Vaugondy placed the de la Fuente geography in an inset with a disclaimer: "We have traced them . . . without adopting them."*

"sage, judicious, impartial, disinterested critique." (p. 1905). Robert de Vaugondy responded with an article in Antoine Boudet's more sympathetic *Journal Œconomique* in November, 1753, in which he reiterated his views against the narrative. Buache answered in the December issue of the same journal; Vaugondy responded, and Buache made a final reply. This dialogue was published by Boudet in 1754 in two columns, the arguments side by side: *Lettre de M. Buache...sur les Nouvelles Découvertes, etc. / Réponse de M. Robert de Vaugondy.../ Réplique de M. Buache*, (Paris: Antoine Boudet, 1754).

This dialogue added nothing new to the debate, but the tone grew acerbic. Buache used the opportunity to criticise other Vaugondy maps and the young man himself. Robert de Vaugondy had questioned the use by Buache of the apocryphal "Sea of the West" (an invention of Claude and Guillaume Delisle) as support for the de la Fuente discoveries. Buache found this criticism "disrespectful" and felt that Vaugondy's doubts about the existence of this sea was characteristic of a "jeune homme." Vaugondy, in his turn, reminded Buache that Delisle himself had only been thirty-two years old when he presented his paper on the Sea of the West to the Ministry, saying that "one should not be surprised to see a young man speak on a matter that another young man has discussed." Buache accused Vaugondy of only looking for glory for himself "by fighting people of rank in the republic of letters," of only raising doubts without establishing anything of truth, particularly when "this is an important question in a geographical matter which has equally to do with History and Religion." In Buache's eyes, Vaugondy was not only a young *parvenu*, he was also in league with map-selling bookdealers stripping profits from other geographers, and worse, he might even be a heretic. The Jesuit periodical, *Mémoires de Trévoux*, had published only the month before an argument that a land or water connection between North America and Asia would explain how the Americas came to be populated and would be consistent with Biblical accounts of the Flood and sons of Noah.

Robert de Vaugondy's scepticism, and his reliance on reason rather than religious argument, no doubt endeared him to other exponents of the French Enlightenment and to those in foreign quarters as well.

In London, John Green (*alias* Braddock Mead) published *Remarks on the New Discoveries in America* (London, 1753); and in St. Petersburg, G. F. Müller, a German astronomer not on friendly terms with Delisle, published *Un Lettre d'un Officier de la Marine Russienne* (Berlin, 1753). Both joined Vaugondy against the de la Fuente relations.

Nor did other French geographers immediately adopt the de la Fuente discoveries. D'Anville did not show them on any of his North American maps. J.N. Bellin, though he had prepared the map of the discoveries for Joseph Nicolas Delisle, dismissed it on his own "Carte réduite des parties septentrionales du globe situées entre l'Asie et l'Amérique" (1756) with the words, "It is in this region that some geographers have placed the presumed discoveries of Admiral de la Fuente but I have found the story too suspect to use." Vaugondy had a similar warning note engraved on the second state of his father's mappemonde of 1749: "Some authors put here the lands which they say have been discovered by Admiral de la Fuente in 1640. But the lack of proof and the absolute ignorance about this Admiral in Spain strips any credence from his reports. We know that a little while ago the Spanish Jesuits asked for men and money from His Catholic Majesty in order to pursue the matter; we await the result of this next expedition." In this way Bellin and Vaugondy recognized the existence of new information without accepting it.

The rejection of the de la Fuente report did not mean an end to the search for a northwest passage or indeed a northeast passage. The problems surrounding the *illustration* of the narrative, whether true or not, inspired both Buache and Vaugondy to experiment with maps as illustrations of geographical ideas that were not necessarily reality. Buache's own practice of this type of cartography in his *Considérations Géographiques et Physiques* were perhaps an inspiration to Vaugondy.

The Mappemonde in Four Sheets of 1760

Over the next twenty years, Vaugondy experimented with the mapping of northern regions, placing conjecture next to "reality." One of his more unusual results was the four-sheet Mappemonde of 1760. Consistent with his presentations to the Academy, Didier Robert de Vaugondy did not show the de la Fuente discoveries on this world map. More surprising, however, is the presence both of Guillaume Delisle's "Sea of the West," which Didier had rejected as apocryphal, and an unusual land connection between North America and Asia. Further, the usual southern continent or the suggestion of one is missing. The Sea of the West lies between 45° and 49° north; running from it northeast towards Hudson Bay are these words: "One puts the Sea of the West in this place." The continents of Asia and North America are connected by an isthmus of land which joins Siberia at 120° east longitude to a huge polar extension of North America. The isthmus is annotated, "which could give a link to America by way of the polar lands." By using such inscriptions, Robert de Vaugondy distanced himself from the features illustrated.

He had discussed this land link between North America and Asia in the "Préface Historique" of the *Atlas Universal* in the section on the map, "Orbis Vetus": "It would appear probable that towards the pole there is a continent which communicates from Asia to America and which makes up part of Greenland and the lands of Baffin; that the tongue of undefined land situated at 120 degrees longitude and 77 degrees latitude presents an isthmus, which is truly joined to some coasts whose bearing should run from east to west, forming two gulfs of the northern ocean. The land seen to the north of Asia in 1723 would appear to be part of the eastern coast which, in rounding the easternmost Cape of Asia comes to be joined to lands discovered by the Russians and to California. The west coast of the isthmus seems to have to run to the north of Spitzbergen and reaches the coast of Greenland. It is evident that this isthmus would have been sufficient for the juncture of the new world to ours; that the people of the former are descendents of Japhet rather than of Shem."[8] Yet Vaugondy did not show this link on his map "Orbis Vetus" in the *Atlas Universel* nor on any other world map prior to or after 1760. On later states of the four-sheet Mappemonde (1775, 1786), the Asia/North America connection is erased and the polar regions are shown free of any land mass. Vaugondy seems to be experimenting with a map as the representation of a system or hypothesis. Eight years later he turned these same energies to "systematic" maps of the northeast passage.

The Polar Geography of Samuel Engel

Didier Robert de Vaugondy's interest in polar cartography was furthered by the work of Samuel Engel (1702-1784), a Swiss geographer in Berne. Engel's first "northern" paper, published in 1735, developed ideas about a passage from the Pacific to the Atlantic by the Arctic Sea. He further expanded them in his *Mémoires et Observations géographiques et critiques, sur la situation des pays septentrionaux d'Asie et d'Amérique* (Lausanne, 1765). Engel rejected the de la Fuente discoveries, repeating many of Didier's arguments. He believed that the most sensible route from the Atlantic to the Pacific was along the north coast of Siberia. Engel claimed that this coast was free of ice for a good part of the year and that its length was only about 500 leagues. Engel maintained that the breadth of Siberia was considerably less than that shown in the *Atlas Russien*, which had widened Siberia, making the Russian empire seem larger than it was in reality.

Engel's *Mémoires et Observations* contained a map of the northern Pacific similar to those of Joseph Nicolas Delisle and Philippe Buache. However, Engel truncated the east coast of Siberia to 175° (from the Ile de Fer) and pushed out the west coast of North America to 180° in order to accommodate all the Indian reports of lakes, rivers, and mountains in this continent.

Didier Robert de Vaugondy adopted Engel's "Système" and in 1768 published his "Carte Systèmatique des Pays Septentrionaux de l'Asie et de l'Amérique, dressée...d'après les mémoires et Observations Géographiques et Critiques de M*** publiés à Lausanne...1765." [Cat. 451] This map accompanied his *Lettre de M. Robert de Vaugondy à M*** [Engel] au Sujet d'une Carte Systèmatique des pays Septentrionaux* (Paris: Boudet, 1768). Robert de Vaugondy's letter expressed admiration for Engel's work and suggested some small changes on his map, namely that the east coast of Asia (Siberia) should end at 185°, not 175° longitude (in accordance with d'Anville's maps for Du Halde's *Atlas de Chine*), and that the west coast of North America (the Côte d'Anian) end at 190° instead of 180°. Vaugondy closed his letter with his explanation of a 'systematic' map: "I desire nothing so much as to see the system take the place of reality: but our information is subject to too much uncertainty. We would be deliberately allowing ourselves to be carried away to wish to present speculations, however well founded, as mathematical truths." [p. 8] Didier's "Carte Systèmatique" and his aptitude for illustrating hypotheses must have caught the attention of Charles Joseph Panckoucke, publisher of the *Supplément* of the *Encyclopédie*. The *Supplément* consisted of four volumes of text, one of plates, and a *Table Alphabétique* for the whole of the *Encyclopédie*. The first seventeen volumes of the *Encyclopédie* had been criticised for their cursory treatment of geography. When Panckoucke took over its publication in 1768, he promised that "the discoveries made in the last 25 years and those previously omitted would be added; other omissions and errors would be corrected."[9]

Panckoucke employed Samuel Engel to write articles about the northern regions and Didier Robert de Vaugondy to prepare a series of ten maps to illustrate them. They specifically accompanied the articles "Amérique Septentrionale," "Asie," "Californie," "Mer de l'Ouest," and "Passage par le Nord." The maps, found in the *Suite du Recueil des Planches, sur les Sciences, les Arts Libéraux, et les Arts Méchaniques, avec leur explication* (1777), are dated 1772 and 1773. Didier signed six of them, though it is likely that he oversaw the engraving of the other four, which are copies of maps from Buache's *Considérations Géographiques* and Captain Cluny's *American Traveller*; another is a composite view showing changing ideas about the coast of California, drawn from various cartographers. [Cat. 402, 406, 452-455, 460, 462, 473, 474]

The duc de Croÿ and the 'Hémisphère Australe'

As Vaugondy was preparing these northern maps, his interests turned south when he met the duc de Croÿ (1718-1784), Maréchal de France, in 1772. De Croÿ represented the segment of eighteenth-century French nobility that personified the Enlightenment in the range of his intellectual pursuits. A soldier of distinction (his rank of Maréchal rewarded his defence of the Pas de Calais), he was well travelled and well read. An amateur scientist, he had an observatory installed in his chateau at Châtillon and for seven years assembled notes for his own *Histoire Naturelle, considérée en physicien, chimiste et naturaliste.*

TERRES POLAIRES ARC
BAYE DE BAFFIN
MER DU NORD
Cercle polaire arctique
MER DU SUD OU PACIFIQUE
OCEAN
OCCIDENTAL
Tropique de Cancer
GOLPHE DE MEXIQUE
Antilles
MER NOIRE
Ligne equinoctiale ou Equateur
OCEAN
MERIDIONAL
Tropique de Capricorne
Cercle polaire antarctique

Didier Robert de Vaugondy's Mappemonde of 1760 [Cat. 10]. This large, four-sheet map displays some unusual features: the connection of North America, Europe, and Asia to an arctic land mass is the most striking, as is the complete absence of an antarctic continent which had been a regular feature on world maps since the sixteenth century. (BN: Ge.A.1229) (Photo courtesy of the Bibliothèque Nationale, Paris)

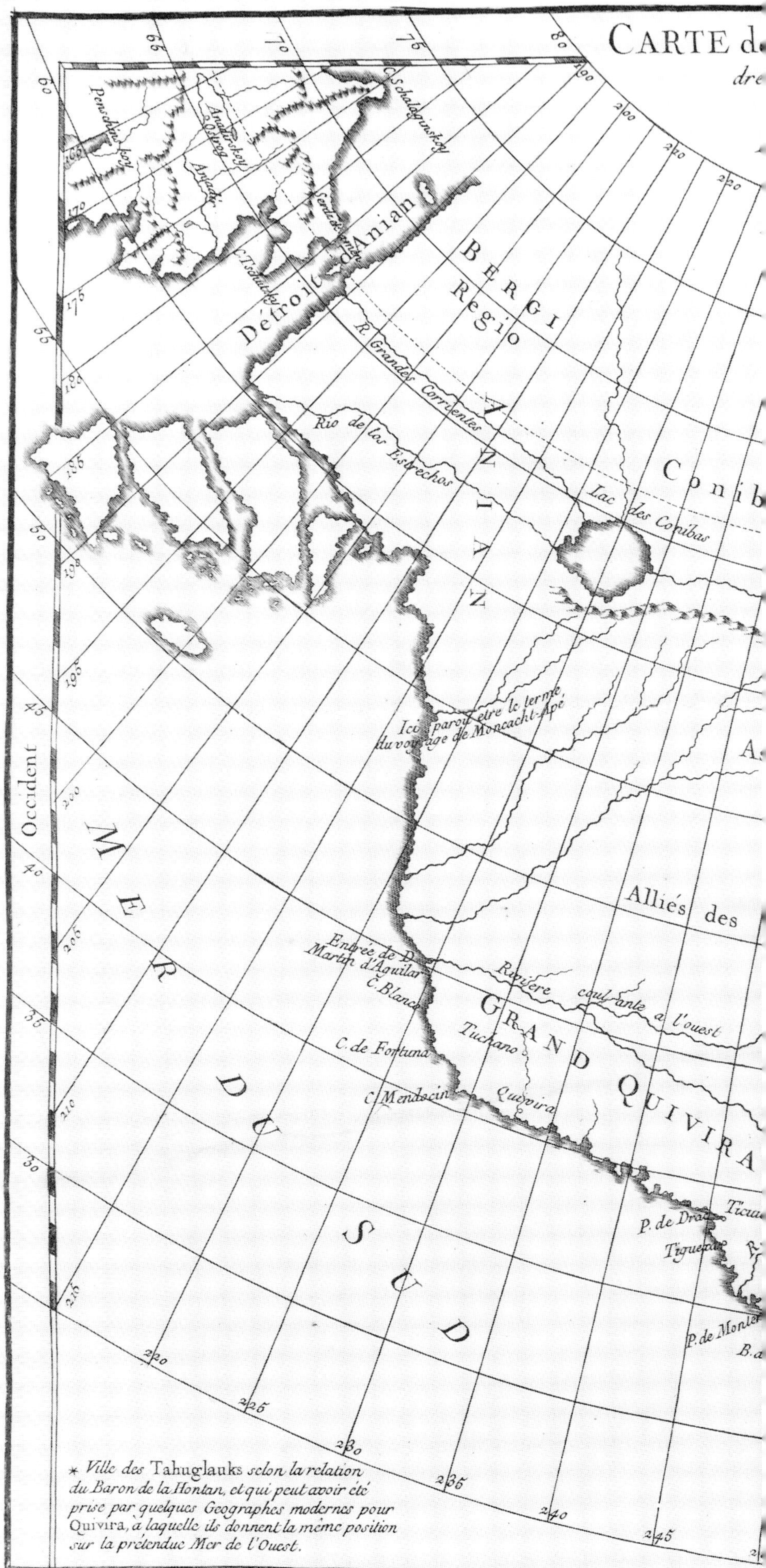

Didier Robert de Vaugondy's rendition of Samuel Engel's map of 1764. [Cat. 453] It was published as part of Charles Panckoucke's **Supplément** *to the* **Encyclopédie** *in 1772. Vaugondy altered slightly Engel's Asian and American coasts. (Photo courtesy of the William L. Clements Library, University of Michigan)*

Septentrion

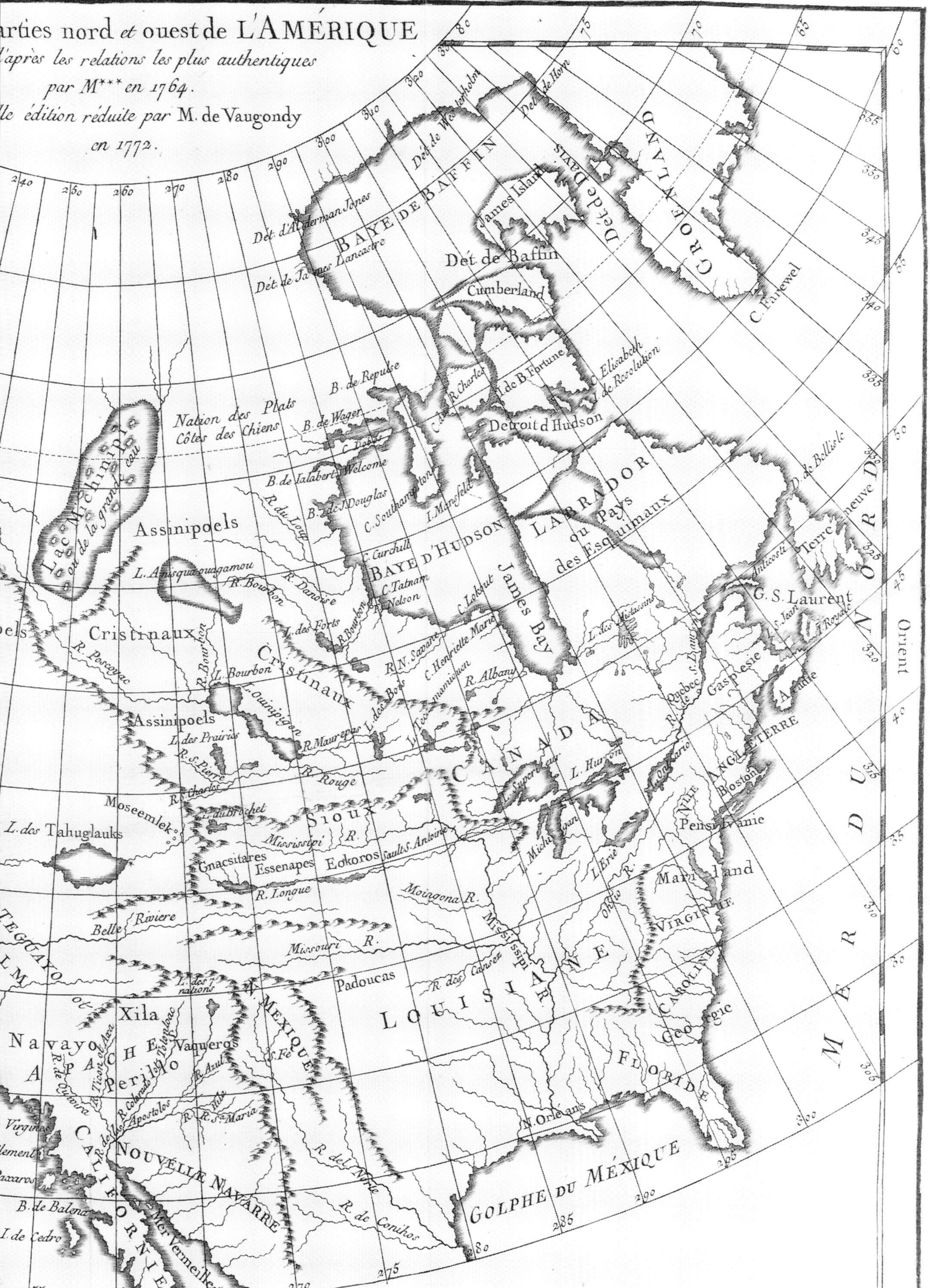

His large map collection was not just limited to the interests of a military man. Exploration was one of his passions which he combined with an intense patriotism. The discoveries made by Bougainville and Cook on their first voyages persuaded him that France ought to undertake more exploration in the southern hemisphere, and he gave financial support to the second voyage of Kerguelen in search of the 'Southern Continent' (1773). In an effort to illustrate for the King and his administration both Kerguelen's route and to make the case for southern exploration, de Croÿ set about to learn as much as he could of the geography of the southern hemisphere. He made the acquaintance of Bougainville and met several Paris geographers, but "found in M. Robert de Vaugondi [sic] the one who has the best globes and who understands them the best."[10]

De Croÿ owned a ten-inch Vaugondy globe for which he designed a new mounting that eliminated the meridian circle. To replace it Didier Robert de Vaugondy designed a copper instrument of three hinged quarter circles, which he called an azimuthal. By setting this device on the horizon, two of its branches opened to form a meridian and could be moved in any way on the globe while the globe itself could be moved in any direction in its mounting. Being able to manipulate his globe in this more flexible way led the duc to conceive a distinctive map of the southern hemisphere drawn on an oblique projection which would show places not usually visible on polar projections. Didier Robert de Vaugondy helped him prepare this unusual "Hémisphère Austral ou Antarctique..." of 1773. [Cat. 27][11]

The oblique projection allowed ports north of the equator to be included on the map, places such as Madras, Pondicheri, Canton, that were important to navigators. The duc also wanted the shape of Europe and its cities to be drawn on the map as the Antipodes, a device both Gilles and Didier had employed on earlier maps. This would show the King the relationship of places and give some idea of scale in the faraway south Pacific. So careful was Vaugondy in preparing this map for publication that he traced the projection onto the copperplate himself "in order to avoid the errors of tracing which, combined with the effects of printing, can cause some faults in precision."[12]

Didier Robert de Vaugondy had already demonstrated his skill as an engraver on his celestial maps of 1762 (see Chapter Six) and in supervising the engraving of the maps for the *Atlas Universel*. On this map, he used line gradations to distinguish places in the south Pacific: a heavy line for those with certain astronomical positions; a thin line for "less certain" places with no observed longitude, and a dotted line for dubious locations for which neither longitude nor latitude had been determined. Such distinctions allowed him to move New Zealand, a "less certain place," eight or nine degrees from its usual position to match the outline of Spain.

Robert de Vaugondy worked steadily with de Croÿ on the map from January 1773, bringing proofs for correction every fifteen days while the duc assembled the research on the voyages and checked geographical information. Vaugondy presented the map to the Academy of Sciences in March 1773, but advised his audience that it was unfinished as the duc de Croÿ wished to add more discoveries to it as they occurred. The duc continued his researches until February 1774 when he gave his corrections to Vaugondy for engraving.

"Hémisphère Austral ou Antarctique...dressé sous les yeux de M. le duc de Croÿ, par le Sr. de Vaugondy...1773" [Cat. 27]: State 2, with the voyages of Cook. The slightly tilted oblique projection allows areas of southeast Asia and India north of the equator to be shown. The map was designed to encourage the French to more exploration in the southern hemisphere. (Photo courtesy of the William L. Clements Library, University of Michigan)

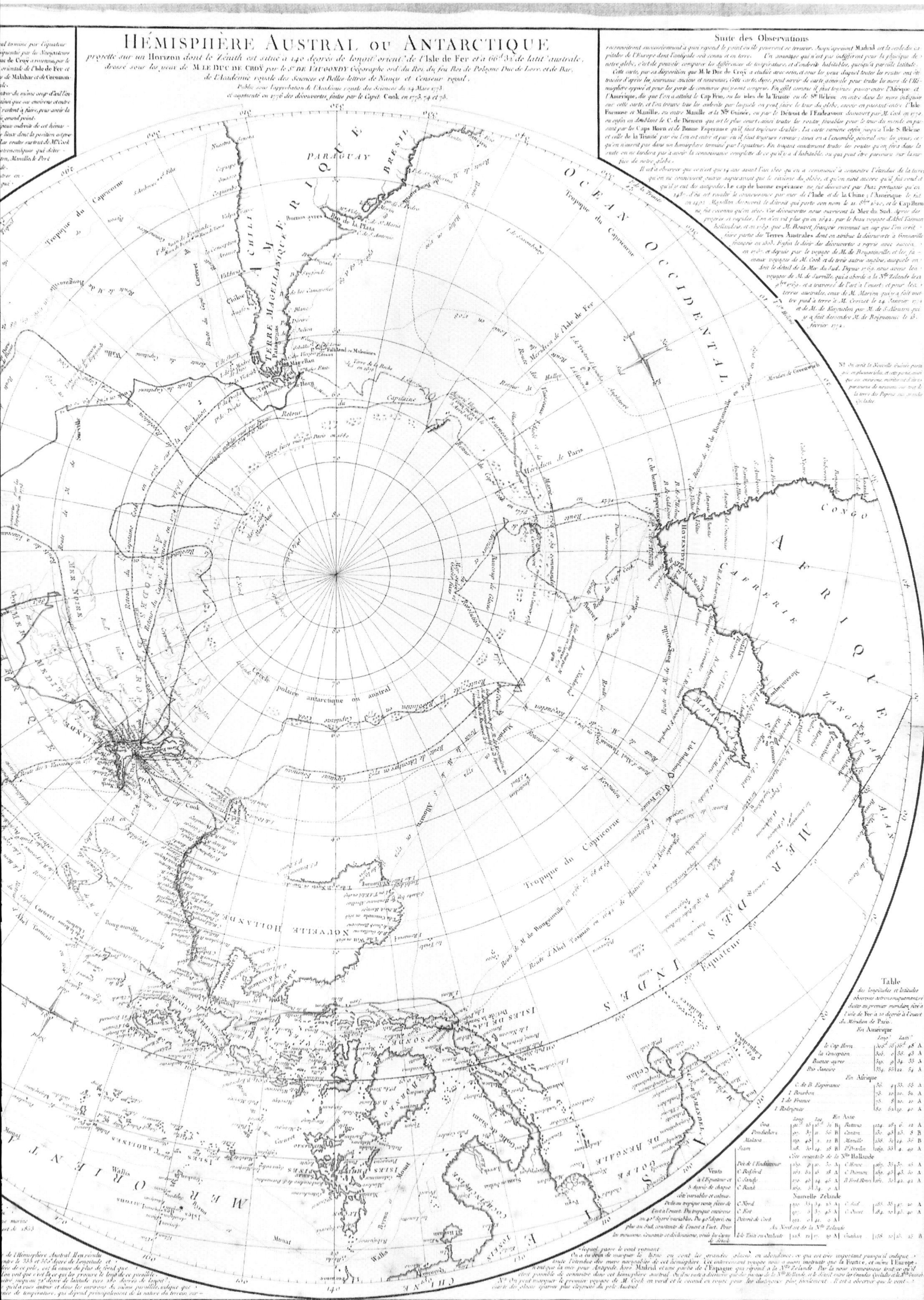
HÉMISPHÈRE AUSTRAL ou ANTARCTIQUE
projetté sur un Horizon dont le Zénith est situé à 140 degrés de longit. orient. de l'Isle de Fer et à 66.d 32' de latit. australe,
dressé sous les yeux de M. LE DUC DE CROŸ par le S.r DE VAUGONDY Géographe ord. du Roi, du feu Roi de Pologne Duc de Lorr. et de Bar,
de l'Académie royale des Sciences et Belles-lettres de Nancy et Censeur royal,
Publié sous l'approbation de l'Académie royale des Sciences du 24 Mars 1773
et augmenté en 1776 des découvertes faites par le Capit. Cook en 1773, 74 et 75.
Suite des Observations
PARAGUAY
BRESIL
OCEAN OCCIDENTAL
AFRIQUE
CAFRERIE
CONGO
MER DES INDES
NOUVELLE HOLLANDE
Tropique du Capricorne
Table

When the map was finally completed, de Croÿ brought both Vaugondy and the map to Versailles. Late in the evening of 15th April, they presented the map to the Minister of the Navy, M. de Boynes. The duc was somewhat anxious lest de Boynes or his "jealous Bureau des Plans" should take umbrage and think the map was encroaching on naval territory. De Boynes, however, was so enthusiastic about the map that he approved its presentation to the King Louis XV. It was then midnight; Vaugondy and de Croÿ had to wait until nearly two in the morning before being admitted to the royal presence. De Croÿ held back, allowing Vaugondy to make the presentation, during which the King was much taken by the display of the European Antipodes, though to de Croÿ's mind he did not really apply himself to the whole *ensemble* of the map.

From the King, they went to show it to the three princes, the future Louis XVI, Louis XVIII, and Charles X. The Dauphin was at breakfast before going hunting. Recorded de Croÿ: "We spoke of hunts and tumbles, then I presented Vaugondy and the map. The Dauphin understood the important things completely, but the hunt called and objects of science not being fashionable, [the interview] was short." Leaving the princes, they met the duc d'Aiguillon, Minister of Foreign Affairs, who examined the map "like a connoisseur." Although Didier was not used to such all night visits ("il n'avait jamais fait de telles séances"), de Croÿ was well pleased: "Our presentation succeeded well and gave us every liberty to make this curious piece of geography public, which will be for a long time the basis of this part [of the world]."[13]

A first state of the map was given the approbation of the Academy of Sciences on 24th March 1773, but may be dated 1774. De Croÿ supervised a later edition of his map in 1776, published in 1777, which shows the discoveries of Captain Cook between 1773 and 1775. By this time, however, relations between the duc and the geographer had grown strained: "...Vaugondy enrages me...I have to have great patience...", "...he is accustomed to making haste slowly...", "...he annoys me often."[14] It was probably Didier's painstaking, deliberate nature which irritated him, a man impatient to finish his map and turn to other projects. Nonetheless, they corresponded and Didier continued to send de Croÿ copies of his maps, asking for comments.[15]

The Map of the North Pole

For all his efforts on behalf of the de Croÿ vision of the southern hemisphere, Vaugondy did not abandon his own ideas about polar exploration. After his March presentation to the Academy of the de Croÿ map and explanatory paper, he presented another in August 1773: the "Essai d'une Carte Polaire Arctique," [Cat. 25].

This paper, which remained unpublished, concerned the polar lands and the means of navigating in the northern sea. The map, drawn on an orthographic polar projection, uses the technique of line gradations for certain and uncertain coasts as on the "Hémisphère Australe." As it is a navigational map, it shows a series of dotted spiral lines representing wind rhumbs. It also incorporated a good deal of geography which Vaugondy had previously rejected: the lac de Fonte (here named Ouinipigouchin) and the Lac Bernarda (here, Lac Michinipi). He adopted the islands from the Japanese map of Hans Kaempfer which had been used by Buache to support Russian discoveries in the north Pacific. He filled the Bering Straits with the archipelago of Aleutian islands (Iles Alexiennes) and copied his north coast of North America between 110° and 150° longitude from Captain Alexander Cluny in *The American Traveller* (London, 1769). All these features formed part of the maps Didier Robert de Vaugondy was preparing for the *Supplément* to the *Encyclopédie.*

Robert de Vaugondy's presentation to the Academy discussed three possible routes across the north pole to to the Pacific. His map illustrated what he thought was the most feasible of the three: a free and open sea which could be traversed during the summer, taking advantage of the perpetual daylight. He outlined the problems related to the mapping of an expedition of this kind, such as the kind of compass to take, the method for counting days (since there would be no night), the measurement of winds and currents, and what to do if the winds and currents were opposed to the proposed routes, as well as the type of projection to employ.

The "Essai d'une Carte Polaire," 1774, by Didier Robert de Vaugondy. (BN: Ge.D.12865) [Cat. 25] Vaugondy presented a paper to the Academy of Sciences to accompany this map which proposed three hypothetical routes across the north pole to the Pacific. The map illustrates the favoured hypothesis: an open polar sea, free of ice, that could be crossed during the summer months. Compare this open polar area to Didier's 1760 Mappemonde with its polar mass. (pp. 80-81) (Photo courtesy of the Bibliothèque Nationale, Paris)

The "Nouveau Systême Géographique"

The Academy approved both map and paper for publication, though only the map was printed, in 1774. In that year, Vaugondy continued his "northern" presentations to the Academy with the *Mémoire sur les Pays de l'Asie et de l'Amérique situés au Nord de la Mer du Sud* (Paris: Boudet, 1774). As a *supplément* to his previous work, this paper discussed current knowledge of northeast Asia and north west America and was illustrated by a "Nouveau Systême Géographique par lequel on concilie les anciennes connoissances sur les Pays nord-ouest de l'Amérique avec les nouvelles découvertes des Russes au nord de la Mer du Sud" [A New Geographical System in which older information on the northwest of America is reconciled with the new discoveries by the Russians in the north Pacific] [Cat. 456] Vaugondy called it a "Nouveau Systême" to differentiate it from his "Carte Systèmatique" of 1768. In fact, the two maps are quite different, especially regarding the Aleutian islands and the land of Yeço.

Like the "Essai d'une carte polaire," this map did not pretend to show a geographical reality but attempted to reconcile different geographical hypotheses, what geographers in the twentieth century might call a model. Vaugondy stressed that his map was a "system" by which discoveries, whether real or not, might be illustrated; it was not a depiction of the true nature of the land or sea. To emphasize this caveat, he opened his treatise with Cicero: "Nos sequimur probabilia, nec ultra id quod verisimile est progredi possumus, et refellere sine pertinaciâ et refelli sine iracundiâ parati sumus," (We follow things which are probable; we can not proceed beyond that which is likely to be true; and we are ready to refute without stubborness and be refuted without anger.) [*Tusculanae Quaestiones*, Book 2] On the "Nouveau Systême" Robert de Vaugondy showed the coast of North America as discovered by Bering and Tchirikow in 1741 and the rivers described by the Indian Moncacht-Apé from Le Page du Pratz's *Histoire de la Louisiane* (1768). He also allowed himself to include the discoveries of Admiral de la Fuente, explaining that even the de la Fuente discoveries might be shown without accepting their reality: "I have no difficulty in showing them on my map, as if they were real, without any pretence of deciding in their favour. All those who have defended them have done so on the grounds of probability which collapse upon reading the narrative...But supposing that they were real, I hope that the use one makes of them is only as an accessory or *hors d'oeuvre*, and that the manner in which they are enshrined, so to speak, in among those about which we have more certitude, might satisfy the authenticity which they lack."[16]

Didier Robert de Vaugondy's willingness to present the de la Fuente discoveries in this new system was an extension of the positive geography which formed the maps of the *Atlas Universel*: comparing sources and reconciling differences. When faced with regions about which no information is certain, the geographer can only try to co-ordinate equally uncertain accounts. The de la Fuente story, though in itself improbable, did in fact coincide with earlier explorers' accounts. Indeed, this was the basis on which Philippe Buache had accepted the de la Fuente narrative twenty years earlier. Thus, reasoned Buache then and Vaugondy now, any truth in de la Fuente's explorations would be confirmed by the more secure authenticity of others who described similar geography.

On the other hand, Vaugondy continued to reject the Sea of the West because it did not coincide with earlier accounts or with the more recent reports from the Indian Moncacht-Apé published by Le Page du Pratz. The Indian reported not a sea in the west but a great river, La Belle Rivière, that flowed from the centre of the continent into the Pacific. In addition, the Indian also described a salt lake, 300 leagues long and 30 leagues wide, similar to that reported by Baron de La Hontan at the end of the seventeenth century, shown by Guillaume Delisle on his map of Canada in 1703. Thus not only location but existence depended on the reconciliation of sources.

Didier Robert de Vaugondy's most important act of reconciliation, however, concerned the longitude of the Kamchatka peninsula in Asia and the region of California in North America. Their correct location determined the breadth of both of these continents. Robert de Vaugondy based his outline of Kamchatka on the astronomical observations made at Avatcha (modern Petropavlovsk) by Bering in 1728 and 1729. These had placed Avatcha at 171°45' from the Ile de Fer. Robert de Vaugondy rejected the measurements taken in 1741 by Louis de la Croyère, brother of J. N. Delisle (a rejection that would later prove troublesome); he added to Bering's observations those taken by the astronomer Islenief in 1761. This placed the easternmost point of Siberia at 191° longitude, much further east than shown on previous maps.

The shape of California proved an even more complicated task. The two fixed points for this coast were Cap Lucar at the southern end of the Baja peninsula (not shown on the "Nouveau Système Géographique") and Cap Fortuna in the north, with an intermediate checkpoint at Cap Mendocino. There was enough agreement between maps old and new to establish the latitude of these points, with Mendocino at 41-42° north latitude and Cap Fortuna at 50°. No astronomical observations had been made along this coast, so Didier resorted to an interesting, if ultimately erroneous, logic. Acting on the principle that most east-west longitudinal bearings given on pre-eighteenth century maps had been too wide, he determined that the longitude given on older maps had to be reduced by at least ten degrees. This placed Cap Fortuna at 220° and Mendocino at 230° from the Isle de Fer. This "reduction" in longitude in fact so expanded the breadth of North America that the Vaugondy California coastline bulges considerably westwards, like the outline of California favoured by Engel. This bulge allowed enough room in North America to contain the most recent knowledge about the western regions of Canada — that the distance between Lac Ounipigouchin (Lake Winnipeg) and the California coast had to be at least 40 degrees of longitude in order to accommodate the great rivers which flowed to the west: the Rivière de l'Ouest, la Belle Rivière, and la Grande Rivière. Vaugondy also included Captain Cluny's coastline of northern Canada as well as Cluny's connection between Repulse Bay and the Sea of Tartarie.

By such operations, Didier Robert de Vaugondy worked towards a system which would illustrate a current state of 'probable' knowledge by reconciling all sources. Unfortunately, Didier's adjudicators at the Academy were not persuaded by his results: "He [Vaugondy] opposes conjectures with conjectures, arguments with arguments, relations which are hardly decisive with relations only a little more conclusive. An unbiased mind would leave this discussion a little more uncertain than it was before."[17]

Robert de Vaugondy's intent, however, was to stress that a northwest passage was highly conjectural and fraught with problems, and that the polar route he advocated offered a greater chance of success. The arguments in the *Mémoire sur les pays septentrionaux* very much reflect the sentiments of Engel found in the *Encyclopédie*. Vaugondy quotes his correspondence with Engel throughout the *Mémoire*, and it is probably safe to attribute the logic supporting the northeast passage to Engel rather than to Vaugondy, whose concern was to illustrate the idea.

Jean Nicolas Buache de la Neuville

Didier Robert de Vaugondy's polar presentations, like his papers on the de la Fuente discoveries twenty years earlier, did not pass without retort from a Buache. On this occasion it was not Philippe, who had died in January of 1773, but his nephew and heir, Jean Nicolas Buache de la Neuville (1741-1825). The time was particularly ripe for a clash, since the death of Buache had left the position of *adjoint géographe* in the Academy of Sciences vacant, as well as that of *premier géographe du roi*.

Buache de la Neuville had just completed a mappemonde which differed considerably from Robert de Vaugondy's "Nouveau Système Géographique." He also published an accompanying *Mémoire sur les Pays de l'Asie et de l'Amérique situés au Nord de la Mer du Sud pour servir de supplément aux Considerations Géographiques et Physiques de feu Mr Buache sur les nouvelles découvertes au Nord de Cette Mer.* This paper not only justified his own map but also defended the name of his uncle, Philippe Buache.

In a reprise of his uncle's complaints of twenty years earlier, Buache de la Neuville felt Robert de Vaugondy was once again slighting a Delisle. Vaugondy's rejection of Louis de la Croyère's astronomical observations was seen as an alliance with the German astronomers at St. Petersburg with whom J. N. Delisle had never seen eye to eye. "M. Vaugondy supports himself with a passage by M. Gmelin on which he insists very much and some remarks by M. Delisle, the astronomer, over which he passes very lightly."[18]

As well as criticizing Robert de Vaugondy's arithmetic and use of sources (criticisms Vaugondy was to hear repeated a year later), Buache de la Neuville particularly objected to the westward bulge of North America. When Vaugondy reduced the longitude in North America, he did it in the wrong direction. In fact, he should have reduced the size of North America and increased the size of the Pacific which "Dampierre, Frondat, and other famous navigators have said is much larger than the maps of their time show."[19]

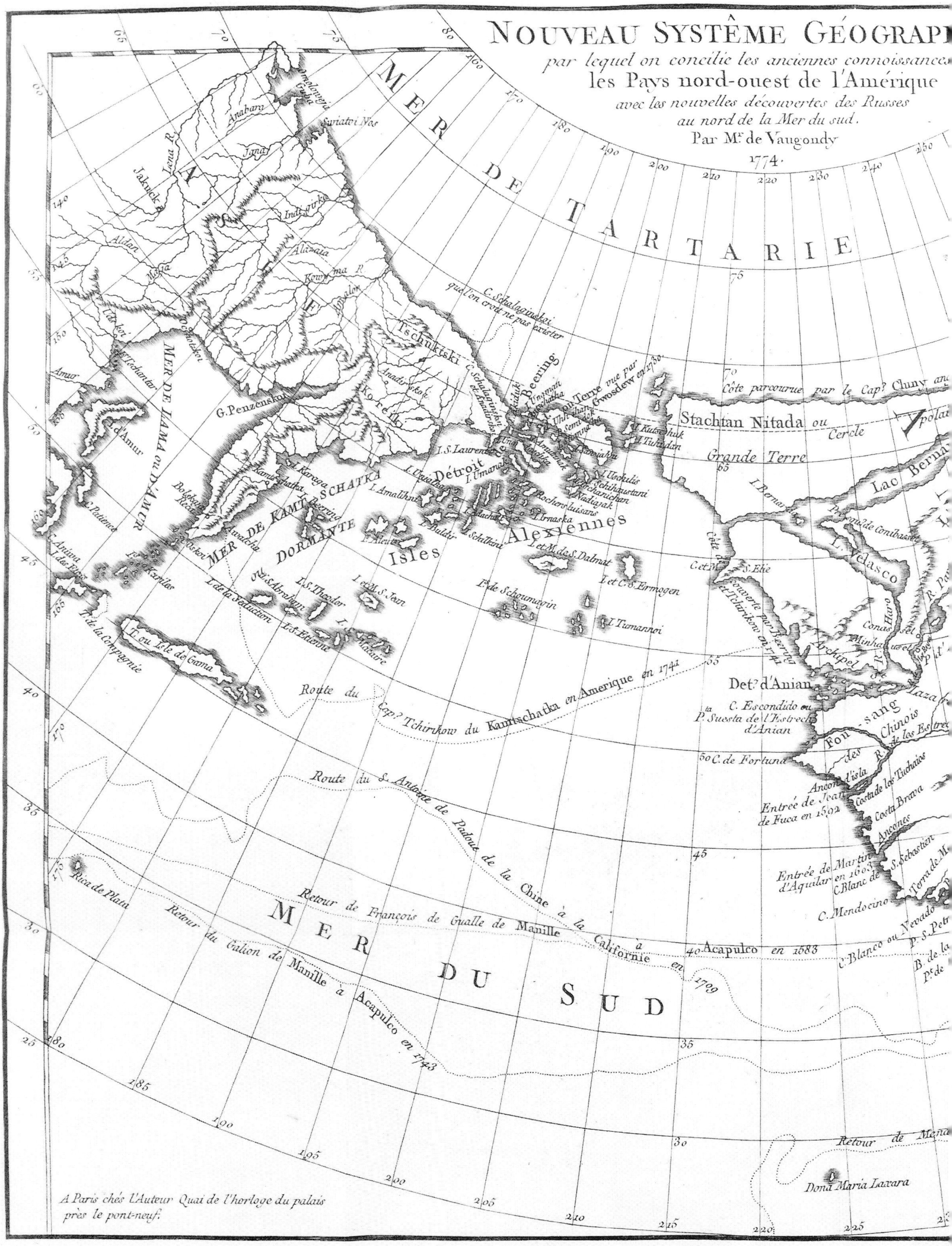
NOUVEAU SYSTÊME GÉOGRAPI
par lequel on concilie les anciennes connoissances
les Pays nord-ouest de l'Amérique
avec les nouvelles découvertes des Russes
au nord de la Mer du sud.
Par Mr. de Vaugondy
1774.
MER DE TARTARIE
Swiatoi Nos
Anabara
Jana
Indigirka
Aldan
Alasaia
Kowyma R.
Tschukitski
C. Schalaginskoi qu'l'on croit ne pas exister
Bering
Côte parcourue par le Cap. Cluny
Stachtan Nitada ou
Cercle
polai
Grande Terre
Lac Bernard
Velasco
G. Penzenskoi
MER DE LAMA ou D'AMUR
MER DE KAMTSCHATKA
DORMANTE
Isles Alexiennes
Détroit
I. S. Laurent
Kurilles
Ile de Schoumagin
I. Tumannoi
T. ou Isle de Gama
T. de la Compagnie
Route du Capt. Tchirikow du Kamtschatka en Amerique en 1741
Det. d'Anian
C. Escondido ou P. Suesta de l'Estrecho d'Anian
Fou-sang
C. de Fortuna
Entrée de Jean de Fuca en 1592
Entrée de Martin d'Aquilar en 1603
C. Blanc de
C. Mendocino
Route du S. Antoine de Padoue de la Chine à la Californie en 1709
Retour de François de Gualle de Manille à Acapulco en 1683
Retour du Galion de Manille à Acapulco en 1743
Rica de Plata
MER DU SUD
Dona Maria Laxara
A Paris chés l'Auteur Quai de l'horloge du palais près le pont-neuf.

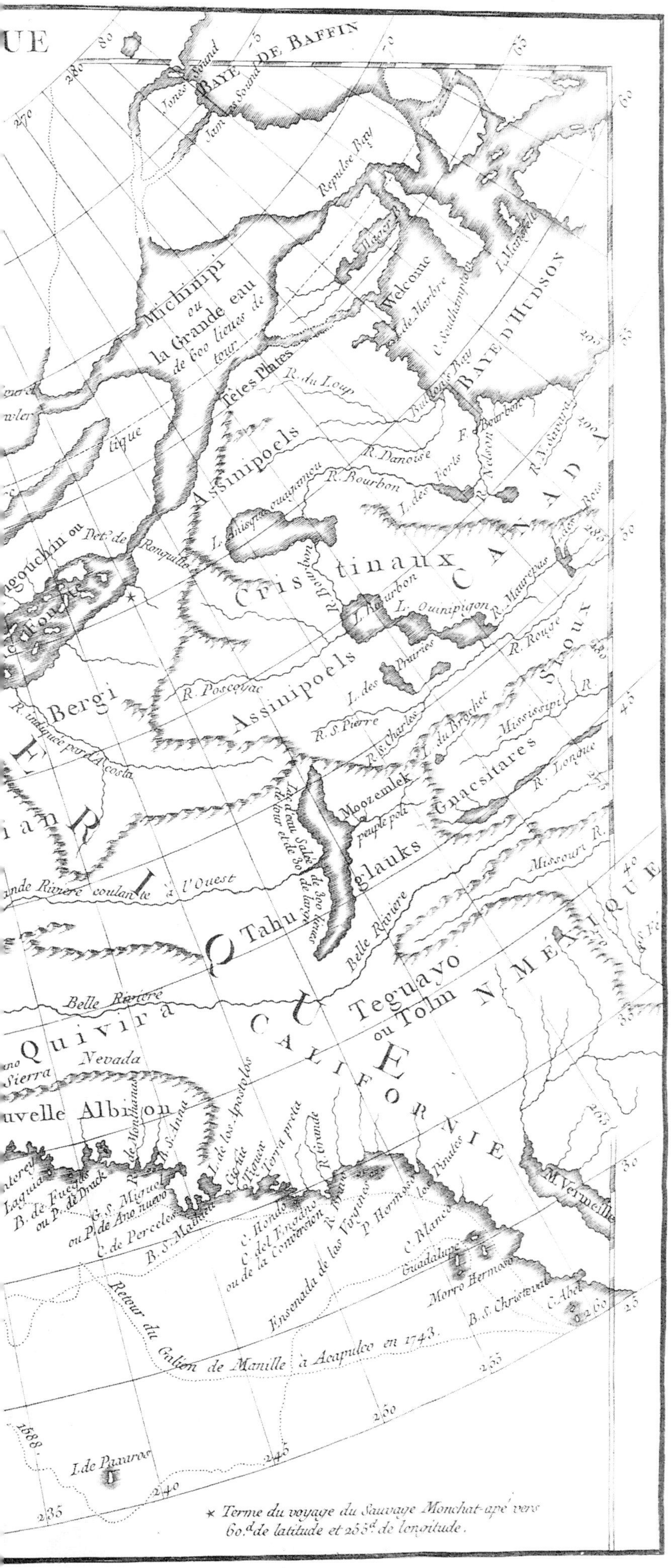

The 'Nouveau Système Géographique,' by Didier Robert de Vaugondy, 1774 [Cat. 456], published with his Mémoire sur les Pays de l'Asie et de l'Amérique situés au Nord de la Mer du Sud *(Paris, Boudet: 1774). It is not a map of reality but a conjectural system reconciling various accounts of the geography of northwest America. (Photo courtesy of the William L. Clements Library, University of Michigan)*

Buache de la Neuville read his paper to the Academy of Sciences, which approved it for publication in March 1775 and applauded him for determining the "true geometrical position of Kamchatka" and for having succeeded as conclusively if not better than Engel and Vaugondy. Vaugondy did not refute Buache's paper on the northern regions but advertised it next to his own, just as he had published Philippe Buache's arguments along with his own in 1754.

The progress of Didier Robert de Vaugondy's cartography of the northern Pacific regions and his thoughts on the de la Fuente discoveries demonstrates his increasing interest in maps as a means of analysis. Cartography could become a useful tool for verifying conflicting data and for outlining possible explorations, a development typical of Enlightenment thought. As Numa Broc has pointed out: "The *géographe de cabinet* intervenes where the explorer stops; argument takes the place of observation and 'systèmes' become all the more necessary as the parts of the world to which they apply become less accessible."[20]

Didier Robert de Vaugondy's systematic geography did not win him the chair in the Academy of Sciences — this went to d'Anville — nor did it endear him to the younger Buache. Nonetheless, he had made a mark with the Academicians who described him as working with "intelligence and sagacity...animated by a love of the truth...a worthy heir of the famous Sanson, he is industrious, precise, and enlightened."[21] He was to apply these talents to the more prosaic realm of utilitarian maps and atlases.

References

1. The Latin reads: "totam Novam Scotiam quoque sive Acadiam totam limitibus suis antiquis comprehensam...": "all Nova Scotia also or all Acadia comprehended by its ancient boundaries..."
2. This advertisement and the concerned reaction of the English boundary commission are recorded in the conference manuscripts and letters of William Mildmay, one of the English commissioners. William L. Clements Library: Mildmay Papers.
3. T. Jefferys, *The Conduct of the French with Regard to Nova Scotia...* (London, 1754).
4. The best summaries are by H.R. Wagner (in English) and Numa Broc (in French). H.R. Wagner, "The Delisle-Buache Fantasy," *The Cartography of Northwest America* (Berkeley: University of California Press, 1937): pp. 158-163; N. Broc, "Autour de la Mer de l'Ouest," *Géographie des Philosophes*, pp. 152-166. More recently, an excellent summary and new contributions to the history of this cartography are found in Lucie Lagarde, "Le Passage du nord-ouest et la Mer de l'Ouest dans la cartographie française du 18e siècle, contribution à l'étude de l'oeuvre des Delisle et Buache," *Imago Mundi* 41 (1989), pp. 19-44.
5. *Monthly Miscellany or, Memoirs for the Curious*, London, 1708, April, pp. 123-126; June, pp. 183-186. Some have suggested that it was written by a James Petiver or Daniel Defoe: Percy G. Adams, *Travelers and Travel Liars (1660-1800)* (Berkeley: University of California Press, 1962; reprinted, New York: Dover Press, 1980): p. 65.
6. See George Kish, "Early Thematic Mapping: The Work of Philippe Buache," *Imago Mundi* 28 (1976), pp. 129-136.
7. J.N. Delisle gives this man's name as l'Abbé Pic de la Mirandole, though in the manuscript of the mémoire presented to the Academy, Vaugondy asked that Mirandole's name be witheld. Delisle's comment is in AN:2JJ 79 (14).
8. Didier Robert de Vaugondy, "Préface," *Atlas Universel*, p. 20a.
9. K. Hardesty, *The Supplément to the Encyclopédie*, (The Hague, 1977), p. 3.
10. le Vicomte de Grouchy et Paul Cottin, ed., *Journal inedit du duc de Croÿ (1718-1784) publié, d'après le manuscrit autographe conservé à la Bibliothèque de l'Institut...* (Paris: Flammarion, 1906-1907), Vol. 3, p. 3.
11. Marie-Pierre Dion, "Itinéraire Intellectuel et Réussite Nobilaire au siècle des Lumières: Emmanuel de Croÿ (1718-1784)," (unpublished thesis), (Paris: Ecole des Chartes, 1984). This projection was not entirely new: in 1753 Nicolas Boullanger published a mappemonde in two hemispheres tilted at a slight angle, cited by Dion, p. 152, note 96.
12. Académie des Sciences, *Procès-Verbaux*, mercredi 24 mars 1773, p. 76.
13. The visit to Versailles is described in the *Journal du Duc de Croÿ*, Vol. 3, pp. 78-79.
14. Dion, *op. cit.*, p. 252; *Journal du Duc de Croÿ*: Vol. 3, pp. 41, 308.
15. Dion, *op. cit.*, p. 256, n. 119, letter, Vaugondy to de Croÿ.
16. Didier Robert de Vaugondy, *Mémoire sur les pays de l'Asie et de l'Amérique* (Paris: Robert de Vaugondy and Boudet, 1774): pp. 14-15.
17. Bibliothèque de l'Institut (Paris): MS 2311, Report of Pingré and Jeurat, 29 January 1774.
18. Bibliothèque de l'Institut: MS 2314 (10), Buache de la Neuville, "Mémoire sur les Pays de l'Asie et de l'Amérique...: pour servir de supplément aux Considérations Géographiques et Physiques de feu M. Buache...," p. 5. Gmelin and Müller had been invited to Russia also to participate in the surveying of the kingdom. Buache de la Neuville's paper was published in 1775.
19. *ibid.*, p. 7.
20. Broc, *La Géographie des Philosophes*, p. 269.
21. Pingré, Jeurat, and de Lalande, Report to the Academy of Sciences, 13 August 1774.

CHAPTER FIVE

Maps that Illustrate and Teach

Didier Robert de Vaugondy's "industry, precision, and enlightenment" made him a prolific cartographer throughout the late 1750s and into the 1760s. They were characteristics which also attracted several major authors of the Enlightenment to use his skills. Vaugondy prepared maps as illustrations for Buffon, Montesquieu, and de Brosses, and applied his pedagogical instincts to a new atlas for students.

Like his father, Gilles Robert, Didier Robert de Vaugondy often prepared maps to illustrate the texts of others. His earliest work as an illustrator was for George Louis Le Clerc, le comte du Buffon (1707-1788). Didier prepared the two maps in the first volume of Buffon's monumental *Histoire Naturelle* (Paris, 1749-1804, 44 volumes): the "Carte de l'Ancien Continent" and "Carte du Nouveau Continent" [Cat. 16, 18]. They show the hemispheres oriented along their longest diameter rather than along a north-south axis, demonstrating Buffon's idea of the balance between the Old and New Worlds: land offset by water.

Buffon's idea was expanded into a call for South Pacific exploration by his great friend and fellow Burgundian, the Président Charles de Brosses, whose work, the *Histoire des Navigations aux Terres Australes* (Paris: Durand, 1756) summarized voyages to the south Atlantic and Pacific to support the search for a Southern Continent. Like Buffon, de Brosses needed a geographer to make maps for his book. While Buffon thought well of Vaugondy (see Chapter Seven), he introduced de Brosses to Joseph-Nicolas Delisle. Delisle was not a geographer *per se* but an astronomer, and his experience had not been in the southern hemisphere but in Russia and Siberia. Nonetheless, he was Buffon's colleague in the Academy of Sciences. It was clear that Delisle made a very distinct impression on de Brosses at their first meeting in 1754: "There emerged from my [ante-] chamber a large, florid man, announced on behalf of M. Buffon. He said to me, 'Monsieur, I come from Kamchatka. I know Siberia by heart, my name is Delisle, brother of the great Guillaume. I bring you my works and my maps. Whatever you might say to me, the Northwest Passage is true, the Admiral de la Fuente is no less true and did not lie in a single word. God will forevermore bless you for having written the *Terres Australes*. I am asking you the favour of entrusting me with the maps. I have all the manuscript maps, papers, and letters of my brother, which I will communicate but on the condition that you believe that the de la Fuente report is true...' "[1]

Portrait of George Louis le Clerc, comte de Buffon (1707-1788), author of the Histoire Naturelle, *an immensely influential work of the Enlightenment, for which Didier Robert de Vaugondy prepared two maps. (Photo courtesy of the University of Michigan Library)*

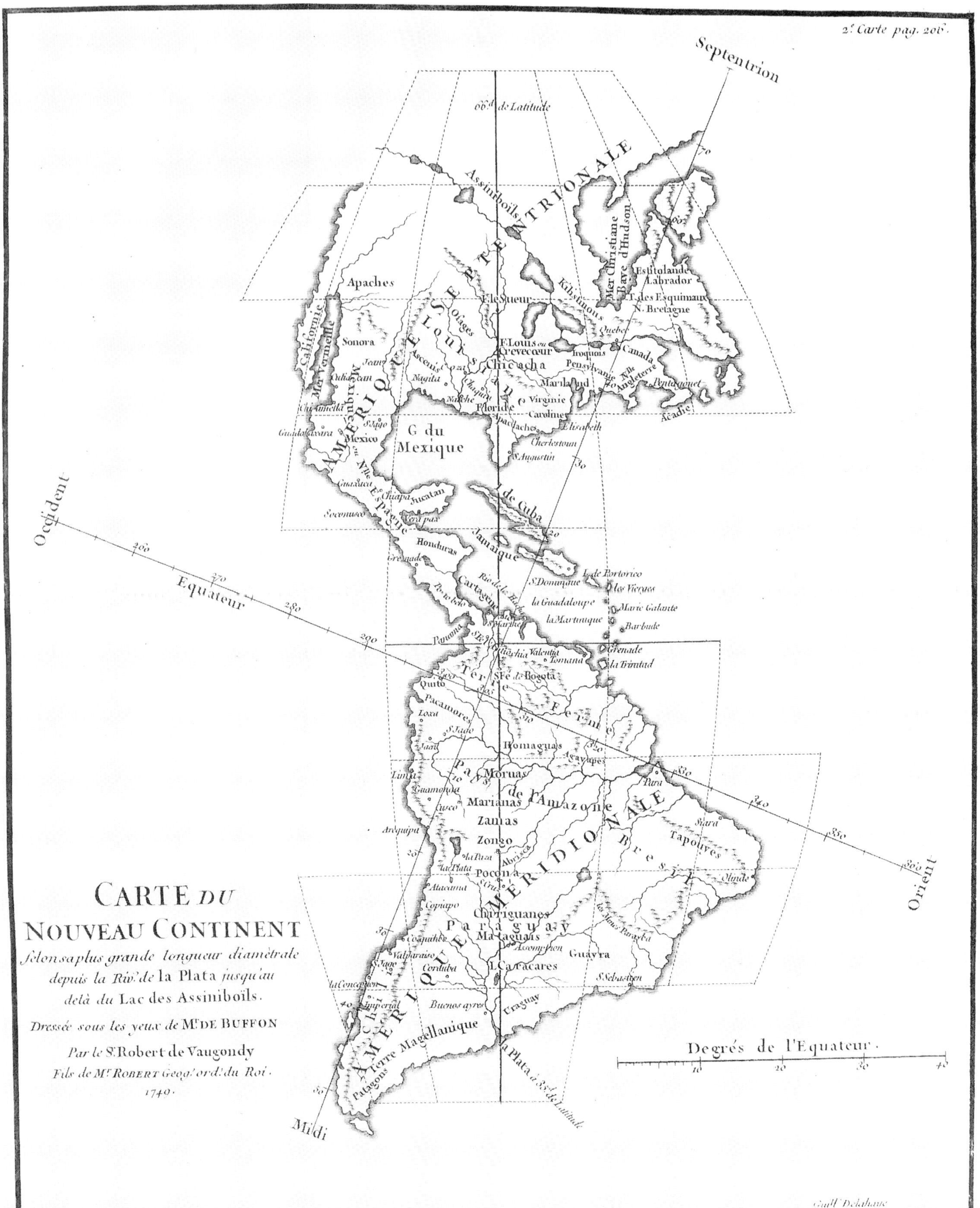
2e Carte pag. 206.
Septentrion
Occident
Orient
Midi
Equateur
66d de Latitude
AMERIQUE SEPTENTRIONALE
AMERIQUE MERIDIONALE
Assiniboils
Mer Christiane
Baye d'Hudson
Estotilande
Labrador
T. des Esquimaux
N. Bretagne
Kilistinons
Quebec
Canada
Iroquois
Pensylvanie
Nlle Angleterre
Pentagouet
Acadie
Apaches
Californie
Mer Vermeille
Sonora
Culiacan
Chiametla
Guadalaxara
Mexico
Nlle Espagne
Guaxaca
Chiapa
Jucatan
Soconusco
Vera pax
Honduras
Grenade
F. le Sueur
F. Louis ou Crevecoeur
Louisiane
Chicacha
Marilând
Virginie
Caroline
Floride
Apalaches
Elisabeth
Cherlestoun
S. Augustin
G du Mexique
I. de Cuba
Jamaique
S. Domingue
I. de Portorico
les Vierges
la Guadaloupe
Marie Galante
la Martinique
Barbude
Grenade
la Trinitad
Porto belo
Panama
Cartagene
Rio de la Hacha
Valentia
Tomana
S. Fé de Bogota
Terre Ferme
Quito
Pacamores
Loxa
S. Jago
Jaen
Lima
Guamonga
Cusco
Arequipa
Omaguas
Morúas
Pays de l'Amazone
Marianas
Zamas
Zongo
la Paza
la Plata
Pocona
Atacama
Copiapo
Chiriguanes
Paraguay
Mataguais
Assomption
Guayra
Valparaiso
S. Iago
Corduba
L. Caracares
S. Sebastien
la Concepcion
Imperial
Chili
Buenos ayres
Uraguay
Terre Magellanique
Patagons
la Plata à 35d de Latitude
Para
Siara
Tapouyes
Bresil
Olinde
Degrés de l'Equateur.
CARTE DU NOUVEAU CONTINENT
selon sa plus grande longueur diamètrale
depuis la Riv.re de la Plata jusqu'au
delà du Lac des Assiniboïls.
Dressée sous les yeux de Mr. DE BUFFON
Par le Sr. Robert de Vaugondy
Fils de Mr. ROBERT Geog.e ord.re du Roi.
1749.
Guill. Delahaye

Delisle's dogged insistence on the truth of de la Fuente should have warned de Brosses that their working relationship would not be easy. Initially they had agreed that Delisle would be in charge of the geographical portion of de Brosses' book, making the maps and tables with notes, "not [with] that intolerable continuous discourse."[2] However, if Delisle's discourse was intolerable, he himself became even more so: "This Delisle doesn't leave me; he pesters me and makes me lose an infinite amount of time. When he works on the maps, he enrages me even more, for he won't work!"[3]

In exasperation, de Brosses renounced his collaboration with Delisle and turned instead to Buffon's cartographer, Didier Robert de Vaugondy, well known to Durand, who was to publish the *Histoire des navigations* for de Brosses. Vaugondy prepared the seven maps for de Brosses' book, showing the southern hemisphere, various parts of the South Pacific and Indonesia, and the Straits of Magellan [Cat. 26, 417-421, 495]. Like those for Buffon, they were engraved by Guillaume Delahaye, an indication that the engraver was not discouraged by the troubles over the *Atlas Universel*.

Following Buffon, de Brosses argued for the existence of land in the South Pacific because it would be "capable of holding the globe in equilibrium in its rotation, serving as a counterweight to the mass of northern Asia."[4] Vaugondy's maps were designed more to emphasize this point than to be "exact" depictions of geography. De Brosses entertained no illusions about the accuracy of the maps he had had Vaugondy put together. When he learned that there was indeed a strait between the Falkland Islands (Iles Malouines) as shown on the map in his book, he commented, "My map is better than I hoped, which gives me great pleasure." He was no more confident of the general map: "[it] is without any exactitude as to that part of the Southern Sea visited by Roggewin."[5] Yet Cook himself praised the maps in de Brosses: "I have compared the part of the coast that I have visited with the maps found in the French work entitled *Histoire des Navigations aux Terres Australes.* I have found them quite exact."[6]

Above *Portrait of Charles de Secondat, baron de Montesquieu (1689-1755). Robert de Vaugondy drew the maps for the posthumous edition of his* **l'Esprit des Lois.** *(Courtesy of the British Museum)*

Left *'Carte du Nouveau Continent' from the* **Histoire Naturelle** *[Cat. 18]. The continents of North and South America are tilted slightly in order to be oriented along their longest diameter. The axis for the 'Nouveau Continent' runs from the mouth of La Rivière de la Plata to Lac des Assiniboïls, west of Lake Superior in North America. The maps are engraved by G. Delahaye. (Photo courtesy of the University of Michigan Library)*

The cartographic work done for Buffon and de Brosses must have affected Didier Robert de Vaugondy's own notions on polar exploration and the search for a north west passage to the Pacific. While Buffon and de Brosses were writing to much larger purpose — the former to classify and explain the natural world, the latter to encourage colonisation of Australasia — their systematic and analytical approach may have impressed Vaugondy. Their process of assessing the possibilities of discovery by setting out previous explorations, their arguments for and against ice at the poles, and their descriptions of types of navigation necessary were absorbed into Robert de Vaugondy's arguments in his "polar" presentations to the Academy of Sciences in 1773 and 1774.

A further influence on Didier's work was

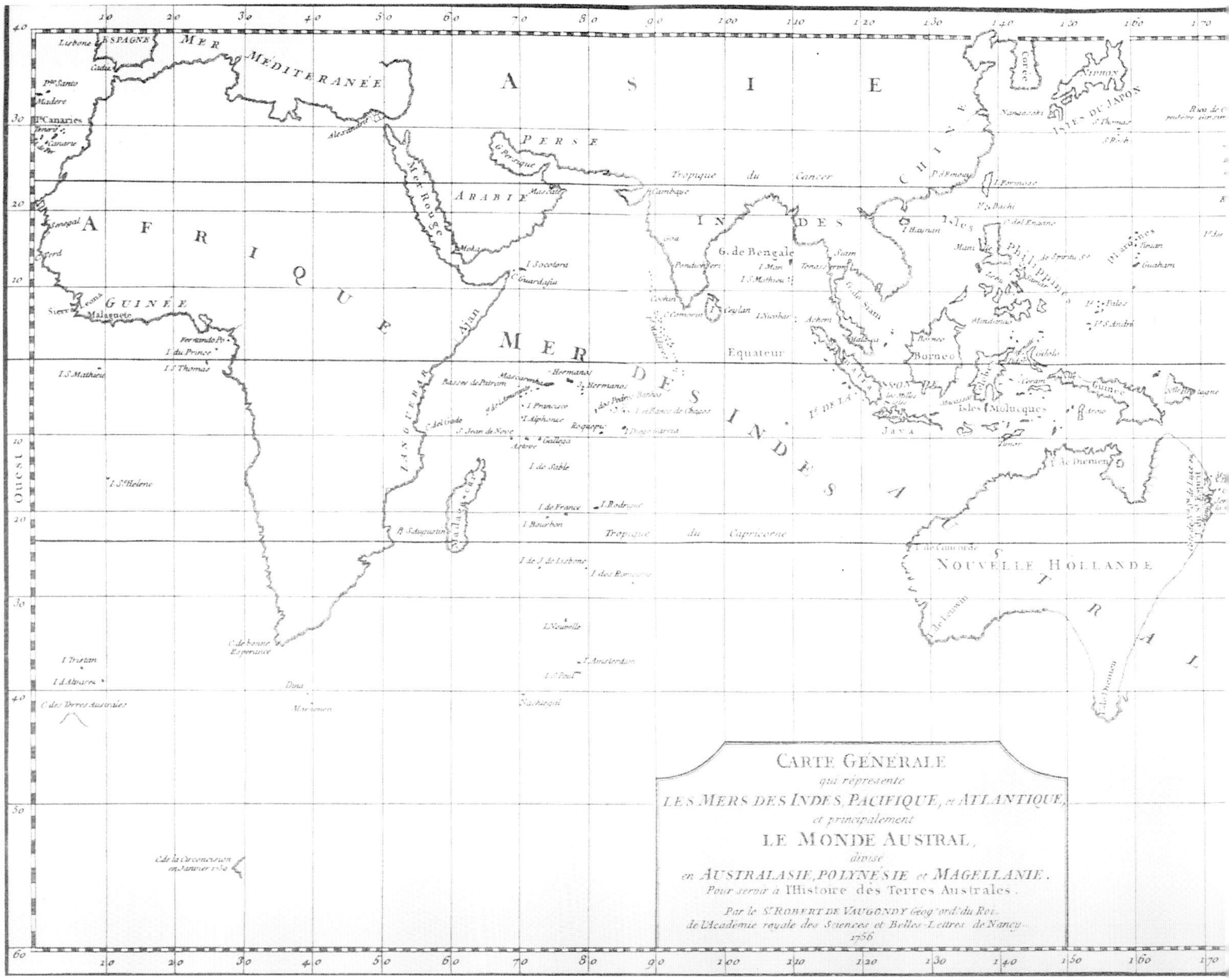

"Carte Générale qui représente les Mers des Indes...et le Monde Australe," [Cat 26], by Didier Robert de Vaugondy, one of the maps accompanying de Brosses' **Histoire des navigations aux Terres Australes.** *This map and de Brosses' work were used and admired by James Cook. (Photo courtesy of the University of Michigan Library)*

Charles de Secondat, baron de la Brède et de Montesquieu (1689-1755), for whom he prepared the two maps in the posthumous edition of *l'Esprit des Lois*: a world map and one of Europe, the Mediterranean, and Asia Minor, again engraved by Guillaume Delahaye [Cat. 9, 33]. The maps, which locate places described in Montesquieu's narrative, are inserted in Book XXI, a history of commerce with special emphasis on the ancient world and the modern world after the discovery of the Americas. Montesquieu's interest in the relationship between climate and man's moral and physical character influenced Didier Robert de Vaugondy's exploration of similar questions in his textbook, *Institutions Géographique* of 1766. In his atlas designed for students, the *Nouvel Atlas Portatif* (1762), he included three thematic world maps which show population distribution by religions, by skin colour, and by facial type. These are among the earliest examples of ethnological maps.[7]

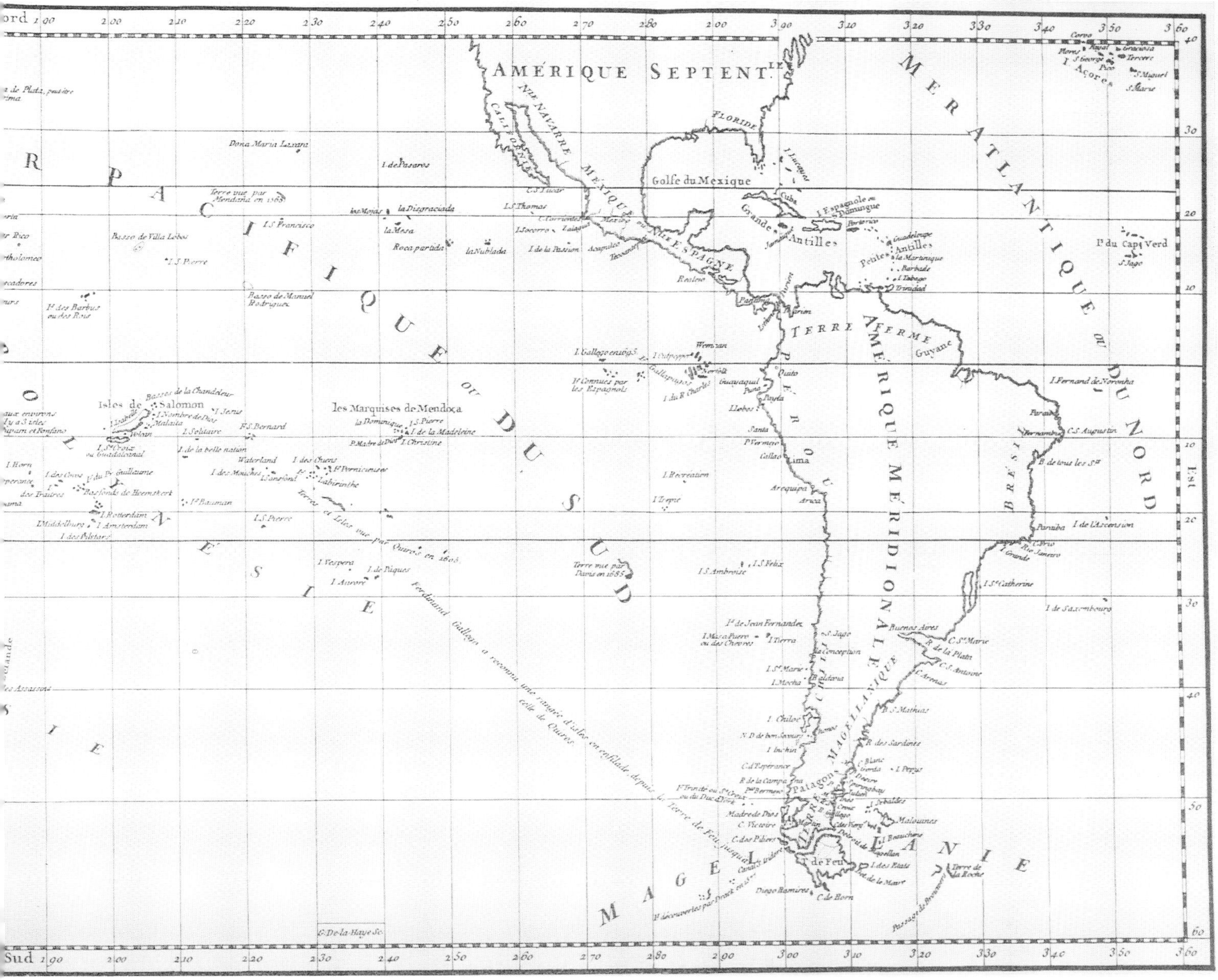

The *Nouvel Atlas Portatif* and Teaching Geography

The *Nouvel Atlas Portatif* of Didier Robert de Vaugondy was his only independent atlas venture, published without apparent financial support from anyone else. It was designed primarily for the education of the young, and as a preface, he wrote a "Discours sur l'étude de la Géographie" in which he outlined his programme for a child's geographical education.

By the middle of the century, geography was a changing discipline and no longer a simple exercise in nomenclature. Didier Robert de Vaugondy pointed out that "if one wants to assess Geography by the variety of means one can think of at the moment which explain its principles, one would have to say that it is the most difficult science of all...[for it] embraces everything that constitutes the physical and moral existence of our globe."[8] Vaugondy considered geometry and astronomy to be necessary to understand the arrangement of parts on the surface of the earth; physics, mechanics, botany and minerology he considered essential to explain different aspects of natural phenomena encountered on earth; and history was vital to an understanding of the peoples of the world with their different mores and customs.

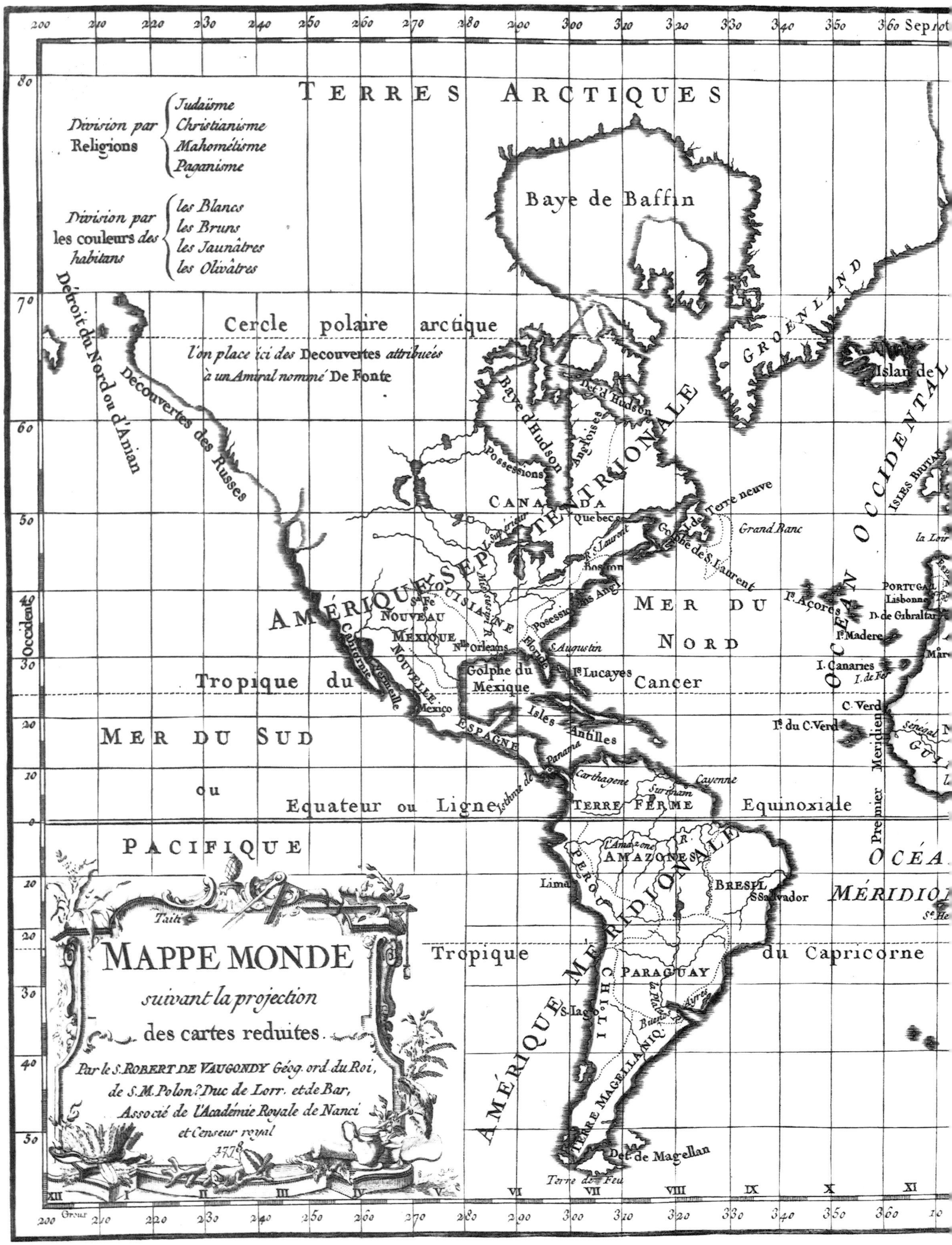

MAPPE MONDE
suivant la projection
des cartes reduites.
Par le S. ROBERT DE VAUGONDY Géog. ord. du Roi,
de S.M. Polon.e Duc de Lorr. et de Bar,
Associé de l'Académie Royale de Nanci
et Censeur royal
1778
Division par Religions
Judaïsme
Christianisme
Mahométisme
Paganisme
Division par les couleurs des habitans
les Blancs
les Bruns
les Jaunâtres
les Olivâtres
TERRES ARCTIQUES
Baye de Baffin
GROENLAND
Islande
Cercle polaire arctique
l'on place ici des Decouvertes attribuées à un Amiral nommé De Fonte
Détroit du Nord ou d'Anian
Decouvertes des Russes
Baye d'Hudson
D.t d'Hudson
Possessions Angloises
CANADA
AMÉRIQUE SEPTENTRIONALE
Quebec
Golphe de S. Laurent
Grand Banc
Boston
LOUISIANE
NOUVEAU MEXIQUE
Nle Orleans
Floride
S. Augustin
Golphe du Mexique
Californie
NOUVELLE ESPAGNE
Mexico
I.s Lucayes
Isles Antilles
Panama
MER DU NORD
I.s Açores
PORTUGAL
Lisbonne
D. de Gibraltar
I. Madere
I. Canaries
C. Verd
I.s du C. Verd
OCÉAN OCCIDENTAL
ISLES BRITAN
Tropique du Cancer
MER DU SUD ou PACIFIQUE
Equateur ou Ligne Equinoxiale
Carthagene
Surinam
Cayenne
TERRE FERME
Premier Meridien
AMAZONES
PEROU
Lima
BRESIL
S. Salvador
AMÉRIQUE MÉRIDIONALE
Taiti
Tropique du Capricorne
PARAGUAY
CHILI
S. Iago
Buenos Ayres
TERRE MAGELLANIQUE
Det. de Magellan
Terre de Feu
OCÉAN MÉRIDIONAL
Occident

Gravé par E. Dussy

Vaugondy explained that in order to teach this difficult and complex subject, a variety of "mechanical means" had been invented like the geographical garden, consisting of "zigzags and hieroglyphics" that assumed children to be automatons instead of rational human beings. He pointed out that students were likely to find such methods a hindrance to learning, causing teachers to despair and parents to regret the money they had spent. Trying to teach the complexities of geography in its widest definition simply encumbered students with all the phenomena of nature and overloaded their weak memories with an "indigestible weight of facts" turning them into "walking encyclopedias."

Vaugondy rejected all such methods, reducing geography to its simplest definition: "nothing other than the description of the different parts of the inhabited globe." He felt that if geography were given a "seductive attraction by its application to what is most familiar to us, [then] the very youngest child is susceptible of appreciating its utility and advantages." Stripped of its mysteries, geography could be easily comprehended using logic, memory, and practical application. The Vaugondy geography curriculum therefore followed a simple, "natural" order. The student of either sex should begin at the earliest age with a world map or, even better, a globe, the true model of the earth. By putting an insect such as a wood louse on the globe and watching its movements, the youngster would learn to recognize the shapes of the continents, later tracing them to memorize their contours. Avoiding the question/answer method used by many popular geography manuals would allow the natural curiosity of the student to determine basic questions such as why the days increase and diminish during the course of the year, why the sun does not always rise and set in the same place, why there are different seasons, or why the moon changes shape. With a globe, a sphere, and a book like his own *Usage des Globes* (1752), such questions were easily answered with simple experiments instead of rote responses.

Having played with a globe from early childhood, a young student would soon be ready for the maps of regions whose shapes would now be familiar. To describe these regions, Didier Robert de Vaugondy recommended a text, the *Géographie Moderne* (Paris: Hérissant) by l'abbé Nicolle de la Croix. This work had been republished several times between 1747 and 1762; Vaugondy liked it because it was less boring than most geographies. "The historical points found sprinkled throughout it relieve the monotony that the description of places inevitably brings." Its only fault was that not all the place names mentioned in the *Géographie Moderne* could be found on maps by various authors. To correct this, Vaugondy prepared two indices for the *Nouvel Atlas*: a general table of contents and an index of place names from de la Croix's book keyed to the maps in the atlas. Thus the design of Didier's *Nouvel Atlas Portatif* joined the "useful to the pleasant," just as Gilles Robert's *Atlas Portatif Universel et Militaire* had done for Gordon's *Grammaire Géographique*.

Whether intended or not, the timing of the publication of the *Nouvel Atlas Portatif* could not have been better. The expulsion of the Jesuits from France after 1762 had left a void in secondary teaching in general and in the subject of geography in particular. Furthermore, the cry for more rational systems of teaching, rather than rote memorisation, echoed Vaugondy's own philosophy of learning.[9] Jean-Jacques Rousseau, also arguing for a "natural" approach to learning, published his controversial *Emile* in 1762, the same year as Vaugondy's *Nouvel Atlas*. In Book Three of *Emile* Rousseau advocates teaching geography from experience and observation rather than from books. More extreme in its upheaval of current methods, Rousseau's ideal system would throw away all spheres, globes and maps ("nothing but machines") and teach the child to make his own. The Vaugondy geography curriculum, moving at the pace of the child, simply encouraged the practical application of what had been learned.

Didier Robert de Vaugondy was not the only one to realize the market potential of publishing an atlas to accompany de la Croix's *Géographie Moderne*. In the same year, 1762, the engraver Jean Lattré joined J. Hérissant, the publisher of *Géographie Moderne*, to produce the *Atlas Moderne* which attempted to

---

Right *The title page from Robert de Vaugondy's* **Nouvel Atlas Portatif.** *(Photo courtesy The Newberry Library)*

Preceding page *Mappemonde from Robert de Vaugondy's* **Nouvel Atlas Portatif** *[Cat. 15]. This map is repeated three times and coloured thematically for the distribution of population by religion, by skin colour, and by facial types. (Photo courtesy of the Newberry Library, Chicago)*

NOUVEL ATLAS
PORTATIF
destiné principalement pour l'instruction
DE LA JEUNESSE,
d'après la Géographie Moderne
de feu l'Abbé Delacroix
Par le S. ROBERT DE VAUGONDY Géographe ord.re du Roi,
de S. M. Polonoise Duc de Lorraine et de Bar,
de la Société Royale des Sciences et Belles Lettres de Nancy,
et Censeur Royal.

A PARIS
Chéz le S.r Fortin Ing.r Mécanicien du Roy
Rue de la Harpe près la rue du Foin
1778.

fulfil this same purpose. According to Lattré's 'Avertissement:' "M. l'abbé Nicolle de la Croix who, in his *Géographie Moderne*, has principally applied himself to give details relative to the reading of history, desires an Atlas whose exactitude, clarity, and size might satisfy the Public, and facilitate the study of geography for youth. It is in part in order to fulfil his views that we have laboured." The Lattré atlas in its first edition consisted of thirty-six maps, prepared by Rigobert Bonne, Janvier, and G. A. Rizzi-Zannoni. Though it was less extensive than Vaugondy's atlas of fifty-four maps, Lattré's format was slightly larger. The Lattré cartouches were slightly more opulent, many of them engraved by Choffard and Arrivet, who had designed and engraved the cartouches for Vaugondy's atlas too.

Lattré's work was cheaper than Vaugondy's, as the chart below shows:

	Vaugondy Nouvel Atlas Portatif	Lattré Atlas Moderne
Binding:		
paper binding	21 *livres*	19 *livres* 10 *sous*
calf binding	24 *livres*	24 *livres*
thin paper	36 *livres*	30 *livres*
coloured *à la Hollandoise* [10]		

Competition with Lattré did not seem to diminish the success of the *Nouvel Atlas Portatif*. It continued to be published well into the nineteenth century by Vaugondy's successor Delamarche, under whose imprint it was adopted by the Ecole Impériale Militaire.[11]

Teaching Geography

The educational theories Didier Robert de Vaugondy set out in his "Discours sur l'étude" stemmed from his own teaching experience. His catalogue of 1777 tells us that his mornings were spent *en ville* teaching, and afternoons working on his own geography at home on the quai de l'Horloge. Announcements of his mathematics and geography classes appeared in the *Journal de Paris*, such as this one dated 28th July 1780: "M. de Vaugondy, *géographe ordinaire du roi*, of the Society of Nancy and Censeur Royal, quai de l'Horloge du Palais, near the Pont Neuf, proposes to open a course of mathematics and Geography for young people. The lessons for this double course will be arranged as three days for Maths, three for geography. Persons who wish to attend are asked to register at the end of August. When twelve have enrolled, they will be told the day the course starts." Didier Robert de Vaugondy's practical approach to teaching geography allowed him to incorporate the work of his students on his own maps, as on his "Plan de Paris" (1760): "I have made use of those quarters of the city which I have surveyed geometrically by having my students apply on the ground the mathematical principles I have taught them."[12]

By 1766, Didier Robert de Vaugondy no longer saw Nicolle de la Croix's *Géographie Moderne* as a complete course. To replace it he published his own teaching text, the *Institutions Géographiques* described in its preface as a tool for learning the 'true principles' of geography. Though he based the *Institutions* on Guillaume Sanson's *Introduction à la géographie* (issued in a fourth edition by Gilles Robert in 1747), Robert de Vaugondy explained that Sanson's *Introduction* was inadequate for teaching. Mere nomenclature was not enough. Echoing his thoughts from the 'Discours' of the *Nouvel Atlas Portatif*, he maintained: "Geography is not, as many think, a science which only demands eyes and ears; otherwise, learning it would consist only in the talent of filling one's memory with a prodigious quantity of names of places and provinces, without understanding the connection these things have with the heavens, with each other, and with history."[13] Geography was not to be studied in a vacuum, and Vaugondy emphasized this opinion with his choice of an opening quotation from Vergil: "Et quo sub caelo tandem, quibus orbis in oris jactemur doceas: ignari hominumque locorumque erramus." (And so may you learn under what heaven, on what shores of the earth we might be thrown; for ignorant of people and places, we go astray.) [*Aeneid* I, 335] The Vaugondy method continued to be child-centred and based on a certain 'natural' order: "My design is to conduct the student as a traveller from one region to another, without any interruption, instead of following the methods used at present, which consist for the most part in having him travel through a region and then transporting him to another one far away, [with the result that] he will not be able to perceive any connection between this region and the one he just left. I will consider the student as a being who, descended onto our globe, wishes to know all the nations which cover its surface. I would tell him that there needs to be order in this enterprise, that the trip takes time."[14]

The order of Vaugondy's enterprise in the *Institutions Géographiques* was broken into two parts: Speculative Geography and Practical Geography. He divided Speculative Geography into astronomical, natural, and historical. The astronomical section was, in large part, an expansion of his earlier works, *Abrégé de la Sphère* (1745) and *Usage des Globes* (1752). Natural geography briefly describes the world's terrestrial and aquatic features. Historical geography he approached with the enlightened tolerance of Montesquieu: "Our globe is in fact covered by a prodigious number of peoples who form among themselves societies of which each has its form of government and its particular laws,"[15] in which light he examined the phenomena of religion, race, and language.

Vaugondy's descriptions follow the divisions made on his thematic world maps in the *Nouvel Atlas Portatif*, the religions of the world being again Judaism, Christianity, Mohammedanism, and Paganism, and race defined by skin colour and facial type: white, brown, yellow, olive, and black (though the *Nouvel Atlas* map omits black) and "figure Européenne, Chinoise, Indienne, Americaine, Laponne, Nègre." Each type, except the European, was given a detailed description; for example, the "figure Chinoise usually [has] a flat face, flat nose, oval eyes, ending in a point."[16] The *Nouvel Atlas Portatif* map shows the same distribution based on four rather more simplified categories: European, Chinese, Laplander, African.

In explaining causes for different races, Didier did not accept that climate caused such differences or encouraged slavery; he preferred instead to invoke an inscrutable Deity. "One might really think that the swarthy colour of some of these people comes from their living in the torrid zone. But why are the Africans, situated in the same latitude as the inhabitants of South America, the only blacks and the only ones of whom is made a commerce so shameful to humanity? It is an enigma of nature difficult or perhaps even impossible to resolve, if only to say that such has been the will of the Supreme Author, of which we only see effects, without being able (nor ought we dare) to penetrate its depths."[17] Vaugondy also touched on the origins of language. His list of fifteen mother tongues from which all languages spring reflects the beginnings of linguistic research which would not flower fully until the nineteenth century.[18]

The second half of the *Institutions Géographiques* — Practical Geography — deals with map projections. As an elaboration of the summary work found in the *Essai sur l'histoire de la Géographie* (1755), it provides a practical guide for choosing the most appropriate projection and constructing maps on it.

Such were Didier Robert de Vaugondy's interests and activities in the teaching and learning of geography in the 1760s. While the work on books and the new atlas appear to be his alone, the support and collaboration of his father, Gilles Robert, cannot be forgotten. When his father died in April, 1766, Didier was moved to append a eulogy to the Preface of the *Institutions Géographiques*: "The loss which I have just felt of a father so respectable does not allow me to impose a silence on my regrets, although I have the consolation of seeing them shared with everyone who knew and had contact with him. Candour, probity, and an enlightened sense of duty guided all his actions. The uniformity in his style of life would have perhaps prolonged his days had he not spent too much time in his study because of his love of reading and work. Chosen at the age of forty-two by the last of the Sansons to succeed to the complete geographic stock of that famous name, he made it a point to respond to the intentions of his predecessor. As I naturally was to succeed him, I had the good fortune at the end of my classical schooling to have him alone for a master. Since then I always shared his work with him, and I have not published works except those whose undertaking and execution were directed by his counsels. Recovered in 1738 from a terrible illness, he had no need of medicine except in the last three or four years."[19] The loss of his closest colleague and adviser may account for the decline in production of maps by Vaugondy after this period. No more atlases appeared, and Didier gave more attention to specialized work: the map for the duc de Croÿ, celestial maps, his presentations to the Academy of Sciences and "systematic" geography, and a final dispute with another geographer over accuracy in maps. However, he did not forget the cartographic principles of the beautiful and the useful which he had formed with his father, and he applied them on perhaps his most useful maps of all: the plans of Paris.

References

1. A.C.Taylor, *Le Président de Brosses et l'Australie* (Paris: 1937): p. 53.
2. *ibid*, p. 54.
3. Letter of Président de Brosses, 24 avril 1754, quoted by Taylor, *loc. cit.*
4. Charles de Brosses, *Histoire des Navigations aux Terres Australes* (Paris: Durand, 1756): Tome I, p. 47; quoted by Broc, *Géographie*, p. 183.
5. A.C. Taylor, *op. cit.*, p. 63
6. Numa Broc, *La Géographie des Philosophes*, p. 305, quoting Hawkesworth, *Relations des Voyages entrepris par ordre de sa Majesté Britannique...* (1774): Volume IV, p. 156.
7. A later example of anthropological cartography is: "Esquisse d'un Tableau général du genre humain où l'on apperçoit...les religions et les moeurs des differents peuples...et les prinicipales variétés de forme et de couleur de chacun d'eux, par Mlle Le Masson le Golft du Cercle des Philidelphes." (c. 1786) BN: Ge.C.8674. This large map (675 x 500 cm.) is a more complex rendering of man's cultural features than Vaugondy's small ones. It has been summarily analysed by George Kish, "Une mappemonde anthropologique au XVIIIeme siècle," *Scritti geografici in onore di Aldo Sestini* (Firenze: Società di Studi Geographific, 1982).
8. "Discours sur l'étude de la Géographie", *Nouvel Atlas Portatif* (Paris, 1762): pp. 3-7, from which come all the following quotations.
9. Mireille Pastoureau, "Les Atlas scolaires Français: 16e-18e Siècles," (paper written for the symposium, Images of the World: The Atlas through History, Library of Congress, October 25-26, 1984). Methods of teaching geography which were fashionable at this time were card games and puzzles; another was *La Parterre Géographique* in which one's garden was arranged with coloured pyramids to represent continents, countries, and capitals. De Bouys, *La Parterre géographique et historique, ou nouvelle méthode d'enseigner la géographie et l'histoire* (Paris: 1737), re-edited in 1753.
10. Didier Robert de Vaugondy, "Catalogue des Ouvrages," (1777), BN: Ge.D.7808. Lattré's prices from his list of works in the *Atlas Moderne* (1762), WLCL: Atlas E-3-B.
11. Delamarche, "Ouvrages Géographiques," (Paris:1788): BN: Ge.F.2443 pièce.
12. Didier Robert de Vaugondy, *Mémoires sur les differens accroissements de la ville de Paris* (Paris: Robert de Vaugondy and Boudet, 1760): p. 3.
13. Didier Robert de Vaugondy, *Institutions Géographiques* (Paris: 1766): p.xxi.
14. *ibid* , pp. xxii-xxiii.
15. *ibid* , p. vii.
16. *ibid* , p. 242.
17. *ibid* , p. 243.
18. According to Robinson, Leibniz proposed using maps for marking language areas; Gottfried Hensel published four maps of continents in 1741 entitled "Europa (Africa/Asia/America) Polyglotta" which use colours and labels to distinguish language divisions. A.H. Robinson, *Early Thematic Mapping in the History of Cartography* (Chicago: The University of Chicago Press, 1982): p. 54, Fig. 55.
19. Didier Robert de Vaugondy, "Préface," *Institutions Géographiques* (Paris: Boudet & Desaint, 1766): p. iv.

CHAPTER SIX

The Accuracy of Maps

While Didier Robert de Vaugondy's *Nouvel Atlas Portatif* and the *Institutions Géographiques* tapped the growing pedagogical market in Paris in the 1760s, he met the growing tourist demand by producing a series of maps and guides of Paris. His first "Plan de la Ville et des Faubourgs de Paris," published in 1760 [Cat. 119], was intended as an addition to the *Atlas Universel.* To accompany it, he wrote the *Mémoires sur les différens [sic] Accroissements de la Ville de Paris, depuis César jusqu'à Présent...pour servir d'intelligence au nouveau plan de cette ville* (Paris: Robert de Vaugondy and Boudet, 1760).

The *Mémoires* reveal Didier Robert de Vaugondy not just as the *géographe de cabinet* he has seemed

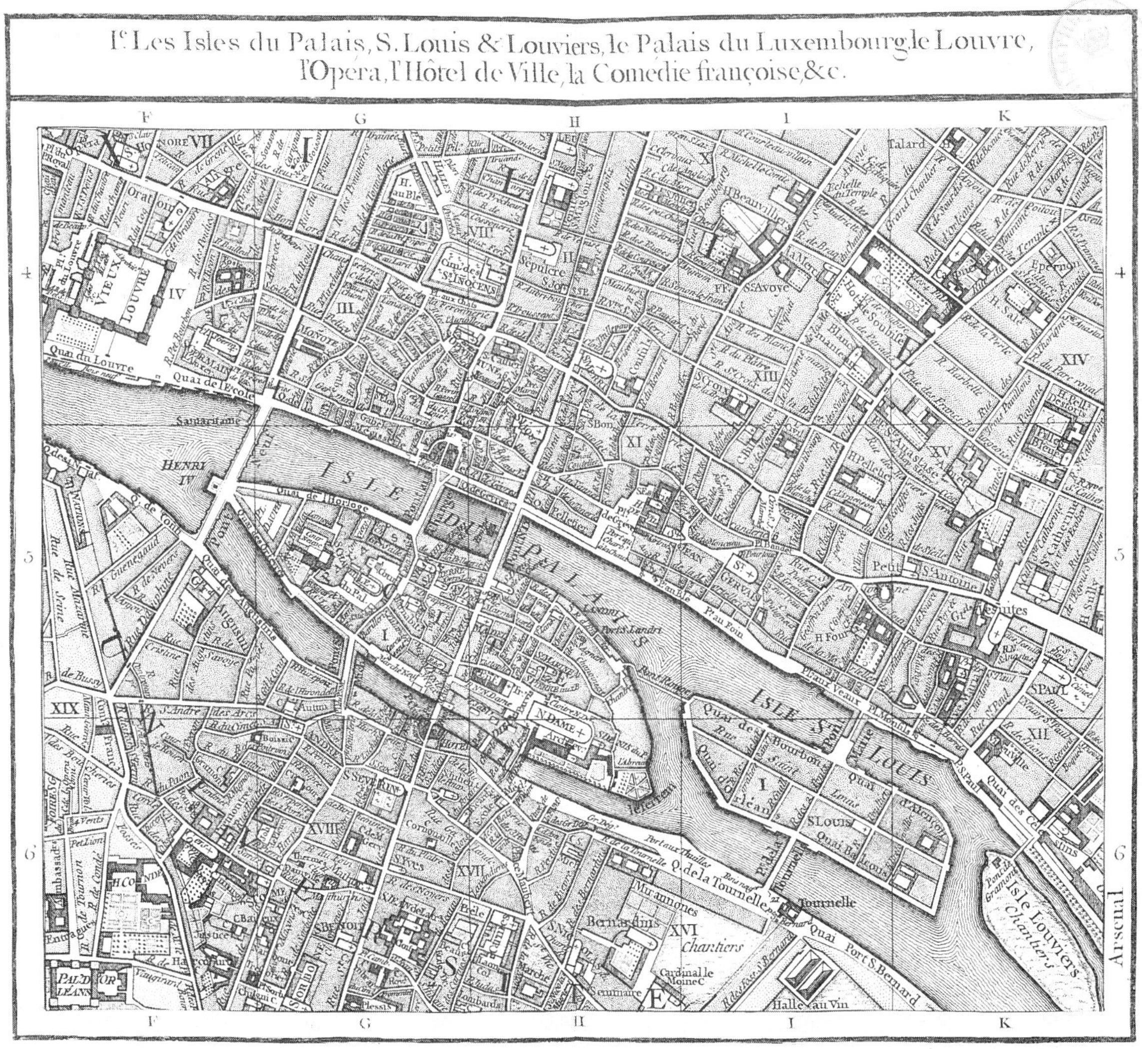

One of the pages from Didier Robert de Vaugondy's **Tablettes Parisiennes** *(1760), the portable form of his large plan of the city of Paris, also first published in 1760. [Cat. 119] (Photo courtesy of the Bibliothèque Nationale, Paris).*

Place de Louis le Grand
Statue d'Henri IV
Place des Victoires
Le Palais Bourbon
Chateau de Versailles
TABLE DES RUES
NOUVEAU PLAN ROUTIER De la Ville et Fauxbourgs de PARIS
Avec ses Principaux Edifices et Nouvelles Barrieres
Année 1789
Nouvelle Eglise de la Madeleine
Eglise de St. Sulpice
College des Quatres Nations
Hotel de Ville
La Sorbonne
MONTMARTRE
SEINE
RIVIERE
DE
QUAI DES TUILERIES
CHAMPS ELISÉES
ECHELLE DE CINQ CENT TOISES
Vue de l'Ecole Militaire du Coté de la Cour
Hotel des Monnoyes
Vue du Chateau des Tuileries du Côté du Jardin

La Place de Louis XV
Le Palais Royal
Ecole de Chirurgie
Statue Equestre de Louis XV
La Place Royale
Nouvelle Eglise de Ste Genevieve
Eglise de St Eustache
Eglise Cathedrale de Notre Dame
Le Val de Grace
Eglise de S. Gervais
Le Palais du Luxembourg
Hotel Royal des Invalides
Galerie du Louvre du Côté de la Riviere
Colonnade du Louvre
SUITE DE LA TABLE DES RUES
REMARQUES
RIVIERE DE SEINE
QUAI SAINT BERNARD
RUE DU FAUBOURG ANTOINE

thus far, but as a surveyor in the field. He credits his students for their work in surveying different parts of Paris under his guidance. He also describes the help received from his brother. "The original of [this map] has been designed by my brother, who has aided me greatly in research about antiquities as well as the survey of many places."[1] Martin Robert [Vaugondy] de Savigny, an engineer-architect, appears occasionally in financial documents of the Vaugondy household and in the *Atlas Universel* lawsuit, where he is depicted supervising the tracing of maps onto copperplate, examining proofs for errors, and transporting maps to and from the Delahaye workshop. The map of Paris is the only instance where he is given credit for helping to make a map. Just as Gilles Robert became the silent partner of Didier Robert de Vaugondy in the later years, so Martin may also have played a larger role in the Vaugondy map business than is apparent.

The surveying of Paris stimulated Didier Robert de Vaugondy's interest in the city's history and archaeology. He noted the vestiges of ancient walls "which enclosed the city in times long gone by and which are known by few people."[2] He marked these walls on the map using a single line for remains and a double line for walls which still existed. He complained that few city maps showed these historical details and recommended that proprietors of houses located in areas suspected of having ancient walls be careful in their re-building programs and that they should make notes of the excavated foundations.

The *Mémoires sur les différens Accroissements*, while only sixty-four pages long, rapidly surveys the history of Paris and the placement of the six walled enclosures from the Romans to Charles le Chauve to Louis XIII. As a *supplément* to the *Mémoires*, Didier bound in a plan of the Hôtel Royal de l'Ecole Militaire [Cat. 121], which showed in great detail the new building opened in 1756 for the school founded in 1753. The *Mémoires* also had an index to streets, including their older names, open squares, quays, bridges, colleges, convents, remarkable places, and the boundaries of the *quartiers*.

The "Plan de la Ville de Paris" went through several editions in Robert de Vaugondy's lifetime: 1766, 1771, 1772, 1776, each with changes in streets and buildings; after 1771, the plan shows parish boundaries too. It continued to be published after Vaugondy's death by his successor Fortin and then by Delamarche with new editions as late as 1818.

In 1760, Didier Robert de Vaugondy published the "Plan de la Ville de Paris" in a different format, as the *Tablettes Parisiennes / qui contiennent le Plan/de la Ville.../ avec une dissertation sur les Rues, Quais, Ports, Places publiques, Colleges, etc./...*. This was a combination of the *Mémoires sur les différens Accroissements* and the "Plan de la Ville de Paris" which was cut up and pasted onto pages with printed frames and titles, followed by an alphabetical index. "When I published the 'Plan de Paris' I wanted to make it as useful as possible and in a portable form. Thus the *Tablettes Parisiennes*."[3]

As he had done with his globes, Didier offered this portable map in various luxurious bindings:

in cardboard . . . . 8 *livres*
bound in calf . . . . 9 *livres* 12 *sous*
on *tablettes* of black or blue morocco,
with copper trim . . . . 21 *livres*
with silver trim . . . . 24 *livres*
on *tablettes* of red morocco,
with gilt trim . . . . 27 *livres*[4]

After 1760, Didier did not offer the *Tablettes Parisiennes* again in these morocco bindings. The 'Plan de la Ville de Paris' appears in the 1777 catalogue in a less luxurious version of the *Tablettes Parisiennes*:

cut-up, on linen, with a case . . . . 6 *livres* 4 *sous*
in-octavo, glued on taffeta . . . . . 3 *livres*[5]

Encouraged by the success of his Paris maps, Didier Robert de Vaugondy published another touring map, the "Environs de Paris / Divisés par Pays..." in 1761 [Cat. 117]. Prepared according to l'Abbé LeBeuf's *Description du Diocèse de Paris* it was accompanied by a small guide book, *Les Promenades des Environs de Paris en Quatre Cartes avec un Plan de Paris*. The maps included a general plan and four maps of the outer environs of Paris: St. Denis in the north, Bourg de la Reine in the south, St.Cloud

---

This attractive road plan of Paris ("Nouveau Plan Routier") is dated 1789. It gives some idea of the extent of the city in the time of the de Vaugondy family. (Courtesy of the Bibliothèque Nationale)

in the west, and Vincennes in the east. Vaugondy's classical quotation on the title page set the tone for using his *Promenades*: "Cantantes licet usque minus via laedet eamus" (May we yet go singing on our way, it makes the road less irksome.) [Virgil, *Eclogues* 9, 64] The text describing each promenade contains directions for leaving Paris, a description of interesting features to be seen along the way, and an historical description of the suburb — surely an early Baedecker.

Like the "Plan de Paris," this work continued to appear in the Vaugondy catalogues from 1764 until Delamarche's catalogue of 1803. Such continuous publication suggests consistent sales for Didier, who remarks that the map was adopted by the police, "for distribution to commissioners and inspectors."[6] Particularly satisfying for him must have been the high praise given to the plan of Paris and the *Mémoires* by the *Journal des Scavans* of November 1760 (pp. 729-733): "Robert [de Vaugondy] has prided himself on giving us a work as perfect as it can be. The author has joined beauty of execution to great precision, and although he only regards this advantage as an accessory, we can assure the public that it has hardly ever seen maps and plans so handsomely executed...We can only praise the zeal of M. de Vaugondy and the care he has taken for the exactitude of his work."

Celestial Maps

Didier Robert de Vaugondy applied that same zeal and care to a pair of maps of the heavens, the "Hémisphère Céleste Arctique...," and the "Hémisphère Céleste Antarctique...", published in 1764 [Cat. 1 and 2], which were joined by his text, *Uranographie ou Description du Ciel en deux Hémisphères, calculés et construits pour l'année 1763* (Paris: Robert de Vaugondy and Boudet, 1764). These maps of the northern and southern skies are very large (628 x 665 mm.) and exhibit a striking example of printing in two colours: black for the stars themselves and red for the figures of the constellations. Colour printing from copper plates was growing popular in the eighteenth century, and Paris was a centre of experimentation between 1740 and 1790. Didier Robert de Vaugondy had probably seen some of the sea charts produced in two colours by Jacques Nicolas Bellin of the Dépôt de la Marine.

Not only did Didier Robert de Vaugondy use this novel device of colour printing, he also gave the maps a unique projection: instead of representing the celestial sphere as a concavity, as if looked at from earth below, Didier showed it as a convexity, as if the eye were opposite the hemisphere projected. In order to obtain the precision and exactitude he desired, he traced the projection on the copper plate himself, just as he had done on the "Hémisphère Australe" for the duc de Croÿ.[7] The advantage of the convex projection was that it could be oriented on a table or desk in an ordinary way without reversing the point of view. This projection was praised by the astronomer de Lalande who, as, Censeur Royal, approved the book and maps in 1764, recommending Didier's planispheres as the best available because of this unique construction.[8]

Accuracy and Maps: The Question of the Flattened Earth

"Precision and exactitude": the words of praise in the *Journal des Sçavans* for his Paris map and the approval of de Lalande for his celestial maps must have been balm to Robert de Vaugondy after the harsh criticisms of the *Atlas Universel* ten years earlier, some of which had also appeared in the *Journal des Sçavans*. But were these the accolades Didier sought? While he obviously was concerned with precision, he had his own views on the importance of "accuracy" of maps. Clarity and resolution of conflict were of more interest to him than the more elusive "accuracy," as witnessed by his methods for reconciling conflicting data employed for the maps of the *Atlas Universel*, his interest in the cartography of hypothetical regions, and his understanding of the problems of projections and their distortions. The question of accuracy and the problem of the projection of an Earth flattened at the poles were the focus of his last paper presented to the Academy of Sciences in July, 1775: *Mémoire sur une question de géographie pratique: Si l'applatissement de la terre peut être rendu sensible sur les cartes géographiques* [Mémoire on a question of practical geography: if the flattening of the earth can be shown on geographical maps].[9]

The question required both mathematical and technical answers. A mathematical formula was needed which would translate the increasing dimension of the degrees of latitude towards the poles onto a projection. Once these dimensions and the resulting projection were plotted, they had to be transferred with the utmost accuracy onto a manuscript map, then drawn again onto the copperplate, requiring extremely accurate measuring, drawing, and engraving instruments. In the process, allowance had to be made for paper shrinkage as the map dried after printing. Even though such formulae could be worked out mathematically, for "ordinary maps" the differences in the size of degrees towards the poles was hardly perceptible to the human eye.

For Vaugondy, there was no point in going through the calculations necessary to show an increase of one-twelfth of a *ligne* (less than a millimetre) on a map of ordinary size (c.500 by 600 mm.) in the latitudes of Europe. He carefully pointed out the incompatibility of achieving mathematical accuracy with the practical problems of printing such an accurate map. He made the analogy that one can determine mathematically how thin an ounce of gold needs to be to cover 146 square feet; the difficulties lay in pounding the gold to such a thinness and building an instrument to measure it.[10]

Practical problems aside, Vaugondy also pointed out that there was no universal agreement about the absolute proportion of flattening because all measurements were based on human observations which themselves differed. He quoted de La Condamine, whose purpose in his expedition to Peru was to determine the shape of the earth: "Inasmuch as theory and actual measurements agree on making the earth a spheroid flattened at the poles, so do they differ about the *amount* of its flatness..."[11] This amount varied from 1/132 to 1/303 depending on the formula used. If accuracy were the issue, Vaugondy felt it was better to accompany a map with a table explaining the increase in the size of latitudes instead of straining to show it graphically.

Vaugondy rejected the argument previously advanced by the geographer Rigobert Bonne that unless geographers paid attention to this flattening they risked making serious errors on their maps which would have "troublesome consequences." He maintained that the degree of error was barely perceptible on a map and one should not hold a geographer responsible for this degree of accuracy. He concluded: "Let us not seek to dazzle the eyes of the public with an imposing display of calculations which...add nothing to graphic precision. It is overly judicious to demand in practice the same precision presented to us in theory, the employment of which our sense organs reject."[12] On this basis, Vaugondy felt that maps could be useful and beautiful without being completely subservient to the perfecting impulse of mathematics. In his view, the only type of map requiring such precision was the marine chart where the accurate plotting of longitude and latitude was paramount to navigators. His opening quotation from Horace summarized his thoughts: "Utilitas, justi prope mater et aequi" (Utility, godmother of correctness and balance) [*Satires* Book I, 1]

Didier Robert de Vaugondy was very sceptical of the so-called "accuracy" of maps that claimed to take into account the earth's flattening. In the *Institutions Géographiques* he indicted the authors of such maps as guilty of false advertising, considering them as guilty as those plagiarists who copy the work of others and after altering a date or decoration, display the work as their own. These authors, pretending to show the earth's flattening and "wishing to parade a scrupulous exactitude, announce to us maps on which they make perceptible the flattening of the earth towards the poles. These are charlatans who only cause a science that has deserved the research of famous men to fall into an all too real degradation."[13]

Didier Robert de Vaugondy must have been specifically thinking of Rigobert Bonne as one of these "charlatans." Bonne's map of the Mediterranean of 1762 was constructed on a projection with increasing latitudes, advertised as showing the flattening of the earth. Vaugondy's 1775 *Mémoire sur une question de géographie pratique* continued a debate begun by Bonne in 1765 when he had published his *Réfutation d'un ouvrage de M. Rizzi Zanoni [sic]* (Paris, 1765). Bonne's barbed invective was aimed at the *Dissertation sur les differents points de Géographie* by Giovanni Antonio Rizzi-Zannoni in which the Italian geographer had criticized those Bonne maps which claimed to show the Earth's flattening.

The Vaugondy *Mémoire* drew immediate fire from Bonne in his publication *Examen de Mémoire sur une question de Géographie pratique de M. Robert de Vaugondy* (Paris: Lattré, 1776). Published by Lattré, for whom Bonne had made maps for the *Atlas Moderne* (see Chapter Five), it appeared at the same time with a different imprint designed to take a further jab at Vaugondy's work: A Londres, Chez

Binding of the **Atlas Portatif et Militaire** *with the coat of arms of the marquise de Pompadour, who would become a subscriber to the Vaugondy* **Atlas Universel.** *(Courtesy of the Royal Geographical Society)*

M. Peevish à la Bévue Perpetuelle [London, at Mr. Peevish's at the sign of the Perpetual Blunder]. Bonne's tone was bilious and personally insulting from start to finish. He matched Vaugondy's Horace with his own selection: "Sumite materiam vestris qui scribitis aequam viribus et versate diu quid ferre recusent, quid valeant humeri." (You who write, take on a subject which is equal to your powers and think for a long time about what your shoulders will refuse to bear, and what they are strong enough to carry.) [*De arte poetica* 39]

Bonne's opening sentence made it clear that this was not merely a mathematical discussion: "It is with an extreme repugnance that I respond to the mémoire of Robert de Vaugondy." Bonne regarded Vaugondy's paper as a personal affront of large proportions, exclaiming, "Voilà — thus is war declared; M. Robert has made the first hostilities with every possible advantage." Bonne was uncomfortable about Vaugondy resuscitating an eleven-year-old work about which Bonne no longer felt entirely confident: "Our weapons are all rusty and one knows that I am not as I was eleven years ago, in form and ready. Thus my defeat is probably assured; nevertheless I will begin defending myself."[14]

Bonne's defence was a mixture of objective rebuttal and personal invective, claiming that Vaugondy did not understand mathematics, particularly the calculations determining the earth's flattening. Bonne saw this as a fault which stripped all credibility from Vaugondy's arguments: "his frequent mistakes lead me to think that the titles with which he is decorated suppose some knowledge but don't give any...One can forget about believing M. Robert when it concerns [mathematical] calculations."[15] According to Bonne, not only was Vaugondy adverse to "extrême précision," but his arithmetic showed that he was against any precision at all.

Bonne did not focus on Vaugondy's larger point: that showing the flattening of the earth on most maps was an exercise in futility and that maps should not be judged as to their accuracy in this regard. For Bonne, all maps, whatever their scale, should show the flattening of the earth, including "a large [scale] geographical map of the tiny seigneurie of Vaugondy, be it only a league square."[16] Didier Robert de Vaugondy did not respond to Bonne's paper, as there was little point in pursuing an argument of arithmetic when the larger issue of what constituted precision was not addressed. Nonetheless, he did list Bonne's rebuttal in his *Catalogue des Ouvrages* of 1777, just as he had done for Buache de la Neuville's response to the *Mémoire sur les pays de l'Asie et l'Amérique*.

The question remains as to why Didier presented his *Mémoire sur une question pratique* to the Academy at all; what had precipitated a shift of interest away from the mapping of polar areas, his subjects of 1773 and 1774, to this issue of accuracy as to the shape of the earth? He cannot have been manoeuvring for any official position: the chair of *adjoint gèographe* at the Academy had just been filled by d'Anville; Bellin (who died in 1773) had been replaced at the Dépôt de la Marine by Bonne; Vaugondy himself had been appointed to Bellin's post as Censeur Royal for matters geographic. The impetus may have come from Bonne's presentation of a paper in 1774 on projections and other "means for perfecting geography." Although Didier clearly states in his "Avertissement" that he wished to address the mémoires of Bonne and Rizzi-Zannoni of 1763 and 1764, he may have been spurred by Bonne's more recent paper to make his own case against overly emphasizing exactitude.

Didier Robert de Vaugondy's reviewers in the Academy, the astronomers de Lalande and Bailly and the geographer d'Anville, approved the *Mémoire sur une question de géographie pratique* with words which would also assuage Bonne: "One cannot exclude [from maps] a rigorous precision to which the Academy aims ceaselessly in its own works; but since in the present case, this precision is more metaphysical than practical, as it can be destroyed by the inevitable error of observations, we think that in applauding the efforts of geographers who try to account for the flattening of the earth, the Academy can continue to regard as good maps whereon this flattening is neglected..."[17] With this approbation, Vaugondy must have felt the victory of the map that is beautiful and useful over the slave to accuracy.

References

1. *Mémoire sur les différens Accroissements* (Paris: 1760), p. 6.
2. *ibid.*, p. 4.
3. "Avertissement,"*Tablettes Parisiennes* (Paris: 1760).
4. *ibid.*
5. Didier Robert de Vaugondy, "Catalogue des Ouvrages" (1777).
6. Didier Robert de Vaugondy, "Notice Raisonnée," *Mémoire sur les pays septentrionaux* (Paris: 1774).
7. "Avertissement," *Uranographie ou Description du Ciel en deux Hémisphères, calculés et construits pour l'année 1763* (Paris: Robert de Vaugondy and Boudet, 1764): p. vi.
8. J.N.de Lalande, "Cartes Célestes," *Supplément* to the *Encyclopédie*, Volume II, p. 253.
9. Academy of Sciences: Pochettes des Séances 26 juillet 1775. *Mémoire sur une question de Géographie pratique* (Paris: Boudet, 1775), BN: G. 3212.
10. *ibid.*, section 4.
11. *ibid.*, section 29.
12. *ibid.* Bonne also quoted Vaugondy as saying "que l'extrême précision n'a presque d'autre usage que de contenter l'ésprit philosophique." *Examen de Mémoire sur une question de Géographie pratique de M. Robert de Vaugondy* (Paris: Lattré, 1776): p. 22.
13. *Institutions Géographiques*, pp. xxiv-xxv.
14. Bonne, *Examen*, p. 5.
15. *ibid.*, pp. 14, 22.
16. *ibid.*, p. 28.
17. Robert de Vaugondy, *Mémoire sur une question*, p. vii.

CHAPTER SEVEN

"A man of exceptional simplicity and honesty, overcome by many misfortunes."[1]

If we judge the work of Didier Robert de Vaugondy by his own standards of combining the beautiful with the useful, and not on the modern scale of accuracy and exactitude, we must consider him a success. Continued esteem as measured by the publication of Vaugondy maps (albeit by Delamarche) well into the nineteenth century confirms this. However, our acclaim and the perpetual roll of the press do not translate into success in his time. Didier Robert de Vaugondy saw the last ten years of his life marred by professional disappointment, financial ruin, and great personal loss.

The Academy of Sciences

The titles which both Didier and his father Gilles enjoyed as *géographes du roi* carried no pensions. Only the *premier géographe du roi* received money amounting to 1000 *livres* per year for his title from the Crown. The only other government supported place for a geographer, outside of the army and navy, was the position of *adjoint géographe* (assistant geographer) in the Academy of Sciences. While the appointment of a geographer to the Academy was to some extent at the discretion of the Perpetual Secretary and/or the King, it was subject to the vote of the regular members of the Academy. A study of the *procès-verbaux* of the Academy during those periods when it was necessary to elect a new *adjoint géographe* shows that in the period prior to the election, geographers who were potential candidates presented papers to the Academy. Since the atmosphere of the Academy was often fraught with personality clashes and petty rivalries that could interfere with scientific discourse, such presentations were not always congenial affairs. From his own experience there, the duc de Croÿ described one meeting of the Academy as bedlam: "It is a pity that it is a bit of a bear garden and that the personal rivalries have more effect on it than those of science; they hardly listen to each other, so much is each member preoccupied with his own goals and affairs, only coming there because it is fashionable or out of self-interest."[2]

In 1773, after the death of Philippe Buache, the candidates nominated for the position of associate geographer were Jean Baptiste

Premier géographe du roi: Jean Baptiste Bourguignon d'Anville (1697-1782) (Photos courtesy of Map Collector Publications)

CARTE
DE LA LORRAINE et DU BARROIS
dans laquelle se trouvent
LA GENERALITE DE METZ et AUTRES ENCLAVES,
Dressée pour la lecture du Mémoire de M. Durival l'ainé,
ET DEDIÉE
AU ROY DE POLOGNE
DUC DE LORRAINE ET DE BAR.
PAR LE S. ROBERT DE VAUGONDY
Avertissement
Lorraine
Barrois
en verd
Pays Messin
Verdunois, et Toulois
en jaune
Metz
Nancy

Bourguignon d'Anville, Rigobert Bonne, Jean-Nicolas Buache de la Neuville, Didier Robert de Vaugondy and an unknown named Huguin.[3] D'Anville's election, with Bonne second, was clearly based on merit. At seventy-six, he had already enjoyed a long and illustrious career as member of the sister Academy of Inscriptions and Belles-Lettres. Although his stature as an historical geographer was unrivalled, his character was prickly, as admitted by his colleague Dacier in his eulogy of d'Anville in 1782: "Criticism was to him a kind of sacrilege against the focus of his cult; swept away by a holy anger, he would sometimes cry: 'You profane all antiquity!'"[4]

Given d'Anville's sensitivity, it was probably a mistake for Didier Robert de Vaugondy to point out to him a small error he made in his calculation of the difference between the latitudes of London and Berwick. Vaugondy wrote politely to d'Anville on 31st March 1777 but received no reply; he wrote again on 3rd April. "I return once again, Sir, to the charge and with all due respect wish you to observe ... that it should be **339** instead of **139** [miles] on p. 45, ll. 15 and 20..." [referring to d'Anville's book, *Considérations Génèrales sur l'étude et les connaissances que demand la composition des ouvrages de Géographie*]. To this charge, d'Anville answered coldly, calling the mistake a printing error, though it occurred twice and d'Anville had repeated erroneous calculations based on the error. To emphasize his disdain for Vaugondy's presumption, the *premier géographe du roi* did not even sign his name to his letter, but Vaugondy did not let d'Anville's icy tone and haughty manner pass unremarked. In a reply of 17th April he wrote: " I have the honour of being known and esteemed by many members of the Academy of Belles-Lettres who do me the courtesy of treating me very differently than you do. I know how to read and write and I have not described you as the *premier des géographes* **ordinaires.** It seems to me that it is a waste of time to bring errors of calculation to your attention...your last letter makes me see that you value no one but yourself. Truly, Monsieur, I would have received your servant better than you have received me...I will no longer point out errors to you since you take it so badly...and although you don't deign to sign your letters, nothing will stop me from acting with all the consideration you deserve and call myself your very humble and obedient servant." Vaugondy could not

Didier Robert de Vaugondy's map of Lorraine and Bar from the **Atlas Universel** *[Cat. 129], dedicated to Stanislas Leszczynski, King of Poland, Duke of Lorraine and Bar. (Photo courtesy of the William L. Clements Library, University of Michigan)*

resist a P.S.: "Anyone who wants to learn an egotistical style need only read your works."[5]

Didier Robert de Vaugondy's propensity for pointing out error, begun with the de la Fuente controversies of the 1750s, and his temerity with d'Anville cannot have furthered any potential career in the Academy of Sciences. When d'Anville died in 1782, Vaugondy was again nominated for the Academy's position of geographer along with the familiar competition of Bonne and Buache de la Neuville in addition to Edmonde Mentelle, a student of Philippe Buache. This time Buache de la Neuville was elected with Bonne once again in second place.[6]

Vaugondy must have been a strong candidate for the position. He was nearly sixty, enjoyed a solid reputation among *Académiciens* like Buffon, and had a champion at Versailles in the person of the comte de Vergennes, Minister of Foreign Affairs. But the second place rival, Rigobert Bonne, enjoyed the protection of Joseph-Jérôme de Lalande, an astronomer of considerable renown and political power in Academy circles. Yet both these senior geographers were defeated by a relative youngster, the forty-one year old Buache de la Neuville, nephew of Philippe Buache. He had served the Court with his uncle by tutoring the Dauphin, now Louis XVI, and his brothers, the future Louis XVIII and Charles X, for which he had already been awarded a pension of 500 *livres*.[7] The young Buache de la Neuville wasted no time in petitioning for all d'Anville's positions and perquisites. D'Anville died on 28th January 1782 and on 5th Februrary, Buache was already asking for his place in the Academy, his lodgings in the Galeries du Louvre, and his title of *premier géographe du roi*.[8] He received them all. Didier Robert de Vaugondy seems not to have indulged in such petitioning, nor did he even make a presentation to the Academy in 1782 as he had done prior to the election of 1772. Perhaps he was confident of his chances or perhaps he felt it was useless to try. Either way, we are told by Vaugondy's daughters that losing the election to Buache de la Neuville was a bitter disappointment to him.

Despite his failure to gain a position in the Paris Academy, Vaugondy was not without Academic connections. The Royal Academy of Sciences and Belles-Lettres at Nancy was founded in 1750 by Stanislas Leszczynski (1677-1766), King of Poland, Duke of Lorraine and Bar and father of Marie Leszczynski, wife of Louis XV. The institution boasted a lending library and 3000 *livres* a year to spend on books and manuscripts, and it offered two prizes of 600 *livres* each year in science and the arts to works of "evident utility." In April of 1756, Didier Robert de Vaugondy presented a map of Lorraine and Bar from the *Atlas Universel* [Cat. 129] to the Academy at Nancy and he was elected *associé étranger* on 8th May. He received the title of *géographe du roi Stanislas* with which he signed nearly all his maps thereafter: "géographe du roi et du duc du Lorraine et du Bar, membre de l'Académie des Sciences et Belles Lettres à Nancy."

Didier Robert de Vaugondy also attempted to become a correspondent of the Royal Society of London. He wrote to Thomas Birch, Secretary of the Society, in November of 1753, attaching the first instalment of maps from the *Atlas Universel* and offering to come to London himself to present the rest. He closed his letter: "Happier still will we be if we can earn the suffrage of your illustrious society. For if I dare to be ambitious for the honor of being in correspondence with it, I would regard this advantage as a recompense capable of repaying the work I have put into [the atlas.]"[9] Birch did not invite Vaugondy to London; he merely thanked him on behalf of the Society for the maps from the atlas and praised and encouraged his continued work.

Vaugondy finally received government approbation by his appointment as Censeur Royal in 1773, replacing Jacques Nicolas Bellin, who had died in 1772. In this position, he was responsible for approving books on geography, navigation, and voyages. In one instance, he approved a work with his own map in it: de Fréville's *Histoire des Nouvelles Découvertes faites dans la Mer du Sud en 1767, 1768, 1769 et 1770* (Paris, 1774). [Cat. 416] But even this position carried with it disappointment, for Vaugondy apparently expected the appointment to carry a pension which never materialized.

The title page from the **Atlas Céleste de Flamsteed** *by Jean Fortin, the globemaker who took over the Robert de Vaugondy map and globe business in 1778. (Photo courtesy of the William L. Clements Library, University of Michigan)*

ATLAS CÉLESTE DE FLAMSTÉED,

APPROUVÉ

PAR L'ACADÉMIE ROYALE DES SCIENCES,

ET PUBLIÉ

SOUS LE PRIVILEGE DE CETTE COMPAGNIE.

SECONDE ÉDITION.

Par M. J. FORTIN, Ingénieur-Mécanicien du Roi & de la Famille Royale, pour les Globes & Spheres.

A PARIS,

Chez { F. G. DESCHAMPS, Libraire, rue S. Jacques, aux Associés.
l'Auteur, rue de la Harpe, près celle du Foin.

M. DCC. LXXVI.

The business has to be sold

Financially, the Vaugondy family's only resource came from the sale of maps, globes, and atlases, their work as illustrators, special commissions like that for the duc de Croÿ, and private teaching of geography and mathematics. None of it was regular, reliable income and it must have become more difficult to sustain after the death of Gilles in 1766. In spite of the productivity of the 1750s and 60s, Didier Robert de Vaugondy was unable to meet financial demands on him at home. His wife became mentally ill and needed to move to the country, so, according to a later account by Didier's daughters, he sold his map business in 1778 in order to buy her a country home.

The buyer was Jean Baptiste Fortin (1740-1817), a globe maker, perhaps related to the well known scientific instrument maker, Nicolas Fortin. Fortin was so well regarded for his construction of spheres and globes that he had particularly requested from the Minister of the Booktrade in 1772 a warrant or patent which would allow him to do his work mounting "globes and spheres...without being bothered by the community of papermakers and that of makers of mathematical instruments who threaten him and trouble him in the exercise of his art."[10] The request was forwarded to the Academy of Sciences, where it received support. His reviewers, de Lalande, the astronomer, and Grandjean de Fouchy, the Perpetual Secretary, described Fortin's work as very special: "Fortin is now the only one in Paris who concentrates on this kind of work. He is noted for his intelligence and exactitude of execution. He has executed a sphere of Copernicus of a particular type...The astronomers of the Academy only recognize and employ Fortin."[10]

If indeed Fortin was the only person in Paris making globes and spheres of the quality applauded by the Academy of Sciences it is likely that Robert de Vaugondy not only knew Fortin's work but that they had collaborated on globes. The terms of the sale of Vaugondy's business to Fortin are not known, but their relations after the sale seem to have been friendly enough. When Fortin published the catalogue of the Vaugondy stock in 1778, he added a note describing Vaugondy's continuing availability for geographic work: "Although M. Robert de Vaugondy has discharged himself from the sale of his geographical works, he will always continue to look after their corrections and additions relative to the new discoveries that might be made in the future and augment them with some works of interest.

Those persons who wish to make a specific collection of maps may still address themselves to him. He dares, from knowledge he has acquired since tender youth, to promise to deserve their confidence in choosing good works, French as well as foreign....He still resides in his house on the quai de l'Horloge du Palais where one should address letters to him."[11]

After the sale to Fortin, Didier's fortunes took a turn for the worse. A "friend" whom he had left in charge of his household of wife and children in the country absconded with the funds entrusted to him. By 1782, Didier had to abandon his house on the quai de l'Horloge, leaving his creditors to sell his goods. We learn this from a manuscript note inside the cover of a Peter Bertius atlas that had originally belonged to Nicolas Sanson. The atlas was bought by an André Mercier in 1782 at the sale of Vaugondy's household effects. Mercier seems to have been so moved by the provenance of the book and the circumstances of its sale that he inscribed the volume in Latin. Translated, it read: "This book belonged to Nicolas Sanson, the King's geographer; afterward it was Robert de Vaugondy's, King's geographer, who [*illegible*] by many creditors, abandoning all his domestic goods, has escaped and fled, and from the public sale of the goods, this same book has come to me Andreas Charles Francois Mercier of La Rochelle, in 1782. This sale was carried out on the quai des Morfondus [quai de l'Horloge] in the deserted house of Robert de Vaugondy. He was a man of exceptional simplicity and honesty, who had been overcome by many misfortunes. Death had taken away his two infant sons who had offered great hopes of natural talent, which pierced his wife with such grief that she went insane, hence the collapse of his business."[12]

In the year of the sale, 1782, Didier had again missed being elected to the Academy of Sciences and named *premier géographe du roi*. Where he went to live from the quai de l'Horloge (trusting M. Mercier's account of events) is a mystery. He did not disappear entirely from the geographical circle in Paris because he collaborated with Nicolas Le Clerc in 1784 on the eight foot "Globe de Bergevin" described in Chapter Two, preparing the two huge hemisphere maps with the geography for the globe. He also supervised the engraving of Le Clerc's two maps of the Baltic Sea and the Gulf of Finland and helped Le Clerc defend the maps against critical attack in 1785 by appearing as a witness in a tribunal before the Maréchal de Castries, Minister of War.[13]

The Death and Succession

Vaugondy died in 1786 of causes unknown and in a state "bordering on poverty."[14] His family's near penury was momentarily relieved nine years later, when the Revolutionary National Convention awarded him 1500 livres posthumously, which his widow, Marie Dangy, was to have received. However, instead of being given directly to her, the money was entrusted to Charles-François Delamarche, who had purchased the Vaugondy stock from Jean Fortin in 1786, the year of Vaugondy's death.

Delamarche (1740-1817) was a lawyer by training; how he became interested in geography is unknown, but his background recalls that of Jean Frémont, Gilles Robert Vaugondy's original partner, a lawyer with strong interests in geography. Delamarche knew Jean-Baptiste Fortin, Robert de Vaugondy's successor, either from "kinship, friendship, or [mutual] interests."[15] His erstwhile service to the Vaugondy family may be read in the *Procès-Verbaux du Comité d'Instruction Publique de la Convention Nationale*.[16]

The Convention's Committee for Public Instruction recognized Robert de Vaugondy for his work on the Bergevin globe and placed him on the "supplementary list of *savants* who have a right to national munificence." The list was formed in January, 1795, and a figure of 1500 *livres* was determined and decreed on 16 April. Five days later, Vaugondy's widow, Marie Louise Rosalie Dangy, petitioned the committee for the 1500 *livres* to be paid to her. However, on the first of May, Charles-François Delamarche intervened and the Committee decided to give him the 1500 *livres* on the widow Vaugondy's behalf. The published record does not contain his arguments, but in June, Delamarche announced the death of the widow Vaugondy to the committee and told them that she had left four [sic] daughters who were in the greatest need. The committee then turned the matter over to the Executive Commission. During the Commission's deliberations another geographer named Robert — François Robert of Dijon — objected to the presence of Robert de Vaugondy's name on the list of *savants* deserving pensions. In a printed letter to the Committee, he claimed that he was the Robert intended by the committee and attached a brief biography and annotated bibliography.[17] His view of Robert de Vaugondy as a geographer was not flattering: "It was an error. Robert de Vaugondy, under whose name are some atlases and maps, has been dead for nearly 34 years. His son, who was nothing in geography, has been dead for about ten years, without a male heir. The encouragements accorded by the Convention for the advancement of Sciences do not concern either of them. Nor do they concern the widow of Vaugondy *fils*, who is a complete stranger to this matter, and who moreover was not on the list. All the same she presented herself and received the gratuity that was awarded to me by the Convention."

There is some irony in François Robert's complaint: had Gilles Robert begun his career using his father's last name of Vaugondy, such confusion would not have occurred, then or now.

Persuaded by François Robert's dossier and without questioning his view of Didier Robert de Vaugondy as "nothing in geography," the Committee decided in August to give 1500 *livres* to François Robert. They instructed Delamarche to give the children and any other heirs of Vaugondy the gratuity which he was holding, after making a deduction of 77 *livres* 7 *sous* 6 *deniers* "dépense imputable." The Vaugondy daughters were therefore to receive 1423 *livres* 12 *sous* 6 *deniers*. Whether or not they did receive this money is unknown, but a later series of documents in the Archives Nationales suggests that they probably did not.

From 1807 to 1825, Vaugondy's three daughters regularly petitioned the Ministry of the Interior for financial help, and their pleas did not always fall on deaf ears.[18] The comte de Lacepède, Grand Chancellor of the Légion d'Honneur, supported their request for aid: "I knew their father very well. [He was a man] for whom Buffon and d'Aubenton had a particular attachment."[19] The baron Walckenaer, a geographical writer and enthusiast, equally supported their case. In spite of these champions, the Ministry was not generous, allowing each of the daughters only 100 francs, and not upon every request. One of their longer petitions, a letter written in 1812, tells a sad story of their father's difficult last years and their present situation. Addressed to the Secretary of the Minister of the Interior and dated 20th

August 1812, it is signed by three of the sisters but is in the hand of Paule, the youngest. Its poignancy and the rarity of such a document about the life of a geographer make it worth printing in full:

"The daughters of the late Robert de Vaugondy, geographer to the King, take once more the liberty of recalling to Your Excellency their unhappy situation. We dare to believe that the name of our father is favourably known to His Majesty, who will find it difficult to believe that a man who had worked so hard for so many years had never received any reward from the Old Regime. In fact, the works commanded by Louis XV were never paid for. During the reign of his Successor, [our father] certainly claimed payment for them, but his repeated requests were not heard. Then with the suppression of the *tontines*, he lost all that had come to him from the inheritance from his father.[20]

Some years later our mother became deranged; [our father] made every sacrifice for her that his heart told him to; he never wanted to be separated from her. Country air being prescribed for her, he sold the geography business in order to send us with her some leagues from Paris. My father, never imagining that one man would trick another, entrusted his money to someone he thought to be his friend, someone who exploited him and placed himself at the head of the new household my father had established, since his work did not permit him to leave Paris. My sisters at this time were very young and I was only four years old. This so-called friend in whom my father had the greatest confidence, from whom he had never asked any account of his monies for fear of wounding his feelings, this friend took off with the money, leaving us without resource. This is the moment at which my sisters were obliged to work for a living. My father had no remaining income except for that from students, which he shared with the family.

This was our father's and our existence since 1778. The death of M. d'Anville some years later gave him hope for the place of *premier géographe*, but the rival candidate was lucky and my father, although superior in knowledge, did not get it.

Having been named *Censeur Royal* in 1774, twelve years later he should have received the pension attached to this title, in 1786, which was precisely the year in which he died. Fate even refused him the consolation of seeing his wife placed in a *pension*, a thing for which he had taken so many steps. Some months before his death he had again solicited those brilliant aquaintances among whom his reputation gave him a free range, but he died, a soul torn by leaving behind so many unfortunate ones.

At the death of my father, we found ourselves the possessors of a manuscript work in three volumes, for which he had published a prospectus some years before. A subscription had been opened but there were not enough subscribers to print the work.[21] This work as well as a great number of maps were taken from us by a man whose name I won't mention.[22] Under the specious pretext of placing our mother in a *pension*, where did he place her? At *la Salpêtrière*.[23] It was there that our mother ended her days. When M. de Vergennes learned of my father's death, he wrote us a letter filled with concern for our loss and he even had the goodness to send some help; he should have spoken to the King about our situation, but death took him from us and deprived us of the one person really attached to our father who could render us less unfortunate.

Monseigneur, you see that unhappiness has not ceased pursuing us. One can hardly believe that my father, who enjoyed such a great reputation and whose name was so esteemed, left children without money. When the name strikes the ear of people we meet, it surprises them to learn that we are the daughters of the geographer. You yourself, Monseigneur, when you had the goodness to hear me the first time, seemed surprised. It is that which makes me take the liberty of reciting for you our misfortunes.

Monseigneur, deign to cast on us a look of benevolence at this time when His Majesty surely has under his eyes the works of my father. Deign to be our interpreter to him; if my father were still alive, His Majesty would want to share with him the benefices he accords to men of merit. His old age would have been sheltered from want. His unfortunate children, without kin, without any resources, nearly beyond being able to work, dare to ask you.

We have the honour of being the very humble servants of Your Excellency

Marie Rosalie}
Marie Adelaide} De Vaugondy
Paule }"

While we must remind ourselves in reading this that Paule Vaugondy was making a case for help, and that her memory of events as a four-year-old were no doubt reinforced by family stories necessarily tinged with bitterness, we cannot help but be struck by the description of Didier Robert de Vaugondy's bad luck.

Paule's eloquence did not succeed in procuring any permanent pension for the sisters. The Vaugondy dossier contains a series of petitions for money from 1812 until 1824. Some years the sisters received two hundred francs (1808, 1821), some years three hundred (1811), and in at least one year, nothing (1818). By November of 1824, the sisters were "reduced to a state near indigence." Two of them — Rosalie and Adelaide — aged sixty-four and sixty-five, overcome by infirmities and no longer able to work, petitioned for admission to a *hospice des incurables* or some other shelter.

In 1825, only two of the daughters were still alive. Paule wrote now to the recently founded Société de Géographie, once again describing their plight. The Minister of the Interior had only been able to offer her older sister a place at La Salpêtrière, where she had been for one month. But "the state in which one lives at that place, as regards food and many other details, makes the youngest Vaugondy daughter take the liberty of setting this exposé under your eyes and appealing to your benevolence."[24] Paule, then fifty-one, asked for some annual help which her sister could use as long as she lived and for something for herself so that when her own infirmities grew to the point she was unable to work, she would have some comfort. No evidence has been found to show that the sisters received further assistance.

We should not leave Paule's story, however, without asking why Didier Robert de Vaugondy did not train his daughters as geographers? After all, he advocated children of 'both sexes' to study geography ("Discours sur l'étude de la géographie," *Nouvel Atlas Portatif*). In the archives of the Academy of Nancy he is credited with presenting a discourse entitled "Les femmes doivent s'occuper des lettres" (though the actual manuscript has not been seen, so it cannot be ascertained if this was by Vaugondy or a secretarial error.) He encountered and employed women as engravers: Mme Delahaye, the sisters Elizabeth and Marie Catherine Haussard. He must have known of the map of the eclipse of the sun produced by the amateur astronomer Mme Le Paute, engraved by Mme Tardieu and Mme Lattré which had drawn the following favourable comment from the *Mémoires de Trévoux* (June 1762, pp. 1530-1534): "Thus here is a new proof that the knowledge of the Sciences and Arts is not an exclusive privilege with access permitted to only one sex, and women will distinguish themselves so in this noble career when they begin to desire to busy themselves usefully or when the abuse of only giving them a frivolous education is felt."

The Hôpital de la Salpêtrière in Paris. An engraving from "Les Délices de Paris" (1753). On the site of a small arsenal where gunpowder was manufactured with saltpeter, Louis XIV in 1656 established a hospital for the poor women of Paris. Over the decades the hospital became a dumping ground for the insane, the infirm, orphans, prostitutes and criminals, who all were subject to the same treatment. (Photo courtesy of the British Library)

Geographers like Didier Robert de Vaugondy perhaps had higher aspirations for their daughters. Geography was not like book or print trades where women followed husbands or fathers into the business. Unlike engravers or publishers, geographers did not think of themselves as artisans or businessman but as intellectuals with a place among the *philosophes* of the Enlightenment. The sons of geographers became geographers, as witnessed by the families Sanson, Delisle, Cassini, Jaillot, Nolin, Robert Vaugondy, and Bellin. Early in the century, there were examples of women inheriting geographical stock: Pierre Duval's daughter, Marie Angélique, at the beginning of the eighteenth century and Guillaume Delisle's widow, who sold his maps until her death in 1745. But neither of them were geographers; rather than producing new maps, they simply acted as publishers for the old. Later in the century, daughters of geographers do not enter the business; they either marry other geographers, as in the case of Guillaume Delisle's daughter and Philippe Buache, or marry into a higher social class, as did one of d'Anville's two daughters who married a government official in charge of the roads and bridges of Paris. Or they did both: the three sisters who inherited the Jaillot stock in 1749 from their brother, Bernard, carried on the family business until one of them married the wealthy Jean-Baptiste Michel Renou de Chauvigné, who took over both the Jaillot name and the business. Others simply led quiet lives in reduced circumstances, like the Vaugondy girls or d'Anville's other daughter who entered a convent. Philippe Buache's daughter died as a child, and he had no children from his second marriage. While this analysis is not exhaustive, it does suggest that women were not encouraged as geographers in the way they might have been as engravers or printers.

Something of the attitude Didier Robert de Vaugondy may have had toward the education of his daughters is expressed in Mercier's Latin inscription in the Bertius atlas: that it was the loss of the infant *sons* which caused the family's grief and ultimate failure. We note that of Vaugondy's two sisters, neither appeared in a geographical context, unlike his brother Martin de Savigny. With little precedent for educating women to a higher level, with no capital to subsidize their entry into circles wherein they could marry well, and with no desire to give them artisan skills for what would be a social drop, it was perhaps inevitable that the Vaugondy sisters should end their days impoverished.

Documents reveal that Didier Robert de Vaugondy received little if any revenue from his cartographic work after the sale to Fortin, about 1778. Similarly, Delamarche did little to help the Vaugondy family after his purchase of the stock from Fortin in 1786. Yet there is no doubt that the Delamarche business was successful. Baron Walckaener's brief biography of Delamarche in the *Biographie Universelle* tells us that "he exploited the Robert stock with an extraordinary cleverness. For nearly thirty years he had almost exclusive possession of the trade of globes and maps in all the colleges and houses of education, which he owed to the low prices at which his books came. In order to establish for himself a personal reputation, he put his name to maps by Robert [de Vaugondy] which he rejuvenated by re-engraving the plates."[25] Delamarche was succeeded by his son Félix who, in partnership with Charles Dien, continued to publish maps and atlases until the 1840s.

It is ironic that the profit made from the name of Robert de Vaugondy did not enrich its inheritors. It marks a considerable change from the family solidarity seen at the inception of the Vaugondy map business with Moullart-Sanson's will in which every effort is made to continue the Sanson name and stock. It also contrasts with Jean Frémont's willingness to risk capital and forgive debts in order to see his friend Gilles Robert Vaugondy succeed in that business. It further leads us to speculate on the extent to which this type of failure in the map trade was typical of eighteenth-century Paris life.

The Map Trade in Eighteenth Century Paris

The failure of the map business for Didier Robert de Vaugondy may be attributed, in part, to his lack of business acumen, a too trusting nature, a naïve faith in the marketplace, perhaps to a personality too caught up in his subject and not enough concerned with the political pursuits of currying favour, wooing patrons, and winning places in the Academy. He may have spread his work too thin, producing every type of geographical work possible: globes of all sizes and types, wall maps, folio maps, small maps, astronomical charts, theoretical treatises; guides and plans of Paris; cartographic illustration of books; and teaching manuals.

Given the capital intensive nature of cartography — the costs of copper, engraving, and printing, not to mention the geographer's time — it was difficult for a geographer to produce new maps and recoup his expenses from their sale. D'Anville described the labour-intensive nature of map production and the need for a Mycenas in his *Analyse géographique de l'Italie:* "The undertaking of a body of general maps demands such great studies in its preparation, of works often repeatedly on the same subjects, such an ample collection of materials, be it specific maps, *mémoires*, instructions in detail, that although for around twenty years there have left my hands a considerable number of pieces or particular attempts, I would still not dare put my hand to a parallel work if the goodness and truly royal munificence of a Great Prince had not fashioned for me the dictates of this enterprise and facilitated its execution...."[26]

It was very difficult for geographers to publish maps on their own. It is significant that when the Vaugondys were simply reissuing Sanson maps, they barely made a profit (see Chapter One). When they set out to create collections of new maps in atlases they relied on the support of book publishers: Pissot and Durand for the *Atlas Portatif* and of course Antoine Boudet for the *Atlas Universel*. Yet even with financial backing, the geographer needed to be concerned about sales, as Didier Robert de Vaugondy wrote to the duc de Croÿ regarding their map of the 'Hémisphère Australe.' In spite of its low price of three and five *livres* , sales had been slow "except for two deliveries of twelve on cloth and thirty on paper...to Brest and Bordeaux. But I hope once the holidays are over, that the additions and corrections you plan to make will renew and arouse the curiosity of amateurs."[27] Those geographers who did not have government-supported posts had to diversify their stock as the Vaugondys did. Didier's contemporaries, Rigobert Bonne and Louis Charles Desnos, were two who also produced a wide variety of maps, globes, guides, and texts. Like the Vaugondys they did not make a success of it. Desnos declared bankruptcy in 1784, and Bonne's ventures were usually protected by his collaboration with the engraver-publisher Jean Lattré. Geographers needed what Philippe Buache had objected to in the *Atlas Universel* lawsuit, namely, the involvement of book publishers and/or print dealers to publish maps.

Buache had argued for a closed market, strictly regulated, like that of the book trade itself, in which only geographers could publish and sell maps. He reasoned that this would reduce plagiarism, maintain quality, and keep profits in the hands of geographers. Unfortunately, the map buyer was not as discerning as Buache would have liked and was easily contented with the cheap copy. The market, though growing, could not support many individual geographers in competition with each other. There was no review mechanism or sufficient general education for the public to distinguish the good from the mediocre. While geographers competed for a limited market, there could be no real success until one took over the stock of another.

In this way Charles Delamarche was able to succeed where the Vaugondys failed: not only did he buy the Sanson-Vaugondy-Fortin stock but he also purchased the major portion of the Lattré stock, which included most of the work of Bonne, Desnos, and Janvier, thus acquiring his chief competitors. Unlike the Vaugondys, Delamarche did not seek to add much new material to his inventory; he corrected and augmented plates and rearranged maps to create new atlases but did not produce many original works. He profited, too, from the course of events after 1789, shifting the attention of his business to a new and lucrative corner of the post-Revolution market: schools. By educating a captive present audience, he insured a future clientele.

The story of the Robert de Vaugondy family illustrates how the method they adopted of map production and distribution could not support much growth. The poignancy of their tale demonstrates sharply that the market, however fickle, was the only resource for a geographer without a patron or institutional connection. Yet despite the precarious economic conditions of the period, both Gilles and Didier Robert de Vaugondy continued to produce maps, atlases, and globes of high geographical and aesthetic standards, testifying to a love of their work, a desire to serve the public, and a commitment to produce maps, globes, and atlases which were 'beautiful and useful.'

References

1. From a Latin inscription written by André Mercier in 1782 in a Peter Bertius atlas that had belonged to Didier Robert de Vaugondy. Mercier bought the atlas at the sale of goods in Vaugondy's house. See p. 12.
2. *Journal de Croÿ*, Volume III, pp. 54-55, quoted by Jean Etienne Martin-Alaric, *Bougainville, navigateur et les découvertes de son temps* (Paris: Presses Universitaires de France, 1964): Volume II, p. 1399.
3. Academy of Sciences, *Procès-Verbaux* 24 mars 1773, p. 76.
4. Bon-Joseph Dacier, "Eloge de d'Anville," *Histoire de l'Académie des Inscriptions et Belles-Lettres* (Paris, 1782): vol. XLV, p. 160. Also quoted in Gabriel Marcel, cited below, n. 5.
5. These letters were in the private collection of Gabriel Marcel, who published them in "Correspondance de Michel Hennin et de d'Anville," *Bulletin de la Section de Géographie* (Paris: Comité de Travaux historiques et scientifiques, 1907): pp. 446-448.
6. Academy of Sciences, *Procès-Verbaux* 20 avril 1782, p. 72.
7. AN: O^1 120, f°277, 27 October 1773.
8. AN: O^1 1916, f°33, 5 February 1782.
9. British Library: Add.MS. 4320, fol. 15, Didier Robert de Vaugondy to Birch, 21 Novembre 1753.
10. Both Fortin's request and the response of de Lalande and Grandjean de Fouchy are in Academy of Sciences, Pochettes de Séances, 29 août 1772. Stevenson (*op. cit.*) gives Fortin's dates as 1750-1832. However, the dossier on Jean-Baptiste Fortin in the Conservatoire Nationale des Arts et Métiers, Doc. 3928, cites his birth as 2nd December 1740 and death on 16 October 1817. Fortin bought property at Bagneux in 1784, where he was the mayor in 1793 and 1794, and where he died, childless, leaving his property to his wife, Marie Guériot. The purchase of the Bagneux property leads one to speculate on the date of the sale of the Vaugondy stock to Delamarche. No document recording this sale has been found; the date of 1786 comes from the entry in the *Biographie Universelle* by Walckenaer. Fortin's purchase in Bagneux might suggest a date of 1784. Fortin's will is in the Archives des Hauts des Seine, CXXVIII, 336, 1er fructidor An 12.
11. BN: Ge.FF.7192 pièce: "Catalogue des Ouvrages qui composent le Fonds de Géographie de MM. Sanson et Robert de Vaugondy...qui se trouvent chez le Sieur Fortin," (Paris: Vve. Hérissant, 1778).
12. This atlas and its inscription came to light in Catalogue Three of J. T. Monckton, Chicago, Illinois, item 29. I would like to thank the Monckton gallery for allowing me to study the atlas and copy the inscription.
13. *Examen impartial de la critique, sans nom d'auteur, des cartes de la mer Baltique et du golfe de Finlande, présentées à M. le Maréchal de CASTRIES, ministre et secrétaire d'Etat de la marine, par LE CLERC, écuyer, chevalier de l'Ordre du Roi et membre de plusieurs Académies.* Paris, 1786, especially pp. 11-13.
14. Weiss, "Robert," *Biographie Universelle*, ed. Michaud, Volume 36, p. 139.
15. Walckenaer, "Delamarche," *Biographie Universelle*, ed. Michaud, Volume 10, p.296. Walckenaer says that Delamarche was aided in making globes to accompany his *Traité de la Sphère* (1790) by Fortin, "with whom he had forged close connections of family, friendship, or interests."
16. M.J.Guillaume, ed., *Procès-Verbaux du Comité d'Instruction Publique de la Convention Nationale*, (Paris, 1907): Volume VI, pp. 76, 90, 119, 155, 274-275, 426, 472, 486, 590. All the following information about the Committee's deliberations is taken from these pages.
17. The printed version of this request exists as *Au Comité d'instruction publique*, (Dijon: P. Causse, An III) BN: 8° Ln27.17598. François Robert (1737-1819) was the author of the geographical sections of Panckoucke's four volume *Encyclopédie Méthodique* as well as a number of other works.
18. AN: F^{17} (1542^2) Dossier Robert de Vaugondy. Série F^{17} is the division of Science and Letters within the Ministry of the Interior, formed after the Revolution as a continuation of the Committee of Public Instruction.
19. *ibid.* , Letter no. 21, 19 février 1809, from comte de Lacépède to M. le comte de Crétat, Minister of the Interior. Louis-Jean-Marie d'Aubenton (1716-1800) was a naturalist and anatomist who worked with Buffon at the Jardin du Roi and contributed to the *Histoire Naturelle* and the *Encyclopédie.*
20. *Tontines* were a form of group life insurance established by the government as a means of borrowing money. The interest increased with the age of the acquirer and the eventual deaths of all the members of the group; ultimately it was very costly for the government. The suppression was probably that of 5th July 1770. Maurice Block, *Dictionaire de l'Administration Française* (Paris, 1862) pp. 1520 ff.
21. This was the *Géographie ancienne* first advertised and approved in 1760.
22. Delamarche? He later published a *Géographie Ancienne* which sounds like Robert's proposed *Atlas de Geographie Ancienne*: "61 maps in folio, par Sanson et Robert, 78 *livres*." Delamarche, *Catalogue des Globes, Sphères, Atlas, Cartes, et autres ouvrages Géographiques et Astronomiques*, (Paris: Delamarche, 1803): p. 12. This would put Delamarche in a rather different light than that shed by the proceedings of the Committee of Public Instruction.
23. The Hôpital de la Salpêtrière was founded by Louis XIV in 1656 as a hospital for the poor women of Paris. Over the decades it took on the role of a prison, and by 1790 it housed 8000 women: indigents, prostitutes, criminals, and the mentally ill.
24. BN: Mss. de la Société de Géographie, Colis No. 19 (3081).
25. Walckanaer, in *Biographie Universelle*, ed. Michaud: Vol. 10, p. 296.
26. d'Anville, *Analyse géographique de l'Italie* (Paris, 1744): p. viii.
27. Dion, "Itinéraire Intellectuel," p. 255, citing Archives du Duc de Croÿ, Dulmen, K 24, letter from Didier Robert de Vaugondy to de Croÿ, 1 Sept 1774. Brest was the site of the Académie de la Marine; Bordeaux was of course another major port city. As with the subscription sales of the *Atlas Universel*, the "Hémisphère Australe" was being sold in quantity, probably to book or mapdealers.

BIBLIOGRAPHY

This bibliography cites the primary manuscript sources used for the biography of Gilles and Didier Robert de Vaugondy, the printed works of the Vaugondys, and a selected group of secondary literature from the twentieth century. Other books and articles have been cited in the notes.

PRIMARY SOURCES
Manuscripts

(Paris, unless otherwise noted)

Archives de l'Académie des Sciences:
Pochettes de Séances
Procès-Verbaux

Archives des Hauts de Seine:
CXXXVIII, 336, 1er fructidor an 12}
CXXXVIII, 367, 20 octobre 1817} concerning J. Fortin

Archives Nationales:
Série F^{17} 1542^{2}, Papiers du Comité d'instruction publique dossier D. Robert de Vaugondy
Série O^{1}, Maison du Roi
O^{1} 78, (294-5): brevet for Gilles Robert Vaugondy
O^{1} 120, (277): pension for Buache de la Neuville
O^{1} 1194 (29 déc. 1751): Louvre, globe project
O^{1} 1671 (128, 129, 130): Louvre, globe project
O^{1} 1672 (165, 220, 221): Louvre, globe project
O^{1} 1907B (1751: 46): Louvre, globe project
O^{1} 1916 (33, 106, 126): Buache de la Neuville
Série 2JJ 79: papers of J.N. Delisle concerning de la Fuente
Série Y, Archives du Chatelet de Paris
Y 927^{A}, 24 mars 1769: estimation de la maison, rue du Bac/rue St. Dominique
Série Z^{1J} 927, Procès-verbaux d'expertise des greffiers des bâtiments
Minutier Central (Notarial archives):
I, 353, 22 mai 1731: transport de droits, Perrier/Frémont, Robert
XX, 561, 19 juillet 1737: transaction, Frémont/Robert
XX, 606, 12 October 1751: dépot de testament, Frémont
LXXIII, 716, 10 décembre 1739: bail, Frémont,Robert/Raillard
LXXVI, 259, 19 décembre 1733: vente et constitution, Mariette/Robert
LXXVI, 270, 13 juin 1738: constitution, Raillard/Robert
LXXVI, 329, 31 mars 1751: quittance, Raillard/Robert
LXXXIX, 410, 29 mai 1732: transaction, Villemarie/Frémont,Robert
XC, 341, 11 avril 1731: bail, Helye/Robert, Frémont, Perrier
XCI, 880, 17 décembre 1751: bail, Chaban/Robert
CXVII, 188, 30 octobre 1702: donation, Villemarie/Moullart-Sanson
CXVII, 373, 22 juin 1730: testament, Moullart-Sanson
CXIX, 267, 27 février 1742: constitution, Robert/Frémont
CXIX, 268, 9 mai 1742: constitution, Robert/Frémont
CXIX, 272, 14 mars 1743: partage, Robert/Frémont
CXIX, 304, 28 mars 1751: constitution, Robert/Dubuisson

Bibliothèque de l'Arsenal:
Archives de la Bastille (Bicêtre): Ms. 11,166: Jean Vaugondy

Bibliothèque de l'Institut:
Ms. 2311 (12): Buache de la Neuville, Examen d'un mémoire de Robert de Vaugondy
Ms. 2314 (10): Buache de la Neuville, Mémoire sur les pays de l'Asie et de l'Amérique

Bibliothèque Mazarine:
Ms. 1542 (ancien 2688), Bergevin Globe

Bibliothèque Nationale:
Manuscrits:
Fichier Laborde:
n.a.f. 12082 (18095): Guillaume Delahaye
12131 (38381): Pierre Laurent
12195 (64711-64714): Gilles Robert Vaugondy
Collection Tours et Anjou 14 (17,18): maps by D. Robert de Vaugondy

Cartes et Plans:
Ge.FF.13373: Buache contre Vallet et Dheulland

Ge.FF.13374: Procès de Géographie entre Boudet et Delahaye, graveur de cartes; écrit en très grande partie par Buache (1754)
Ge.FF.13375: Discussion de Géographie entre Buache et le libraire Boudet (manuscrit de Buache)
Ge.FF.10808: manuscrit de d'Anville
Manuscrits de la Société de Géographie: Colis No. 19 (3081).

Conservatoire Nationale des Arts et Metiers:
Centre de Documentation de l'Histoire des Techniques: Doc.3928, dossier Fortin

Bibliothèque de la Ville de Nancy:
Mss. 920-921, Tableaux Synoptiques et Annuels des Procès-Verbaux de l'Académie de Stanislas, 1750-1795

British Library, London:
Add.Ms. 4320, f°15: Robert de Vaugondy to Thomas Birch

William L. Clements Library, Ann Arbor, Michigan:
Letters to Jefferys and Faden, 1773-1783.
Mildmay Papers, Conference Manuscripts.
Mildmay Papers, Letters.

PRINTED WORKS BY THE ROBERT DE VAUGONDYS

Gilles Robert
Introduction à la Géographie des Sansons, Paris: Durand, 1743.
Géographie Sacrée et Historique, Paris: Durand, 1747.

Didier Robert de Vaugondy
Abrégé des Sphères, Paris: Durand, 1745.
Usage des Globes, Paris: Antoine Boudet, 1751.
Essai sur l'histoire de la Géographie, Paris: Boudet, 1755. Republished as "Préface Historique dans laquelle on traite de l'Origine, des progrès, et de l'Etat actuel de La Géographie" in the *Atlas Universel*, Paris: Robert de Vaugondy and Antoine Boudet, 1757.
Observations Critiques sur les Nouvelles Découvertes de l'Amiral de la Fuente.., Paris: Antoine Boudet, 1753.
Lettre de M. Buache, Premier Géographe du Roi et de l'Académie des Sciences, sur les nouvelles Découvertes, etc. Réponse de M. Robert de Vaugondy, Géographe ordinaire du Roi. Réplique de M. Buache, Paris: Antoine Boudet, 1754.
"Géographie" and "Globes," *Encyclopédie*, ed. D. Diderot and J. D'Alembert, Paris, 1757, Volume VII, pp. 707-711a.
Mémoire sur les différens accroissements de la ville de Paris, depuis César jusqu'à Présent...Pour servir d'intelligence au nouveau plan de cette ville..., Paris: Robert de Vaugondy and Antoine Boudet, 1760.
Tablettes Parisiennes, Paris: Robert de Vaugondy, 1760.
Les Promenades des Environs de Paris, Paris: Robert de Vaugondy and Antoine Boudet, 1761.
"Discours sur l'étude de la Géographie," *Nouvel Atlas Portatif*, Paris: Robert de Vaugondy, 1762.
Uranographie, Paris: Robert de Vaugondy and Antoine Boudet, 1764.
Institutions Géographiques, Paris: Boudet and Desaint, 1766.
*Lettre de M. Robert de Vaugondy à M.*** au sujet d'une Carte Systèmatique des pays Septentrionaux de l'Asie et de l'Amérique*, Paris: Antoine Boudet, 1768.
Mémoire sur les pays de l'Asie et de l'Amérique, situés au nord de la mer du Sud: Accompagné d'une carte, intitulée: Nouveau Système Géographique, par lequel on concilie les anciennes connoissances sur les pays nord-ouest de l'Amérique, avec les nouvelles découvertes des Russes au nord de la mer du Sud, Paris: Robert de Vaugondy and Antoine Boudet, 1774.
Description et usages de la Sphère Armillaire, suivant le Systême de Copernic..., Paris: Robert de Vaugondy and Antoine Boudet, 1771.
Mémoire sur une question de Géographie Pratique: si l'applatissement de la Terre peut être rendu sensible sur les cartes, Paris: Robert de Vaugondy and Antoine Boudet, 1775.

VAUGONDY CATALOGUES

"Catalogue des Cartes...par les Srs. Sanson et Robert," in *Introduction à la Géographie*, 1743.
"Liste des ouvrages Géographiques des Messieurs Sanson, Robert, et Vaugondy," in *Tablettes Parisiennes*, 1760.
"Ouvrages de Mr. Robert de Vaugondy," in *Institutions Géographiques*, 1766.
"Notice raisonnée des ouvrages de l'Auteur," in *Mémoire sur les pays de l'Asie et de l'Amérique*, 1774.
"Catalogue des Ouvrages," 1777, BN: Ge.D.7808.
Fortin, J.-B.,"Catalogue des Ouvrages qui composent le Fonds de Géographie de MM. Sanson et Robert de Vaugondy...qui se trouvent chez le Sieur Fortin", Paris: Vve. Hérissant, 1778. BN: Ge.FF.7192 pièce.
Delamarche, Charles, "Ouvrages Géographiques," Paris:1788: BN: Ge.F.2443 pièce.
——————-,*Catalogue des Globes, Sphères, Atlas, Cartes, et autres ouvrages Géographiques et Astronomiques*, Paris: Delamarche, 1803. LC:161 838.

SELECTED BOOKS AND ARTICLES

Adams, Percy G., *Travelers and Travel Liars 1660-1800*, Berkeley: University of California Press, 1962; reprinted, New York: Dover Press, 1980.

Berthaut, Henri, *La Carte de France 1750-1898: Etude Historique*, Paris, 1898-99.

————-, *Les Ingénieurs géographes militaires 1624-1831*, Paris, 1902.

Berthelot, A., and Ceccaldi, F., *Les Cartes de la Corse de Ptolémée au XIX siècle*, Paris, 1939.

Breitfuss, L., "Early Maps of North-Eastern Asia and of the Lands around the North Pacific: Controversy between G. F. Muller and N. Delisle," *Imago Mundi* 3 (1939), pp. 87-99.

Broc, Numa, *La Géographie des philosophes*, Paris: Edition Ophrys, 1974.

————, *Les Montagnes vues par les géographes et les naturalistes de langue française au XVIIIe siècle*, Paris: Bibliothèque nationale, 1969.

————, "Un Affaire de plagiat cartographique sous Louis XIV: le procès Delisle-Nolin," *Revue Historique et Scientifique* 1970, pp. 141-153.

Chaussinand-Nogaret, Guy de, *La Noblesse au XVIIIème siècle: de la Féodalité aux Lumières*, Paris: Hachette, 1976; reprint Brussels: Editions Complexe, 1984.

Covillant, I.G.A., "Histoire des Archives et de la Bibliothèque du Service Hydrographique de la Marine," typescript, 1979, BN: Cartes et Plans, Usuel, IV-C.

Dainville, François de, *Cartes anciennes de l'église de France*, Paris: J. Vrin, 1956.

————, *Le Langage des géographes*, Paris: J. Picard, 1964.

Darnton, Robert, *The Business of Enlightenment: A Publishing History of the* **Encyclopédie**, 1775-1800, Cambridge, Mass.: Harvard University Press, 1979.

Daumas, Maurice, *Les Instruments Scientifiques aux XVIIe et XVIIIe siècles*, Paris, 1953.

Destombes, Marcel, "De la chronique à l'Histoire: Le Globe Terrestre Monumental de Bergevin 1784-1795," *Archives Internationales d'Histoire des Sciences* Juin, 1977, Vol. 27, No. 100, pp. 113-134.

Dion, Marie-Pierre, "Itinéraire Intellectuel et Réussite Nobilaire au siècle des Lumières: Emmanuel de Croÿ 1718-1784," unpublished thesis, Paris: Ecole des Chartes, 1984.

Drayperon, L., *Les Deux Buache et l'education géographique de trois rois de France Louis XVI, Louis XVIII, Charles X avec documents inédits*, Paris: Delagrave, 1888.

Du Bus, Charles, "Les collections d'Anville à la Bibliothèque nationale," *Bulletin de la Section de Géographie* Volume 41 (1926), pp. 93-145.

Furet, François, ed., *Livre et Société dans la France du XVIIIe siècle*, Paris, 1965.

Grivel, Marianne, *Le commerce de l'estampe à Paris au XVIIème siècle*, Geneva, 1986.

Hahn, Robert, *The Anatomy of a Scientific Institution: The Paris Academy of Sciences, 1666-1803*, Berkeley: University of California Press, 1971.

Heidenreich, Conrad, and Dahl, Ed, "The French Mapping of North America, 1600-1760," *The Map Collector* No. 13, Dec. 1980, pp. 2-11; No. 19, June 1982, pp. 2-7.

Hillairet, Jacques, *Dictionnaire Historique des Rues de Paris*, Paris: Les Editions de Minuit, 1961.

Kish, George, "Correspondence of Continental mapmakers with a London firm in the 1770s and 1780s Jefferys and Faden," *Imago Mundi* 4 (1947), pp. 75-77.

————, "Early Thematic Mapping: The Work of Philippe Buache," *Imago Mundi* 28 (1976), pp. 129-136.

Konvitz, Josef, *Cartography in France, 1660-1848: Science, Engineering, and Statecraft*, Chicago: University of Chicago Press, 1987.

Lagarde, Lucie, "Le Passage du Nord-ouest et la Mer de l'Ouest dans la Cartographie française du 18e Siècle, contribution à l'étude de l'Oeuvre des Delisle et Buache," *Imago Mundi* 41 (1989), pp. 19-43.

Martin, H.J. *et al.*, *Histoire de l'Edition Française: Le Livre Triomphant 1660-1830*, Paris, 1984.

Pastoureau, Mireille, *Les Atlas Français XVIe - XVIIe siècles*, Paris: Bibliothèque nationale, 1984.

————, "Les Atlas scolaires Français: 16e-18e Siècles," paper for the symposium, *Images of the World: The Atlas through History*, Library of Congress, October 25-26, 1984.

————, ed., N. Sanson, *Atlas du Monde 1665*, Paris: Sand & Conti, 1988.

Pelletier, Monique, "De l'objet de luxe au produit de consommation courante: L'evolution de l'édition des globes en France aux XVIII et XIXe siècles," *Revue de la Bibliothèque nationale*, Paris: Bibliothèque nationale, 1986.

————, "Les globes de Louis XIV. Les sources françaises de l'oeuvre de Coronelli," *Imago Mundi* 34 (1982), pp. 72-89.

Perla, G., "La Géographie dans l'Encyclopédie," *Revue de Synthèse* IIIe s. no. 115, Juillet-Septembre, 1984, pp. 299-311.

Portalis, Baron Roger, and Beraldi, Henry, *Les Graveurs du XVIIIe siècle*, Paris, 1880.

Robinson, A.H., *Early Thematic Mapping in the History of Cartography*, Chicago: University of Chicago Press, 1982.

Ronfort, Jean-Nérée, "Science and Luxury: Two Acquisitions by the J. Paul Getty Museum," *The J. Paul Getty Museum Journal*, Vol 17 (1989), pp. 47-82.

Schmidt, Rudolf, "Eine armillarsphäre von Didier (?) Robert de Vaugondy," Internationale Coronelli-Gesellschaft für Globen-und Instrumentenkunde, Information No. 16, Dezember 1989, pp. 4-8.

Schwarz, Seymour, and Ehrenberg, Ralph, *The Mapping of America*, New York: H. Abrams, 1980.

Stevenson, E.L., *Terrestrial and Celestial Globes*, New Haven: Yale, 1921.

Tooley, R.V., *Dictionary of Mapmakers*, New York: A. Liss; Amsterdam: Meridian, 1979.

————, "French Mapping of the Americas: the Delisle, Buache, Dezauche Succession 1700-1830," *The Map Collectors' Circle*, Volume 4, No. 33, 1967, pp. 3-11.

Vallée, Leon, *Catalogue des Plans de Paris*, Paris: 1908.

Verner, Coolie, "The Fry and Jefferson Map of Virginia," *Imago Mundi* 21 (1967), pp. 70-94.

Wagner, H.R., "The Delisle-Buache Fantasy," *The Cartography of Northwest America*, Berkeley: University of California Press, 1937.

"Carte Nouvelle de l'Isle de Corse," 1756 [Cat. 163]. (Courtesy of the William L. Clements Library, University of Michigan)

INTRODUCTION TO THE CATALOGUE OF MAPS BY GILLES ROBERT AND DIDIER ROBERT DE VAUGONDY

This catalogue contains only maps which are original to the Roberts; thus, new editions of Sanson maps (often noted in titles as "revue et corrigée aux observations astronomiques") are included but re-publications of Sanson maps with only change of date and imprint are not, nor are all Delamarche editions listed. The catalogue organizes maps geographically and closely follows the Roberts' own organization of maps in their atlases. A geographical index is in the form of the Table of Contents. The layout of each entry is explained here:

1	[Cat.No.]	Region	Date	Author
2	Title			
3	Imprint			
4	Measurement			
5	Engravers			
6	RF Scales			
7	Comments			
8	Book Title Author			
9			AP: AU: NAP:	
10	Location			

1 The **Region** is the geographical area covered by the map. It is followed by the **Date** of the earliest known state of that plate and the **Author**, abbreviated **R** for Gilles Robert and **RV** for Didier Robert de Vaugondy. Square brackets are used for information supplied from sources other than the map itself.
2 The **Title** of the map has been written in full, with line endings (/) shown whenever possible. Accents are as they appear in the title, subject as always to the engraver's whim, resulting in some accents being omitted and occasional inversions, such as "òu" for "où." Such omissions and inversions have not been corrected.
3 The **Imprint** is the place of publication and publisher as it appears on the map. If the map was produced with a book, the book's imprint has been given.
4 The **measurement** is the height by width in millimeters, along the printed border of the map. When possible, both border and plate mark measurements are given. If not indicated, the border measurement may be assumed.
5 The **Engravers** are recorded with the signature found on the map itself and cartouche. The map engraver may not be presumed to be the cartouche engraver.
6 The **RF** is the **representative fraction**, ascertained, when possible, along the latitude scale. It is, of course, approximate. The **Scales** are those printed on the map.
7 The **Comments** list the different states of each map's plate with a summary of the changes made, though in the interest of space these descriptions have been kept brief.
8 If a map appeared in a **Book**, the title of the book, the author and the publication date are given.
9 If the map is found in one of the three Robert atlases, its number in the atlas is noted next to one of the following abbreviations: AP: *Atlas Portatif Universel et Militaire* (1748/9) AU: *Atlas Universel* (1757) NAP: *Nouvel Atlas Portatif* (1762) In the case of the *Atlas Universel*, the map numbers will not always correspond with the order of maps in any edition of this atlas, because the Vaugondys added five road maps to the original 103 maps in the atlas. These road maps are sometimes interpolated in the atlas, sometimes placed at the end. I have numbered them 104 to 108.

10 If a map was issued separately, the locations where the map has been seen and recorded are noted using the following abbreviations:

FRANCE

AN: Archives Nationales, Paris
BN: Bibliothèque Nationale, Paris
BM: Bibliothèque Municipale, Nancy

ENGLAND

BL: British Library, London
Bod: Bodleian Library, Oxford
CUL: Cambridge University Library
RGS: The Royal Geographical Society, London

CANADA

NAC:NMC: The National Archives of Canada: The National Map Collection

THE UNITED STATES

JCB: John Carter Brown Library, Brown University, Rhode Island
NL: The Newberry Library, Chicago
UM: The University of Michigan, Ann Arbor
WLCL: The William L. Clements Library, Ann Arbor
USM: The University of Southern Maine

Sometimes a map is only known from a contemporary catalogue reference. The following abbreviations are used.

Julien: Roch-Joseph Julien, *Nouveau Catalog de Cartes Géographiques et Topographiques* (Paris: Julien; Nuremburg: Homann Heritiers; The Hague: Gosse & Pinet; London: Andrew Dury, 1763).
L-D: Nicolas Lenglet-Dufresnoy, *Catalogue des Meilleures Cartes Géographiques Générales et Particulières, avec quelques remarques sur la choix qu'on en doit faire*, (Paris: 1742; reprinted Amsterdam: Meridian Publishing Co., 1965).
Lelong: Jacques Lelong and C.-M. Fevret de Fontette, *Bibliothèque historique de la France*, (Paris: J.-T. Hérissant, 1768-1778).
R: Gilles Robert, "Catalogue des Cartes...par les Srs. Sanson et Robert," in *Introduction à la Géographie*, (1743).
RV: Didier Robert de Vaugondy
"Liste des ouvrages Géographiques des Messieurs Sanson, Robert, et Vaugondy," in *Tablettes Parisiennes*, (1760).
"Ouvrages de Mr. Robert de Vaugondy," in *Institutions Géographiques*, (1766).
"Notice raisonnée des ouvrages de l'Auteur," in *Memoire sur les pays de l'Asie et de l'Amérique*, (1774).
"Catalogue des Ouvrages," (1777), BN: Ge.D.7808.
Fortin, J.-B., "Catalogue des Ouvrages qui composent le Fonds de Géographie de MM. Sanson et Robert de Vaugondy...qui se trouvent chez le Sieur Fortin, (Paris: Vve. Hérissant, 1778), BN: Ge.FF.7192 pièce.
Delamarche, Charles, "Ouvrages Géographiques," (Paris:1788): BN: Ge.F.2443 pièce.
——————, *Catalogue des Globes, Sphères, Atlas, Cartes, et autres ouvrages Géographiques et Astronomiques*, (Paris: Delamarche, 1803): LC:161 838.

While the author has attempted to locate and catalogue every Robert de Vaugondy map, inevitably more will surface. An appendix is already in progress, and author and publisher would be grateful for any additions readers wish to contribute.

TABLE OF CONTENTS CATALOGUE OF MAPS

THE CATALOGUE OF MAPS

1 **Celestial Hemisphere, North** 1764 RV

Hémisphère Céleste Arctique ou Septentrional / Calculé et Construit pour l'année 1763 par le Sr. Robert de Vaugondy / Géographe ord? du Roi, de S. M. Polon? Duc de Lorraine et de Bar et de la Société royale des Sciences et Belles-Lettres de Nancy / Avec Privilège du Roi, 1764.

Paris: ches l'Auteur Quai de l'Horloge.

br: 665 x 628 mm.
pl: 678 x 640 mm.

Engravers:

RF: c.1:

Scales: Echelle de parties égales pour déterminer le rayon d'un horizon particulier Grandeurs des Etoiles: 1e...6e..Nebuleuse (Graduated punch used to express different sizes of stars)

Comments: Map printed in two colors: black for stars; red for figures of constellations; note double plate marks

State 1: 1764

State 2: 1779 [per de Lalande, Bibliographie Astronomique]

State 3: [after 1786]: chez le Sr. Delamarche / Rue du Foin St. Jacques, au Collège de Mtr Gervais; changed Title: ... *Vaugondy / Corrigé et augmenté d'un grand nombre d'étoiles et de nébuleuses, observées par la Caille, Méchain et Messier astronomes.*
Additions and new constellation: Le Taureau Royal de Poniatowski (265-280°); Algenib removed (62°, 33°); le Bouclier [of Orion] removed (68°). New nebulas mostly in constellations of Le Lion, Les Levriers, la Vierge, la Grande Ourse (between 148-212°). Names of parts of body removed from constellation figures

Book:

AP:
AU:
NAP:

Location: BN: Ge.C.8767 (1); Ge.C.8766 (Delamarche); BL: K.1.64II

2 **Celestial Hemisphere, South** 1764 RV

Hémisphère Céleste Antarctique ou Austral / Calculé et Construit pour l'année 1763 par le Sr. Robert de Vaugondy / Géographe Ord? du Roi, de S.M. Polon? Duc de Lorraine et Bar et de la Société Royale des Sciences et Belles-Lettres de Nancy / Avec Privilège du Roi 1764

Paris: ches l'Auteur Quai de l'Horloge

br: 659 x 626 mm.
pl: 676 x 645 mm.

Engravers:

RF: c.1:

Scales: Echelle de parties égales pour déterminer le rayon d'un horizon particulier Grandeurs des Etoiles: 1e...6e..Nebuleueses (Graduated punch used to express different sizes of stars)

Comments: Map printed in two colors: black for stars; red for constellation figures; note double plate marks

State 1: 1764

State 2: 1779 [per de Lalande, Bibliographie Astronomique]

2 **Celestial Hemisphere, South** (continued)

State 3: [after 1786]: chez le Sr. Delamarche Geographe / Rue du Foin St. Jacques, au Collège de Mtr. Gervais; changed Title: ... *Vaugondy / Corrigé et augmenté d'un grande nombre d'étoiles et de nebuleuses observées par le Caille, Méchain et Messier astronomes.*
Additions: Le Solitaire constellation (between 212° and 242°); 17 new nebulas in Le Sagittaire and l'Ecu de Sobiesky; new nebulas in La Licorne and la Dorade. New punch marks used for star designations; Parts of the body on red figures removed

Book:

AP:
AU:
NAP:

Location: BN: Ge.C.8767 (2); Ge.C.8766 (Delamarche, date?); BL: K.1.64I; AN: NN 396/2

3 **Celestial globe, gores** [1750] RV

[Globe céleste en douze fuseaux par Robert de Vaugondy]

manuscript

4 sheets, each 340 x 560 mm.

Engravers: (map) [Gobin]

RF: c.1:

Scales:

Comments: see DRV, 'Préface' in Usage des Globes (1751), for reference to Gobin as engraver of celestial globe. (Illustrated on end papers of this volume)

Book:

AP:
AU:
NAP:

Location: BN: Rés. Ge.DD.2967

4 **World** 1743 R

MAPPEMONDE / ou Description / du GLOBE TERRESTRE / où l'on trouve en general le rapport / que toutes les parties ont avec les / Cieux et entr'elles / Par les Srs Sanson / Geog. ord. du Roi / revue augmentée et assujetie / aux Observations / Astronomiques / Par le Sr. ROBERT / Geog. ord. du Roi/

Paris: ches Robert / Geog. ord. du Roi / Quai de l'horloge du Palais / avec priv. du Roi / 1743

br: 464 x 742 mm.
pl: 469 x 747 mm.

Engravers: (map) gravé par Le Tellier
(cartouche) Gobin Sc.

RF: c.1: 60,000,000

Scales:

Comments: hemispheres; decorative ribbon at top of map with biblical quotation: Book of Joshua, Chapt. 18: 4,8,9

4 **World** (continued)

State 1: 1743

Book:
AP:
AU:
NAP: Ill. 12-13

Location: WLCL: 7-E-6

5 **World** 1748 R

MAPPEMONDE / ou Description du / Globe Terrestre / Par le Sr. Robert Géographe / Ordinaire du Roy. / Avec Privilege / 1748

br: 160 x 307 mm.
pl: 167 x 313 mm.

Engravers: (map) Lattré sculp.

RF: c.1:195,000,000

Scales:

Comments: hemispheres

State 1: 1748

State 2: 1768: engraver (map): Dussy Sc.; shows latest discoveries by the Russians along coastline of Alaska and Aleutians [Smith collection, USM]

State 3: 1774: imprint: Dépôt de Géographie, rue Geoffroy Langevin, no. 328

Book:
AP: 2
AU:
NAP:

Location: USM: Smith Collection (1768); Nancy, BM: 263.500 (1774)

6 **World** 1749 R/RV

Mappemonde / dressée suivant les nouvelles relations / et assujettie aux observations astronomiques, / Par le Sr. Robert Geog? ord?/de Sa Majesté. / Avec Privilege. 1749.

Paris: Ches le Sr. Robert.../Quay de l'horloge du Palais,/ proche le pont neuf.

br: 463 x 730 mm.
pl: 476 x 750 mm.

Engravers: (map) Gravé par Delahaye l'aîné
(cartouche) Gobin inv. et Sculpsit

RF: c.1:60,000,000

Scales:

Comments: Hemispheres

State 1: 1749: no 'avertissement' about Admiral de la Fuente; lacks I. Bermudes

State 2: [after 1753]: addition in NW America of Avertissement concerning the discoveries of Admiral de la Fuente, with ref. to the Journal Oeconomique, Juillet, 1753, p. 55. Addition of complete outline of I. Nouvelles [Falklands] (cf. 1752 Mappemonde from AU with east coast of islands only, Cat. 8)

State 3: 1771: as per RV, 'Notice Raisonnée' (1774); re-edition of 1749 map; signed RV; new cartouche by Arrivet

[State 4: 1778: not seen; by parallel with maps of four continents, published by Fortin]

State 5: [after 1786]: Delamarche [1793?]; new title: *MAPPEMONDE / dressée suivant les nouvelles relations /et assujettie / aux observations astronomiques / Par le Sr ROBERT DE VAUGONDY Géog./ corrigée et augmentée / des nouvelles découvertes du Capitaine Cook / Par DELAMARCHE Géogr.//* Imprint: Chez le Cen. Delamarche Géogr? Rue du Foin / St. Jacques, au Collège de Mtr Gervais

Julien Cat. (1763): price, 1#

6 **World** (continued)

Book:
AP:
AU:
NAP:

Location: BN: Ge.C.8504 (1749); Ge.C.8505 (An XII); Ge.C.8506 (1812); Ge.C.16363 (s.d., corr.et aug. par Delamarche); WLCL: W-2-D (Julien, Theatre du Monde: 1); NAC: NMC 47867 (Delamarche, 1804)

7 **World** 1750 RV

MAPPEMONDE / Par le Sr. Robert de Vaugondy / avec Privilége du Roi. / 1750.

br: 142 mm. x 274 mm.

Engravers:

RF: c.1:undetermined

Scales:

Comments: in two hemispheres; "Page 32" outside border, top rt.

Book: Méthode Abrégée et facile pour apprendre la Géographie
AP:
AU:
NAP:

[l'Abbé Le François], Paris: Libraires Associés, 1770

Location: WLCL; C2 1770 LE; NAC: G 125 L4 1781

8 **World** 1752 RV

MAPPEMONDE / ou / Description du GLOBE TERRESTRE;/ dressée sur les mémoires / les plus nouveaux, / et assujettie aux observations / astronomiques, / Par le Sr. Robert de Vaugondy fils.

Paris: chez Robert..Quay de l'Horloge du Palais; chez Ant.Boudet..rue St. Jacques; Avec privilege 1752.

br: 465 x 705 mm.
pl: 486 x 730 mm.

Engravers: (cartouche) C.Cochin filius delin.; P.F.Tardieu Sculp.

RF: c.1:70,000,000

Scales:

Comments: Hemispheres

State l: 1752: shows voyages of Anson 1740; l'Aigle et la Marie 1738-9; Galion Na. Seignora 1743

State 2: [1768]: voyages remain; added off coast of S. Am.,'I de St. Pierre découverte en 1756'; changes in Baffin Bay; 'Galapos' [sic] islands added; Alaskan coastline added; lettering changes (italics to Roman and vice versa); place name changes, e.g. Baye de Baffin to Baffins Bay

State 3: 1776: adds 'Censeur Royal' to title; adds de Fonte discoveries in NW America; Tardieu erased from cartouche, Groux added (suggests a Fortin publication)

State 4: 1783: added to title: 'Avec les Routes, et Découvertes du célèbre Capitaine Cook'; imprint of Antoine Boudet

Design for cartouche in BM: Prints and drawing

Book:
AP:
AU: 13
NAP:

Location: JCB: Atlas Universel (only copy of 1776, State 3, seen); NAC: NMC A/10,000/1752 (1757); A/10,000/1783.

9	**World**	1756	RV

Carte / pour l'intelligence du / Livre Intitulé / de l'Esprit des Loix / Par le Sr. Robert de Vaugondy Géog. ord. du Roi etc. 1756

[Book: London: Chez Nourse]

br: 232 x 467 mm.
pl.: 243 x 481 mm.

Engravers: (map) G. De la Haye

RF: c.1:62,500,000

Scales:

Comments: Mercator projection

BN: Ge.D.12335 (1): without date, place, author or engraver

Book: OEUVRES / DE MONSIEUR DE MONTESQUIEU, / Nouvelle Edition, / révue, corrigée & considerablement augmentée par l'auteur / Tome Second/

AP:
AU:
NAP:

Montesquieu (1767)

Location: BN: Ge.D. 12335 (1); Bod: Vet.E5 d.65

10	**World**	1760	RV

MAPPEMONDE / ou / Carte Générale / du Globe Terrestre / dessinée suivant les regles de la Projection / des Cartes réduites. / Par le S. Robert de Vaugondy Géographe ordinaire du Roi / de S.M.Polon? Duc de Lorraine et de Bar, et de l'Academie royale / des Sciences et Belles-lettres de Nancy./

Paris: Ches l'Auteur Quai de l'horloge du Palais pres le pont-neuf./Avec Privilege

Four sheets: br. 981 x 1132 mm. (assembled) [BN]

Engravers: (map) gravé par E. Dussy
(cartouche) Arrivet inv. & sculp.

RF: c.1:34,000,000

Scales:

Comments: Mercator projection. Part of series of World and Continent maps in 4 sheets produced from 1760 onwards

State 1: 1760: Arctic land mass has a connection to North America; see RV, AU "Préface" p. 20, Orbis Vetus; M. Pedley, *Révue de la BN* (1984) Later states remove this Arctic connection

State 2: 1775: 'Censeur Royal' added to title; east coast of Nouvelle Hollande shown; 'Nouvelle Zelande en 1770'

State 3: 1778: imprint: Chez Fortin Ingr. Mécanicien du Roy pour les Globes et Spheres / Rue de la Harpe près la rue du Foin

State 4: [1786]: not seen

Julien Cat. (1763), no. 12., 4#

Book:

AP:
AU:
NAP:

Location: BN: Ge.A.1229 (1760); Ge.C.8520 (1775); [no cote no., acq. 1985](1786); LC: 11528 (1778, chez Fortin); BL: MAR.I.19.2.11 Tab. (1775)

11	**World**	1762	RV

MAPPE-MONDE / par le S. Robert de Vaugondy Géog? ord? du Roi / du feu Roi de Pologne Duc de Lorraine et de Bar / de la Société royale de Nanci / 1762

A Paris ches l'auteur Quai de l'Horloge près le Pont Neuf

br: 292 x 446 mm.

Engravers: (cartouche) Arrivet inv. & sculp.

RF: c.1:100,000,000

Scales:

11	**World** (continued)

Comments: in hemispheres

State 1: 1762: Arrivet signs cartouche

State 2: 1780: title adds: "et Censeur Royal"; shows de Fonte discoveries, NW America; complete outline of New Zealand, east coast of Australia; "Groux sculp." on cartouche; Fortin imprint: "Chez Fortin Ingr Mécanicien du Roi / pour les Globes et Sphères / Rue de la Harpe près la rue du Foin"

State 3: 1786: shows route of Cook; imprint: Chez le S. Delamarche, Géog? / Rue du Foin St. Jacques, / au Collége de Mtre Gervais.

Book:

AP:
AU:
NAP: 1

Location: BN: Ge.D. 12341 [Fortin]; RGS: 4.c.9 (Delamarche); NAC: NMC 8078 (1786); 8069 (1795)

12	**World, globe gores**	1745	RV

GLOBE TERRESTRE / dressé / sur les cartes de M. Robert. / Géog. ord. du Roi. / Par le Sr. Robert de Vaugondy son fils.

A Paris / Chés l'Auteur, Quay de l'horloge / avec privilege / 1745.

two sheets; 12 gores: 540 mm. circumference
each sheet: 280 x 225 mm.

Engravers: (map) gravé par G. Delahaye

RF: c.1:

Scales:

Comments: first known map by Didier Robert de Vaugondy, preparatory for six-inch globe presented to Louis XV in 1751

Book:

AP:
AU:
NAP:

Location: BN: Ge. D. 12571

13	**World, historical**	1752	RV

ORBIS VETUS / in utrâque continente / juxtà mentem Sansonianam distinctus, / nec non observationibus astronomicis / redactus, / accurante Robert de Vaugondy / Geographo Regis ordinario / 1752 / Cum Privilegio Regis

Lutetiae Parisiorum: apud Autorem in ripa Sequanae / ab horologis palatii dictâ, prope pontem novum

br: 478 x 715 mm.
pl: 496 x 738 mm.

Engravers:

RF: c.1:70,000,000

Scales:

Comments: hemispheres.

Imprint also: Apud Ant. Boudet Bibliopolam et Typographum / regis in viâ Sti. Jacobi; Quotations from ancient authors scattered throughout the seas, most referring to Atlantis: Plato, Seneca, Theopompus, Manilius; Monitum: concerns the placements of Atlantis as North and South America and Hesperides as West Indies; uses ancient geographers and G. Sanson as testimony; divides continents into ten regions named for ten sons of Neptune; cites Chinese sources of 40 A.D. about to be published by 'a man learned in oriental languages.'

Book:

AP:
AU: 1
NAP:

Location:

14 **World, historical** 1770 RV

ORBIS VETUS / ex Cornelio Tacito. / Auctor Robert de Vaugondy Regis Geog? et regiae Academiae Nanceianae Socius.

[Book: Paris: Delatour, rue St. Jacques; London: Paul Vaillant]

br: 237 x 382 mm.
pl: 250 x 394 mm.

Engravers: (map) Guill? De-la-Haye

RF: c.1:

Scales: Europe: 58 mm. = 750 Milliaria Romana; Asia: 58 mm. = 1125 Milliaria Romana

Comments: two maps on one sheet; Ptolemaic projection; Europe on left, Asia on right outside border, top right: Tom. I, p. 5

Book: C.Cornelii / TACITI / OPERA / recognovit, emendavit, /supplementis explevit,/ NOTIS, DISSERTATIONIBUS / TABULIS GEOGRAPHICIS / ILLUSTRAVIT/

AP:
AU:
NAP:

Gabriel Brotier ed. (1771)

Location: BL: 53.g.1-4; Bod: Vet A.5.e.4826 (a copy in English; Dublin: Luke White)

15 **World, thematic** 1761 RV

MAPPEMONDE / suivant la projection / des cartes reduites. / Par le S. Robert de Vaugondy Géog. ord. du Roi, / de S.M. Polon? Duc de Lorr. et de Bar, / Associé de l'Académie Royale de Nanci / 1761.

A Paris ches l'Auteur Quai de l'Horloge près le Pont Neuf

br: 245 x 398 mm.

Engravers: (map) Gravé par E. Dussy
(cartouche) Arrivet inv. & sculp.

RF: c.1:100,000,000

Scales:

Comments: This map is used three times in the NAP, coloured in outline for religions, skin color, and human shape: "Division par Religions: Judaisme, Christianisme, Mahometisme, Paganisme."; "Division par les couleurs des habitans: les Blancs, les Bruns, les Jaunatres, les Olivatres"; "Division par la figure des hommes: les Europeens; les visages plats et les yeux ovales; les visages d'Ours; les Africains"

State 1: 1761

State 2: 1778: imprint removed; "et Censeur Royal" added to title; "Groux" signs cartouche; in NW America: "l'on place ici des Découvertes attribuées à un Amiral nommé De Fonte"; island 'Taiti' [Tahiti] in title cartouche; Fortin publication

Book:

AP:
AU:
NAP: 2,3,4

Location: BN: Ge.D.12336, 12337A, 12337 [1761]

16 **Eastern Hemisphere** 1749 RV

CARTE DE / L'ANCIEN CONTINENT / Selon sa plus grande longeur diamètrale / depuis la Pointe de la Tartarie Orientale, / jusqu'au Cap de Boñe Esperance. / Dressée Sous les Yeux de Mr. de Buffon / Par le Sr. Robert de Vaugondy / Fils de Mr. Robert Geogr. ord. du Roi./ 1749.

[Book: Paris: L'Imprimerie Royale]

br: 216 x 173 mm.
pl: 227 x 190 mm.

Engravers: (map) Guill. Delahaye

RF: c.1:50,000,000

Scales: 60 mm. = 40 Degrés de l'Equateur

16 **Eastern Hemisphere** (continued)

Comments: North-South Axis at oblique angle to continents

Book: HISTOIRE NATURELLE

AP:
AU:
NAP:

Buffon (1749)

Location: UM: QH 45 .B93

17 **Western Hemisphere** 1740 R

LE NOUVEAU CONTINENT / ou l'AMERIQUE / Divisée / en Septentrionale et Méridionale / où l'on trouve / les Climats les Méridiens les differentes Possessions / les Peuples naturels libres / les differentes Religions et les differentes Langues / Par les Srs. Sanson Geog. ord. du Roi./ Revue et corrigée en plusieurs endroits sur les Mémoires les plus recents et assujettie aux Observations Astronomiques/ Par le Sr. Robert Géog. ord. du Roi./

A Paris: Chès l'Auteur quai de l'horloge du Palais, proche la rue de Harlay / Avec Privilege du Roi / 1740

br: 485 x 520 mm.
pl: 502 x 537 mm.

Engravers:

RF: c.1:

Scales: 91 mm.= 1500 Mille pas Geometriques ou Milles d'Italie; 500 Lieues Marines ou d'une Heure du Chemin

Comments: De Leth imprint: between 50-58°S.: 'Isles de Glace de couvertes [sic] par les Francois en 1739' [Lozier-Bouvet expedition]; Sanson projection; geographical table to the left; Time divisions on meridiens Climate divisions along right hand latitude divisions: 1er - 19th climat towards north; 1er - 9th climat towards south

Book:

AP:
AU:
NAP:

Location: BN: Ge.C.7822;
BL: 69810 (44) (imprint Amsterdam: De Leth, dated 1735?);
WLCL: 2-C-2 (De Leth);
NAC: NMC 118289

18 **Western Hemisphere** 1749 RV

CARTE DU / NOUVEAU CONTINENT / selon sa plus grande longeur diamètrale / depuis la Riv. de la Plata jusqu'au / delà du Lac des Assiniboïls. / Dressée sous les yeux de Mr. de Buffon / Par le Sr. Robert de Vaugondy / Fils de Mr. Robert Geog? ord? du Roi/ 1749.

[Book: Paris: L'Imprimerie Royale]

br: 216 x 177 mm.
pl: 226 x 187 mm.

Engravers: (map) Guill? Delahaye

RF: c.1:50,000,000

Scales: 61 mm. = 40 Degrés de l'Equateur

Comments: North-South/East-West axes at oblique angles to continents; UM:QH 45 .B92 1853: copy of 1749 map; measures: 190 x 138 mm. (br), 240 x 157 mm. (pl); different lettering style on title; book date of 1853 suggests it is lithographed

Book: HISTOIRE NATURELLE

AP:
AU:
NAP:

Buffon (1749)

Location: UM: QH 45 .B93 – QH45.892 1853; BN: Ge.F.5078 [1853 edition; see comments]

19 **Western Hemisphere** 1749 R/RV

L'AMERIQUE / SEPTENTRIONALE et MERIDIONALE / divisée en ses principales parties par les Srs. Sanson Géog? ordin? du Roi; / recitifiée suivant les nouvelles découvertes. / et assujettie aux observations astronomiques, / par le Sr. Robert Géog? ord? du Roi / Avec Privilege / 1749

Inset: *Isle de / La Martinique*

Inset: *Isle de / S. Domingue*

Paris: Quay de l'horloge / proche le Pont Neuf

br: 492 x 512 mm. (637 mm. incl. Table)
pl: 523 x 652 mm.

Engravers: (map) Gravé par Delahaye l'ainé
(cartouche) Gobin inv. et sculpsit

RF: c.1:27,775,000

Scales: 5 mm. = 75 Milles pas geometriques de 60 au Degré; 5 mm. = 25 Lieues marines d'Angleterre et de France de 20 au D.; 7 mm. = 35 Lieues marines d'Espagne de 17 au Degré; 4 mm. = 25 Lieues communes de France de 25 au Degré; 5 mm. = 25 Lieues d'une heure de 20 au Degré

Inset (Martinique): 12 mm. = 1 Lieue Marine

Inset (S. Domingue): 7 mm. = 5 Lieues Marines

Comments: Table Géographique along right hand edge; Along longitude marks at top and bottom, Roman numerals: e.g. XIh [sic], VIIIh., IXh, Xh, XIh, XIIh. for time zones; as on 1740 Mappemonde

State 1: 1749: Gobin cartouche

State 2: 1771: (See Cat. [22]) title changed: Amerique /ou / INDES OCCIDENTALES / par le Sr. Robert de Vaugondy/; cartouche: Arrivet inv. et sculpsit; scales moved from top left and placed below outside border; N. Am has western bulge; Grt. Lakes as per Bellin (1755)

State 3: 1776: as per RV, Catalogue (1777)

State 4: 1778: imprint: ches Fortin Ingr. Mécanicien du Roy pour les Globes et Sphères / Rue de la Harpe près la Rue du Foin

Julien Cat. (1763): price, 1#

Book:

AP:
AU:
NAP:

Location: CUL: Maps aa.645.74.1; BN: Ge.D.11635 (1749); Ge.D.20609, Ge.D.11647; (1778); WLCL: W-2-D (Julien, Theatre du Monde: 7 (1749)); 2-C-6 (1771); NAC: NMC 118951 (1771)

20 **Western Hemisphere** 1760 RV

L'AMERIQUE / SEPTENTRIONALE et MERIDIONALE / divisée suivant / SES DIFFERENS PAYS. // Par le Sr. Robert de Vaugondy Géographe ordin? du Roi, / de sa Majesté Polon? Duc de Lorr? et de Bar, et de l'Académie Royale / des Sciences et B.Lettres de Nancy et Censeur royal/.

Inset: Supplement / pour / les Terres Arctiques / ou se trouvent / les Bayes d'Hudson et de Baffin /

[State 3:] Paris: Chez Fortin Ingr.Mecanicien du Roy pour les Globes et Spheres / Rue de la Harpe pres la Rue du Foin / 1778

Four sheets: 970 x 1100 mm. (assembled) [LC]

Engravers: (cartouche) Arrivet inv. & sculp.

RF: c.1:

Scales: 140 mm. = 720 mille pas geometriques de 60 au degré; 120 mm. = 280 Lieues marines d'Angleterre; 110 mm. = 220 Lieues marines d'Espagne; 130 mm. = 300 Lieues communes de France

Comments:
State 1: [1760?]: quai de l'horloge imprint? per RV, Notice Raisonnée, (1774); Julien Cat. (1763), no. 51, 1760, 4#

State 2: 1767

20 **Western Hemisphere** (continued)

State 3: 1778: Fortin Imprint

Book:

AP:
AU:
NAP:

Location: BN: Ge.AA.1195 (1778); Ge.AA.1196; BL: K.118.20.2 TAB (1767)

21 **Western hemisphere** 1762 RV

L'AMÉRIQUE / Par le S. Robert de Vaugondy Géog? / ord? du Roi, de S.M. Polon? Duc de / Lorr? et de Bar, et de la Société / royale de Nanci. / 1762

A Paris ches l'Auteur quai de l'Horloge près le pont neuf

br: 243 x 225 mm.

Engravers: (map) Gravé par E. Dussy
(cartouche) Arrivet inv. & Sculp.

RF: c.1:50,000,000

Scales: 39 mm. = 1200 mille pas géom. de 60 au dégre; 400 Lieues Marines ou d'une heure

Comments:
State 1: 1762

State 2: [1778]: date and imprint removed; "et Censeur Royal" added to title; Fortin publication

NW part of N. America: "Quoique nous n'adoptions point les Découvertes de l'Amiral de Fonte l'on peut consulter la carte suivant où nous les avons tracées." [refers to NAP 43, Cat. 449]

Book:

AP:
AU:
NAP: 42

Location:

22 **Western Hemisphere** 1771 RV

Amérique / ou / INDES OCCIDENTALES / PAR LE Sr. Robert de Vaugondy Geog? ord? du Roi / du feu Roi de Pologne Duc de Lorraine et de Bar / et de l'Academie royale des sciences / et belles lettres de Nanci / 1771

br: 492 x 640 mm.
pl: 515 x 651 mm.

Engravers: (map) Delahaye l'ainé
(cartouche) Arrivet inv. et sculpsit

RF: c.1:27,775,000

Scales: 80 mm. = 1200 Milles pas géometriques de 60 au degré; 400 lieues marines d'Angleterre; 350 Lieues marines d'Espagne; 490 lieues communes de France; 400 lieues d'une heure ou grandes lieues

Comments: a re-edition of 1749: l'AMERIQUE Septentrionale et Méridionale divisée en ses principales parties... see Cat. 19; scales moved outside bottom border. See Cat. 19 for further states.

Book:

AP:
AU:
NAP:

Location: BN: Ge. D.11647, 20609 (1778); WLCL: 2-C-6 (1771); NAC: NMC 118951

23 **Western Hemisphere, historical** 1740 R

Atlantis Insula

Paris

Engravers:

RF: c.1:

Scales:

Comments: cf. WLCL: AF (7): Novus Orbis potius Altera Continens sive Atlantis Insula..., G. Sanson, 1699, 39.8 x 56 cm.

Book:

AP:
AU:
NAP:

Location: BN: Ge.D.11660 (ed. G. Sanson)

24 **Northern Hemisphere** 1773 RV

Carte qui représente les differentes connoissances que l'on a eues des TERRES ARCTIQUES depuis 1650 jusqu'en 1747 ausquelles il faut comparer la Carte suivante Ce 10 Par M. de Vaugondy 1773

[Book: Paris: Panckoucke, Stoupe, Brunet; Amsterdam: M. M. Rey]

br: 291 x 341 mm.

Engravers:

RF: c.1: various

Scales:

Comments: outside border, top rt.: Suppl. Carte 9e.; bot. rt.: 168 ; 4 maps on one: Selon Sanson en 1650; Selon Delisle en 1700; Selon Delisle en 1703; Carte de la Baye d'Hudson parcourue en 1746 et 1747 par Henri Ellis / pour la recherche du passage par le Nord-Ouest

Book: ENCYCLOPÉDIE: Suite du Recueil des Planches, sur les Sciences, les Arts Liberaux, et les Arts Mechaniques, avec leurs explication... ed. Diderot; Supplement, ed. Panckoucke (1777)

AP:
AU:
NAP:

Location: UM Rare Books: AE 25 .E53; NAC: NMC 6173, 6274, 6951

25 **Northern Hemisphere** 1774 RV

ESSAI D'UNE CARTE POLAIRE ARCTIQUE / Construite d'après toutes les connoissances les plus nouvelles pour servir aux Navigations et aux Découvertes à faire dans la mer Glaciale, / presentée à l'Académie Royale des Sciences avec un Mémoire approuvé le 24 Mars 1773 et jugé digné d'être imprimé dans le Recueil des Mémoires des Savans étrangers / par le Sr. De Vaugondy Géographe ordinaire du Roi, de feu Roi de Pologne Duc de Lorraine et de Bar / de l'Academie royale de Nanci et Censeur Royal. 1774

Paris: Ches l'Auteur Quai de l'horloge du Palais / pres le Pont neut.

500 x 480 mm.

Engravers: (map) Gravé par E. Dussy

RF: c.1:

Scales: Lieues marines de 20 au degre pour la latitude: (marked along 90 degr. longitude):

Comments: accompanies Hémisphère Austral, Cat. 27; See Acad. des Sciences, Procès-Verbaux, Rapport, 24 Mars 1773: Pingré et Jeurat approve it for publication

Book:

AP:
AU:
NAP:

Location: BN: Ge.D.12865; BL: K.iv.57; NAC: NMC 10497

26 **Southern hemisphere** 1756 RV

CARTE GENERALE / qui répresente / Les Mers des Indes, Pacifique, et Atlantique, / et principalement / Le Monde Austral, / divisé / en Australasie, Polynésie et Magellanie, / Pour servir à l'Histoire des Terres Australes / Par le Sr. Robert de Vaugondy Géog? ord. du Roi, / de l'Académie royales des Sciences et Belles-Lettres de Nancy. / 1756

[Book: Paris: Durand, rue du Foin, au Griffon]

br: 239 x 634 mm. [reprint]

Engravers: (map) G. De-la-Haye Sc.

RF: c.1:60,000,000

Scales:

Comments: Mercator Projection; See A.C.Taylor, Le President de Brosses et l'Australie (Paris, 1937) for details on the making of these seven maps; de Brosses chose R de V. over J.N.Delisle (see Chapter Five); Captain Cook used this map from de Brosses' book on his first voyage

Book: HISTOIRE DES NAVIGATIONS AUX TERRES AUSTRALES. Contenant ce que l'on scait des moeurs & des productions des Contrées découvertes jusqu'à ce jour; ... Président Charles de Brosses (1756)

AP:
AU:
NAP:

Location: CUL:D.18.35; UM: G 160 .B9 1967 (reprint)

27 **Southern Hemisphere** 1773 RV

HEMISPHERE AUSTRAL ou ANTARCTIQUE / projetté sur un Horizon dont le Zénith est situé à 140 degrés de longit? orient? de l'Isle de Fer et à 66° 32' de latit? australe, / dressé sous les yeux de M. le Duc de Croÿ par le Sr. de Vaugondy Géographe ord? du Roy, du feu Roi de Pologne Duc de Lorr. et de Bar, / de l'Académie royale des Sciences et Belles-lettres de Nancy et Censeur royal. / Publié sous l'approbation de l'Académie Royale des Science du 24 Mars 1773. /

Publié sous l'approbation de l'Académie royale des Sciences du 24 mars 1773

br: 644 x 620 mm. [BN]

Engravers: (map) Gravé par E. Dussy

RF: c.1:

Scales:

Comments: State 1: 1773

State 2: 1777: added to title: " / et augmenté en 1776 des découvertes faites par le Capit. Cook en 1773, 74 et 75."; Explorers' routes shown: Cook, Wallis, Byron, de Bougainville, Abel Tasman, Bouvet, Allouarn, Carteret, Halley.; In the four corners of the map are: [top left] Observations (explains the use of the oblique projection to show the Mer du Sud and places recently discovered) [top right] Suite des observations [bot. left] Problème (how to determine distances between two places using this projection) [bot. right] Table (longitudes and latitudes for places in America, Africa, Asia, east coast of Nlle. Hollande, New Zealand, north east of New Zealand)

Book:

AP:
AU:
NAP:

Location: BN: Ge.DD.2987(139); Ge.DD.2987(140); Ge.B [SH; 1(23-1); Ge.C.8649 (with mss. notes added); BL: 968.(3) (1777); AN: NN2/19 (1777); WLCL: 8-0-12 (1777); NAC: NMC 132941 in G 1100 .C656 1775 H1 Vault; Tooley (MCS2) 72.

28 **Europe** [1739] R

L'EUROPE / Divisée suivant l'étendue de ses /principaux Etats subdivisé en leurs / principales provinces. / Par les Srs. Sanson Geog. ord. du Roy. / Revue et changée en pluseurs endroits / Sur les Memoires les plus recents / et assujetie aux Observations astronomiques / Par le Sr. Robert Géog. ord du Roy.

[BL]: Amsterdam: Chez Henry de Leth/ à l'enseigne du Pecheur.

br: 453 x 545 mm.
pl: 458 x 550 mm.

Engravers:

RF: c.1:

Scales: 51 mm. = 300 Mille pas géometriques ou Milles d'Italie; 240 Milles d'Angleterre; 60 Lieues communes de Danemarc et de Suede; 47 Lieues communes de Norwege; 400 Werstes de Moscovie; 125 Lieues communes de France; 75 Lieues communes d'Allemagne; 100 Lieues communes de Pologne; 90 Lieues communes d'Espagne; 50 Lieues communes de Hongrie; 100 Lieues d'une heure de chemin

Comments: only copy seen in BL: de Leth imprint along left neat line: climates shown

Book:

AP:
AU:
NAP:

Location: BN: Ge C.8175 (dated 1738; Sanson map?); BL: Maps.1030.(7) (no date; De Leth imprint); Robert cat. (1743): dated 1739; L-D, p. 245-247 (1745) dated 1739.

29 **Europe** 1748 R

L'Europe / divisée / suivant ses princìpx. Etats / par le Sr. Robert Geog? / Ord? du Roi / avec Privilege. 1748

br: 161 x 198 mm.
pl: 165 x 251 mm.

Engravers:

RF: c.1:28,800,000

Scales: 28 mm = 150 Lieues d'une heure:

Comments:
State 1: 1748

State 2: 1769: from later publication of the Atlas Portatif by Panckoucke?; dated 1769, but with 3 in top right corner, as on State 1.

Book:

AP: 3
AU:
NAP:

Location: BN: Ge.F.1365 (1769)

30 **Europe** 1749 R/RV

L'Europe / divisée en ses Empires, / Royaumes & Republiques / Par les Srs. Sanson Géogr? ordin? du Roi; / assujettie aux Observations astronomiques / et corrigée pour la Russie, / d'après les Cartes de l'Atlas Russien / Par le Sr. Robert Geographe ord. du Roi. / Avec Privilege / 1749

Table: *Description d'Europe*

Paris: Chés l'Autheur / Quay de l'Horloge

br: 491 x 665 mm.
pl: 497 x 673 mm.

Engravers: (map) Delahaye l'ainé Scripsit
(cartouche) Gobin fec.

RF: c.1:9,750,000

Scales: 45 mm.= 300 milles d'Angleterre de 75 au Degré; 48 Milles de Suede de 12 au Degré; 48 Woerstes de Russie de 105 au Degré; 80 Gr. Lieues de France de 25 au Degré; 60 Milles communes d'Allemagne de 15 au

30 **Europe** (continued)

D.; 80 Milles de Pologne de 20 au Degré; 70 Lieues d'Espagne de 17 au Degré; 240 Milles communes d'Italie de 60 au Deg.; 40 Milles d'Hongrie de 10 au Degré; 240 Milles de Turquie de 60 au Degré; 80 Lieues Marines de 20 au Degré

Comments: Table to right

State 1: 1749: Robert; Gobin cartouche

State 2: [1771?]: conjectured from RV, Notice Raisonnée, (1774), described as re-edit of 1749 map; new cartouche: Arrivet inv. & sculpsit

State 3: 1776; as per RV, Catalogue (1777)

State 4: 1778: imprint: Chez Fortin Ingenieur / Mécanicien du Roy Rue de la Harpe/; in lower rt. corner of map: "Comme les Etats du Grand-Seigneur sont trop resserrés dans les cartes d'Asie et d'Afrique, on a jugé à propos de les détailer dans cette carte, et de les distinguer par une même couleur, ce qui n'a pas encore été fait jusqu'à present."

Book:

AP:
AU:
NAP:

Location: BN: Ge.C.8179 (1749); Ge.D.20604, Ge.C.8191 (1778); Ge.C.8192 (n.d.); Ge.C.8193 (An XI/1803); Ge.C.8194 (1812); WLCL: Julien, Theatre du Monde, I (4) [1749]

31 **Europe** 1750 RV

L'EUROPE, / Par le Sr. Robert / de Vaugondy. / Avec Privilége du Roi. / 1750.

br: 147 x 153 mm.

Engravers:

RF: c.1:40,000,000

Scales: 33 mm. = 200 lieues d'une heure

Comments: "Pag. 57" outside border, top right cardinal points outside border

Book: Méthode Abrégée et facile pour apprendre la Géographie

AP:
AU:
NAP:

[l'Abbé Le François] Paris: Libraires Associés, 1770

Location: WLCL: C2 1770 Le

32 **Europe** 1751 RV

L'EUROPE / divisée / en ses principaux Etats, / par le Sr. Robert de Vaugondy, / fils de M. Robert Géographe / ordinaire du Roy. / Avec Privilege. / 1751.

br: 482 x 571 mm.
pl: 497 x 627 mm.

Engravers:

RF: c.1:9,500,000

Scales: 56 mm. = 375 Milles d'Angleterre de 75 au degré; 60 Milles de Suede de 12 au degré; 530 Woerstes de Russie de 105 au degré; 100 Grandes Lieues de France de 20 au degré; 75 Milles communes d'Allemagne de 15 au degré; 86 Lieues d'Espagne de 17 au degré; 100 Milles de Pologne de 20 au degré; 300 Milles communes d'Italie de 60 au degré; 50 Milles de Hongrie de 10 au degré; 300 Milles de Turquie de 60 au degré; 100 Lieues marines de 20 au degré

Comments: From AU 'Préface,' pp. 21b-22a: This map had been criticised in foreign journals for the Sorlingues [Scilly Isles] missing and Scanie coloured as part of Denmark. Roberts answered: Sorlingues too small to include; too busy to have time to check colouring, and Delisle did the same thing with Hungary in his Etats du Turc 1700; criticisms like these are raised by 'esprits caustiques et désoeuvrés' but whatever the motive we would be guilty if we didn't profit from them.

32 Europe (continued)

Problem of the length of the Mediterranean: Sanson and Delisle made it too wide (Delisle: 13° to 70° long.); ours is reduced to 12° (Gibraltar) to 54° (Antioch)

Book:

AP:
AU: 14
NAP:

Location:

33 Europe 1756 RV

Supplément / pour la Carte du Livre / intitulé / de l'Esprit des Loix / par le Sr. Robert de Vaugondy / Géog. ord. du Roi, de S. Mte. Polonoise / Duc de Lorraine et de Bar / et de l'Acad. Rle, des Sc? et B.L. de Nancy /

[Book: London: Chez Nourse]

br: 228 x 298 mm.
pl: 247 x 335 mm.

Engravers: (map) G. De la Haye

RF: c.1:17,000,000

Scales: 11 mm. = 100 Mille pas Géometriques de 60 au Deg.; 16 mm. = 50 Lieues d'une heure de 20 au Deg.

Comments: BN:Ge.D.12335 (2): without engraver's name, perhaps from the 1749 second edition which added a map? Bod: (E) C1 (513): dated 1764; no engraver; copy of 1756; some lettering changes from italic to Roman; some deletions e.g. Cassiterides (Scilly Isles) removed

Book: OEUVRES / DE MONSIEUR DE MONTESQUIEU, / Nouvelle Edition, / revue, corrigée & considerablement augmentée par l'auteur. / Tome Second./ Montesquieu (1767)

AP:
AU:
NAP:

Location: BN: Ge.D.12335 (2); Bod: Vet.E5d65; (E) C1 (513)

34 Europe [1758] RV

L'EUROPE / divisée / en ses Etats, Empires, / Royaumes et Républiques / par le Sr. Robert de Vaugondy / Géographe ordinaire du Roy / de sa Majesté Polonaise Duc de Lorraine et de Bar, et / de l'Académie Royale des Sciences et Belles-Lettres de Nancy / Avec Privilege / 1767

A Paris Chés l'Auteur Quay de l'horloge du Palais près le Pont Neuf. / Avec Privilege

Four sheets: 987 x 1117 mm. (assembled)

Engravers: (cartouche) Arrivet inv. & sculpsit

RF: c.1:5,250,000

Scales: 15 mm. = 50 Milles d'Angleterre de 75 au degré; 23 mm. = 12 Milles de Suéde de 12 au degré; 14 mm. = 63 Woerstes de Russie de 105 au degré; 11 mm. = 10 Grandes Lieues de France de 20 au degré; 15 mm. = 15 Milles communes d'Allemagne de 15 au degré; 11 mm. = 10 Milles de Pologne de 20 au degré; 9 mm. = 10 Lieues d'Espagne de 26 1/2 au degré; 11 mm. = 30 Milles communs d'Italie de 60 au degré; 11 mm. = 5 Milles de Hongrie de 10 au degré; 4 mm. = 10 Milles de Turquie de 60 au degré; 12 mm. = 40 Milles Marines de la Méditerranée de 75 au degré

Comments:
State 1: [1758]: not seen; cited in Julien catalogue (1763): no. 24, prix, 4#.

State 2: 1767

State 3: [after 1788]: added to title, 'corrigée et augmentée par le Cen. Lamarche' By parallel with other four sheet maps of world, there is probably a 1778 Fortin imprint which would become State 3

Book:

AP:
AU:
NAP:

Location: BL: 4.86.2 TAB (1767); LC:G5700 176-.D4 Vault (corrigée & augmentée par le Cen. Lamarche)

35 Europe [1762] RV

L'EUROPE / par le S. Robert de Vaugondy / Géog? ord? du Roi, de S. M. Pol? / duc de Lorr? et de Bar / et de la Société royale de Nancy /

[A Paris ches l'Auteur quai de l'Horloge près le pont neuf]

br: 242 x 283 mm.

Engravers: (map) Gravé par E. Dussy
(cartouche) Arrivet inv. & sculp.

RF: c.1:20,000,000

Scales: 54 mm. = 600 Mille pas géométriques de 60 dég?; 200 Lieues d'une heure de 20 au dégré

Comments:
State 1: [1762]

State 2: [1778]: [imprint removed]; "et Censeur Royal" added to title; Fortin publication

Book:

AP:
AU:
NAP: 5

Location:

36 Europe, Roman Empire 1741 R

CARTE / DE / L'EMPIRE ROMAIN, / Pour servir à la lecture / de l'Histoire d'Espagne / de Jean de Ferreras, / Dressée par le Sr. Robert / Géographe ord. du Roi 1741

br: 333 x 625 mm.
pl: 341 x 635 mm.

Engravers: (map) Gravé par Lattré

RF: c.1: 11,000,000

Scales: 78 mm. = 600 Milles Anciens; 4800 Stades

Comments: shows Europe, Mediterranean Sea, Near East to Caspian Sea, Persian Gulf; North Africa to Red Sea; north to Baltic Sea. Top right, outside border: 'Tome 1er 2e partie' no cardinal points. The bookdealer de Hondt to Cardinal Passionnei, 14 Nov. 1752: The Histoire d'Espagne of Ferreras, in 4°, 10 vols., 220 florins [BAV: Vat. Lat. 9812 (345)]

Book: Histoire Générale d'Espagne

AP:
AU:
NAP:

Jean de Ferreras (trad. par d'Hermilly) Paris: [Ganeau], 1751

Location: Vatican: R.G. Storia III.115 (1&2); BN: 4 Oa. 80; BL: 178.c.7-16; L-D (1745), p. 245-247

37 Europe, Roman Empire 1742 R

IMPERIUM OCCIDENTIS / AUCTORE N. Sanson / Christianiss. Galliarum Regis / Geographo. / revisum et emendatum / ab Aeg. Robert / Christianiss. Regis Geographo.

Second sheet: *IMPERIUM ORIENTIS / [continues as above]*

Parisiis / Apud auctorem in Ripa Sequanae / Vulgo dicta Quai de l'Horloge / Cum privil. Regis.

br: 486 x 661 mm.
pl: 492 x 664 mm. [same for both sheets]

Engravers:

RF: c.1:

Scales: 68 mm.= 375 Vetera Milliara Romana; 3500 Stadia

Comments: Table to left on western sheet; to right on eastern sheet; gives ancient names with modern equivalents. Julien price: 2#. BL copy: messy engraving— guide lines show, graticule lines doubled. Meridian of Berenice shown between 35 and 40 east long.

37 **Europe, Roman Empire** (continued)

Book:

AP:
AU:
NAP:

Location: BL: Maps.183.d.1 (19,20); L-D (1745), p. 245; Julien cat. (1763): 1394

38 **Europe, Roman Empire** 1752 RV

ROMANI IMPERII / occidentis scilicet et orientis, / TABULA GEOGRAPHICA, / ex Schoedis Sansonianis desumpta, / observationibus astronomicis subacta; / et, quo ad vicinas regiones Romanorum jugo intactas, / Germaniam scilicet, Daciam, Germano - Sarmatiam, etc. / ne deserta videretur / Locis et Populis apud antiquos autores relatis / illustrata;/ Autore D. Robert de Vaugondy Geographo / Regis ordin? 1752. / Cum Privilegio

Inset: *Fortunatae Insulae.*

br: 482 x 655 mm.
pl: 495 x 673 mm.;
inset (bot.left): 98 x 168 mm.

Engravers:

RF: c.1:8,800,000

Scales: 94 mm. = 562 Millaria Romana quorum 75 gradum efficiunt; 4500 Stadia Graeca quorum 600 ad gradum unum requiruntur

Comments:
State 1: 1752

State 2: [1768]: 'Aquilo' removed as north cardinal point; island of 'Thule' added at 60° N. lat.

Book:

AP:
AU: 4
NAP:

Location:

39 **Mediterranean Sea** 1759 RV

[Mer Méditerranée]

2 sheets

Engravers:

RF: c.1:

Scales:

Comments: not seen in any location; known only by RV catalogue citations

Book:

AP:
AU:
NAP:

Location: RV Catalogues: Tablettes Parisiennes (1760); Catalogue des Ouvrages (1777)

40 **British Isles** 1747 R

LES ISLES BRITANNIQUES / où sont / Le Royaume d'Angleterre / divisé en ses quatre Roymes. des Saxons, / trois des Anglois et la Princté. de Galles. / Le Royaume d'Ecosse / divisé en deça et dela le Tay, ou / Royaumes des Pictes et des Scots / Le Royaume d'Irlande / en ses Quatre Provinces etc. / Tirés de Speed, T. Pont, Gordon, à Stralock etc. / Par le Sr. Sanson Géographe ordinaire du Roy. / Revuë et corrigé / Par le Sr. Robert Géographe ord? de sa Majesté /

Inset: *[Shetland and the Faroes]*

A Paris / chés l'Auteur Quay de l'horloge du Palais / 1747

br: 540 x 754 mm.
pl: 550 x 762 mm.

Engravers: (cartouche) Noel Cochin fecit

40 **British Isles** (continued)

RF: c.1:2,050,000

Scales: 101 mm. = 60 Mille pas Géométriques ou Milles d'Italie; 96 Milles d'Angleterre de 48 au dégré; 80 Milles d'Irlande de 40 au dégré; 50 Lieues Communes de France de 25 au dégré

Comments: Table to left of divisions

Book:

AP:
AU:
NAP:

Location: BL: Maps.183.d.1 (13) (Bailleul atlas); Julien Cat. (1763), II: 7

41 **British Isles** 1748 R

CARTE / DES / ISLES BRITANNIQUES / par le Sr. Robert. Géog. ord. du Roi. / avec Privilege 1748 /

br: 161 x 182 mm.
pl: 165 x 240 mm.

Engravers:

RF: c.1:6,250,000

Scales: 34 mm. = 30 Lieues d'une heure; 96 Milles ou Lieues d'Angleterre

Comments:

Book:

AP: 4
AU:
NAP:

Location:

42 **British Isles** 1750 RV

ISLES BRITANNIQUES / Par le Sr. Robert / de Vaugondy / avec Privilége. 1750

br: 150 x 153 mm.

Engravers:

RF: c.1:6,500,000

Scales: 33 mm. = 40 lieues d'une heure

Comments: "Pag. 313" outside border, top right; cardinal points outside border

Book: Méthode Abrégée et facile pour apprendre la Géographie

AP:
AU:
NAP:

[l'Abbé Le François] Paris: Libraires Associés, 1770

Location: WLCL: C2 1770 Le

43 **British Isles** 1754 R

LES ISLES / BRITANNIQUES / qui comprennent / LES ROYAUMES D'ANGLETERRE, / D'ECOSSE et D'IRLANDE; / Par le Sr. Robert Géographe / Ordinaire du Roi. / Avec Privilege / 1754.

Inset: *SUPPLEMÉNT / pour les Isles / ORCADES, SCHETLAND, / et FERO.*

[Paris]: avec privilege

br: 480 x 592 mm.;
pl: 489 x 615 mm.
Inset: 180 x 172 mm.

Engravers: (map) Gravé par E. Dussy
(cartouche) haussard Sc [State 3: Groux]

RF: c.1:2,000,000

Scales: 107 mm. = 120 Milles d'Angleterre de 60 au degré; 100 Milles d'Ecosse et d'Irlande de 50 au degré; 40 Lieues marines ou d'une heure de 20 au degré; 50 Lieues communes de France de 25 au degré.

43	**British Isles** (continued)		

Inset: "l'Echelle est la même / que celle de la carte"

Comments:
State 1: 1754: no islands west of Ireland; "Flanan I."; "Bellfart" or "Bellfas"

State 2: 1768: islands added west of Irlande; "Flanan I." changed to "Flanel I." and moved to west of Isles Westernes; Bellfast

State 3: 1778: Groux cartouche

State 4: after 1789: Delamarche edition: date and "géographe ordinaire du roi" removed

Book:

AP:
AU: 15
NAP:

Location:

44	**British Isles**	1762	RV

LES ISLES / BRITANNIQUES / par le Sr. Robert de Vaugondy Géog? / ord. du Roi de S.M. Polon. Duc de Lorr. et de Bar / et de la Société royale de Nanci. 1762.

A Paris chés l'Auteur Quai de l'Horloge près le Pont neuf

br: 242 x 284 mm.

Engravers: (map) Gravé par E. Dussy
(cartouche) Arrivet inv. & sculp.

RF: c.1:4,750,000

Scales: 48 mm. = 120 Mille pas géométriques de 60 au degré; 38 mm. = 30 Lieues d'une heure de 20 au dégré

Comments:
State 1: 1762

State 2: 1778: date changed; imprint removed; Fortin publication

Book:

AP:
AU:
NAP: 6

Location:

45	**British Isles, roads**	1757	R

Carte / des GRANDES ROUTES / D'ANGLETERRE, D'ECOSSE, / ET D'IRLANDE. / Par le Sr. Robert Géog? ordin? du Roy / Avec Privilege / 1757

br: 472 x 580 mm.
pl: 500 x 603 mm.

Engravers: (cartouche) Mre Cne [i.e. Marie Catherine] haussard Sculp.

RF: c.1:2,075,000

Scales: 105 mm. = 120 Milles d'Angleterre de 60 au degré; 100 Milles d'Ecosse et d'Irlande de 50 au degré; 40 Lieues Marines ou d'une heure de 20 au degré; 50 Lieues communes de France de 25 au degré

Comments:
State 1: 1757: Bellfart

State 2: 1768: Roman lettering changed to italic for many place names; Bellfart changed to Bellfast

[State 3: 1778: Groux cartouche]

Book:

AP:
AU: 107
NAP:

Location:

46	**British Isles, historical**	1750	RV

BRITANNICAE / INSULAE / in quibus / ALBION seu BRITANNIA MAJOR, / ET IVERNIA seu BRITANNIA MINOR / juxta Ptolemoei mentem divisae, / tum in suas majores partes, / tum in populos exhibentur, / a Nicolao Sanson Christ. Regis Geographo, / Revisae, et ad Observationes astronomicas redactae, / accurante Robert de Vaugondy, filio, / Cum Privilegio Regis / 1750

br: 487 x 530 mm.
pl: 497 x 555 mm.

Engravers: (map) Guill? Delahaye Sculpsit
(cartouche) haussard

RF: c.1:2,300,000

Scales: 92 mm. = 150 Millaria Romana quorum 75 gradum efficiunt; 1200 Stadia Graeca quorum 600 gradui respondent

Comments:
State 1: 1750: "Cassiterides" [Scilly Isles] not shown; cartouche: haussard

State 2: 1750: "Cassiterides" [Scilly Isles] added ; 'haussard' removed?

State 3: 1750: "Sini fl. Ostium" [mouth of r. Shannon] and "Sabrinae Aestuarium" [Severn estuary] labelled

State 4: [1768]: road network added; many names changed from Roman to italic lettering; place name changes, e.g. "Sini fl. Ostium" to "Senius fl."

[State 5: 1778: cartouche signed Groux]

[State 6: 1786: published by Delamarche: date removed]

Book:

AP:
AU: 5
NAP:

Location: See Weinreb & Douwma Catalogue 6 (1970) for all states

47	**England**	1748	R

ROYME. D'ANGLETERRE / divisé par Provinces et / Shires ou Comtés / Par le Sr. Robert Geog. du Roi / avec Privilege 1748

br: 160 x 164 mm.
pl: 178 x 172 mm.

Engravers:

RF: c.1:4,300,000

Scales: 39 mm. = 30 Lieues d'une heure:

Comments:

Book:

AP: 5
AU:
NAP:

Location:

48	**England**	1753	R

LE ROYAUME / D'ANGLETERRE, / divisé selon les sept Royaumes, / ou Heptarchie des Saxons, / avec / La Principauté de Galles ; / et subdivisé en / Shires ou Comtés. / Par le Sr. Robert Geog. ord. du Roi. / Avec Privilege 1753

br: 477 x 514 mm.
pl: 492 x 532 mm.

Engravers: (cartouche) E. haussard fecit

RF: c.1:1,400,000

Scales: 77 mm. = 60 Milles d'Angleterre de 60 au degré; 50 Milles d'Ecosse et Irlande de 50 au degré; 20 Lieues Marines ou d'une heure de 20 au degré

Comments: "Bellfast"/"Bellfart": varies

48 **England** (continued)

Book:

AP:
AU: 16
NAP:

Location:

49 **England** 1762 RV

L'ANGLETERRE / Par le S. Robert de Vaugondy / Géog? ord? du Roi, de S.M. Polon? / Duc de Lorr? et de Bar / de la Société royale de Nancy / 1762

A Paris chés l'Auteur Quai de l'Horloge près le Pont neuf

br: 243 x 223 mm.

Engravers: (map) Gravé par E. Dussy
(cartouche) Arrivet inv. & sculp.

RF: c.1:2,800,000

Scales: 40 mm. = 60 Mille pas géom? de 60 au dégré; 75 Milles d'Angleterre de 75 au dégré

Comments:
State 1: 1762

State 2: [1778]: date and imprint removed; "et Censeur Royal" added to title; Fortin publication

Book:

AP:
AU:
NAP: 7

Location:

50 **London** 1759 RV

Carte des / environs de Londres / qui comprend / le cours de la Tamise / et les provinces / qui l'environnent tant au nord qu'au sud, / tirée de la grande carte d'Angleterre de Chr. Browne / par le Sr. Robert de Vaugondy, Geog. ord. du Roi, / de S.M.Polon. Duc de L. et de Bar et Associé / de l'Académie royale des Sciences et Belles-Lettres / de Nancy. 1759

Paris, chez l'Auteur

br: 488 x 675 mm.
pl: 505 x 694 mm.

Engravers: (map) Gravé par E. Dussy
(cartouche) Arrivet inv. et Sculpsit

RF: c.1:424,000

Scales: 87 mm.= 20 mille pas géom. de 60 au dégré; 113 mm. = 30 milles d'Angleterre; 94 mm. = 18 milles communes d'Angleterre; 79 mm. = 6 lieues marines d'Angleterre

Comments: roads shown, some with dotted, some with full lines

Prices: Mémoires de Trévoux (Nov. 1759, p. 2877): 1# 5s; Julien Cat. (1763), no. 762: 1#

Book:

AP:
AU:
NAP:

Location: BN: Ge. DD. 2987 B (2048)

51 **Scotland** 1748 R

ROYAUME / D'ECOSSE / divisé en Provinces / Par le Sr. Robert Géog. du Roi / Avec Privilege 1748

br: 160 x 197 mm.
pl: 175 x 206 mm.

Engravers:

RF: c.1:2,850,000

51 **Scotland** (continued)

Scales: 38 mm. = 20 Lieues d'une heure:

Comments:

Book:

AP: 6
AU:
NAP:

Location:

52 **Scotland** 1751 R

LE ROYAUME / D'ECOSSE / divisé en / Shires ou Comtés, / Par le Sr. Robert / Geographe ordinaire du Roi. / Avec Privilege / 1751.

[Paris]:avec privilege.

br: 487 x 575 mm.
pl: 496 x 630 mm.

Engravers:

RF: c.1:955,000

Scales: 114 mm. = 60 Mille pas Geometriques de 60 au degré; 50 Mille d'Ecosse de 50 au degré

Comments:
State l: 1751: Flanan I. close to Isles Westernes;

State 2: [1768]: Is. Flanel moved to far west; St. Kilda added; Roman lettering of many place names changed to italics.

Book:

AP:
AU: 17
NAP:

Location:

53 **Scotland** 1762 RV

L'ECOSSE / Par le S. Robert de Vaugondy / Géographe ord? du Roi, de S.M. Pol? / Duc de Lorr? et de Bar, et de / la Société royale de Nancy. / 1762.

A Paris chés l'Auteur Quai de l'Horloge près le Pont neuf

br: 243 x 226 mm.

Engravers: (map) Gravé par E. Dussy
(cartouche) Arrivet inv. & sculp.

RF: c.1:2,800,000

Scales: 41 mm. = 60 Milles pas géom? de 60 au dégré; 50 Milles d'Ecosse de 50 au dégré

Comments:
State 1: 1762

State 2: 1778: date changed; imprint removed; Fortin publication

Book:

AP:
AU:
NAP: 8

Location:

54 **Edinburgh** 1759 RV

Carte des Environs d'Edenburg...Tirée de la grande Carte d'Ecosse de Dowet

460 x 584 mm.

Engravers:

RF: c.1:

Scales:

Comments: not seen; sold by Weinreb & Douwma in 1975; otherwise known only from 18th c. catalogues:

54 **Edinburgh** (continued)

Weinreb & Douwma: 'Original outline colour, trimmed to the engraved surface and laid down (in the period). Based on the work of James Dorret (here corrupted to Dowet.) £ 30.' Dorrets map, London 1750.
Mémoires de Trévoux, Nov. 1759, p. 2877: price, 1# 5s.
Julien Cat. (1763): no. 778, price 1#

Book:

AP:
AU:
NAP:

Location:

55 **Ireland** 1748 R

Isle et Royaume / d'IRLANDE / par le Sr. Robert Geog. Ord. du Roi / avec Privilege 1748

br: 167 x 167 mm.
pl: 180 x 175 mm.

Engravers:

RF: c.1:3,250,000

Scales: 35 mm. = 20 Lieues d'une heure:

Comments:

Book:

AP: 7
AU:
NAP:

Location:

56 **Ireland** 1750 R

ROYAUME / D'IRLANDE / divisé en / ses quatre Provinces, / et subdivisés en Comtés / et Baronies / Par le Sr. Robert / Geographe ordinaire du Roy. / Avec Privilege / 1750

br: 487 x 545 mm.
pl: 497 x 626 mm.

Engravers: (map) Guill? Delahaye Sculpsit

RF: c.1:1,080,000

Scales: 99 mm. = 60 Milles d'Italie de 60 au Dégré; 50 Milles d'Irlande de 50 au Dégré

Comments:

State 1: 1750: title includes "et Baronies"; fewer italic place names; no frame around scales; B. de Galloway has no islands; north tip of Anglesey labelled as "Moylroniad"; "Lundye" [island] in Canal de Bristol; "Belfast"; North and Middle ground off east coast

State 2: [1750]: "et Baronies" removed from title; many italic place names added; islands added in B. de Galloway; two grounds added off east coast, names removed

State 3: [1750]: "Belfast" changed to "Bellfart"; "Moylroniad" becomes "I. Moylroniad," island north of Anglesey; Lundye "I."

State 4: [1768]: section of Scotland added (CANTYRE) betw. 12-13° E.; islands added to west; "Bellfast"; "I. Moylroniad" changed to "Skerry I."; "The South Islands" of Arran added W. of Gallway Bay

Book:

AP:
AU: 18
NAP:

Location:

57 **Ireland** 1762 RV

L'IRLANDE / Par le S. Robert de Vaugondy / Géog? ord? du Roi, de S.M. Pol? / Duc de Lo? et de Bar, et / de la Société royale de Nancy / 1762

57 **Ireland** (continued)

A Paris chés l'Auteur Quai de l'Horloge près le Pont neuf

br: 241 x 224 mm.

Engravers: (map) Gravé par E. Dussy
(cartouche) Arrivet Inv. & sculp.

RF: c.1:2,850,000

Scales: 41 mm. = 25 Lieues de 25 au Degré; 50 Milles d'Irlande de 50 au Degré

Comments:

State 1: 1762

State 2: [1778]: date and imprint removed; "et Censeur Royal" added to title; Fortin publication

Book:

AP:
AU:
NAP: 9

Location:

58 **Ireland, historical** 1757 RV

CARTE DE / L'ANCIENNE IRLANDE / dressée pour la lecture / de L'Histoire de ce Pays. / Par le Sr. Robert de Vaugondy / Geog. ordin. du Roy, de S. M. Pol. Duc de Lorraine et de Bar / et Associé de l'Acad. Roy. des Sc. et B. L. de Nancy. / 1757

[Book: Paris: Antoine Boudet, Imprimeur du Roi, rue Saint-Jacques, à la Bible d'Or]

br: 320 x 211 mm.
pl: 335 x 225 mm.

Engravers:

RF: c.1:1,650,000

Scales:

Comments:

Book: HISTOIRE DE L'IRLANDE ANCIENNE ET MODERNE.
Tirée des Monumens les plus authentiques.

AP:
AU:
NAP:

M.l'Abbé Ma-Geoghegan. (Paris: Boudet 1758)

Location: Bod: Gough Ireland 113

59 **Ireland, historical** 1762 RV

Royaume d'Irlande...XII - XVII siecle

[Book: Paris: Antoine Boudet, rue Saint-Jacques, à la Bible d'or]

Engravers: (map) Gravé par E. Dussy

RF: c.1:

Scales:

Comments: This map is tipped into Volume II of the CUL copy of Histoire de l'Irlande; note date later than book's publication date

Book: HISTOIRES DE L'IRLANDE ANCIENNE ET MODERNE.
Tirée des Monumens les plus authentiques.

AP:
AU:
NAP:

l'abbé Ma-Geoghegan. (Paris: Boudet, 1758)

Location: CUL: Hib.4.758.1-3.

60 **Ireland, Connaught** 1757 RV

PROVINCE DE CONNACIE / divisée en Dynasties pour / les premiers siecles du Christianisme / Par le Sr. Robert de Vaugondy / Geog.ord.du Roi, de S.M.Pol. Duc de Lorraine / et du Bar. et associé de l'Academie Rle. / des Sc. et Belles lettres de Nancy. / 1757

[Book: Paris: Antoine Boudet, Imprimeur du Roi, rue Saint-Jacques, à la Bible d'Or]

60 **Ireland, Connaught** (continued)

br: 211 x 185 mm.
pl: 224 x 202 mm.

Engravers:

RF: c.1:

Scales:

Comments:

Book: HISTOIRE DE L'IRLANDE ANCIENNE ET MODERNE.
AP:
AU:
NAP:

l'Abbé Ma-Geoghegan. (Paris: Boudet, 1758)

Location: Bod: Gough Ireland 113

61 **Ireland, Leinster** 1757 RV

PROVINCES DE / MIDIE ET DE LAGENIE / divisées en Dynasties, / Par le Sr. Robert de Vaugondy / Géog. ord. du Roi, de S.M. Pol. Duc de Lorraine / et de Bar, et associé de l'Acad. Rle. des Scien. / et Belles Lettres de Nancy / 1757

[Book: Paris: Antoine Boudet, Imprimeur du Roi, rue Saint-Jacques, à la Bible d'Or]

br: 211 x 177 mm.
pl: 222 x 205 mm.

Engravers:

RF: c.1:

Scales:

Comments:

Book: HISTOIRE DE L'IRLANDE ANCIENNE ET MODERNE....
AP:
AU:
NAP:

l'Abbé Ma-Geoghegan (Paris: Boudet 1758)

Location: Bod: Gough Ireland 113

62 **Ireland, Munster** 1757 RV

PROVINCE DE / MOMONIE / divisée en Dynasties pour les / premiers siecles du Christianisme / Par le Sr. Robert de Vaugondy / Geog. ord. du Roy de S.M.Pol. Duc de Lor- / raine et de Bar et Associé de l'Acad. Royale / de Nancy / 1757

[Book: Paris: Antoine Boudet, Imprimeur du Roi, rue Saint-Jacques, à la Bible d'Or]

br: 214 x 240 mm.
pl: 223 x 270 mm.

Engravers:

RF: c.1:

Scales:

Comments:

Book: HISTOIRE DE L'IRLANDE ANCIENNE ET MODERNE...
AP:
AU:
NAP:

M. l'Abbe Ma-Geoghegan (Paris: Boudet, 1758)

Location: Bod: Gough Ireland 113

63 **Ireland, Ulster** [1757] RV

PROVINCE / D'ULTONIE / Divisée en Dynasties / pour les premiers siecles du Christianisme. / Par le Sr. Robert de Vaugondy / Geog.ord. du Roi, de S.M.Pol. Duc de Lorraine / et de Bar et Associé de L'Acad. Rle. des Sc. et / Belles Lettres de Nancy

[Book: Paris: Antoine Boudet, Imprimeur du Roi, rue Saint-Jacques, à la Bible d'Or]

63 **Ireland, Ulster** (continued)

br: 215 x 248 mm.
pl: 225 x 260 mm.

Engravers:

RF: c.1:1,000,000

Scales:

Comments:

Book: HISTOIRE DE L'IRLANDE ANCIENNE ET MODERNE...
AP:
AU:
NAP:

M. l'Abbe Ma-Geoghegan. (Paris: Boudet, 1758)

Location: Bod: Gough Ireland 113

64 **Scandinavia** 1750 RV

SUEDE ET / NORWEGE. / Par le Sr. Robert de / Vaugondy. / Avec Privilege 1750.

br: 150 x 154 mm.

Engravers:

RF: c.1:12,000,000

Scales: 35 mm. = 80 [unit not given]

Comments: "Pag. 343" outside border, top right cardinal points: Septentrionale only

Book: Méthode Abrégée et facile pour apprendre la Géographie
AP:
AU:
NAP:

[l'Abbé Le François] Paris: Libraires Associés, 1770

Location: WLCL: C2 1770 Le

65 **Scandinavia** [1759] RV

LES COURONNES DU NORD / Le Danemarck, La Norwege, l'Islande / et la Suede./partie de la Russie d'Europe, de la Pologne, la Prusse; / partie de l'Allemagne et des Frontieres de la France./

Inset: *[Iceland]*

[State 2]: Paris: C.F.Delamarche, rue du Foin St. Jacques, au College de Mtre. Gervais

south sheet: br: 451 x 627; pl: 470 x 664 mm.
north sheet: br: 460 x 630; pl: 472 x 660 mm.

Engravers:

RF: c.1:

Scales: 150 mm. = 37 Lieues communes de Danemarck, de 13 au Degré; 28 Lieues communes de Norwege, d'environ 10 au Degré; 28 Lieues communes de Suede, de 10 au Degré; 70 Lieues Françoises de 2500 Toises; 51 Lieues Marines, de 20 au Degré; 230 Wersts communs de Russie, de 85 au Degré; 55 Lieues communes de Pologne, de 20 au Degré; 41 Lieues communes d'Allemagne, de 15 au Degré

Comments: Inset: 175 x 196 mm.
RV, Tablettes Parisiennes (1760); Catalogue (1777)

State 1: 1759: cited by RV

State 2: 1792: Delamarche imprint (only copy seen)
Along top, longitudes given in degrees from Meridian of Paris; along bottom, from Meridian of Isle de Fer

Book:
AP:
AU:
NAP:

Location: BL: Maps.34375.(35): Delamarche (1792)

66	**Sweden**	1762	RV

LA SUEDE / Par le S. Robert de Vaugondy / Géog? ord? du Roi, de S.M. Pol? / Duc de Lorr? et de Bar / et de la Société royale de Nancy. / Avec Privilege 1762.

A Paris chés l'Auteur Quai de l'Horloge près le Pont neuf

br: 241 x 194 mm.

Engravers: (map) Gravé par E. Dussy
(cartouche) Arrivet inv. & Sculp.

RF: c.1:7,500,000

Scales: 38 mm. = 150 Mille pas géom? de 60 au dégré; 30 Lieues de Suede de 12 au dégré

Comments: State 1:

1762

State 2: [1778]: privilege, date, and imprint removed; "et Censeur Royal" added to title; Fortin publication

Book:

AP:
AU:
NAP: 10

Location:

67	**Sweden and Norway**	1748	R

Royaumes de / SUEDE et DE NORWEGE / Par le Sr. Robert, Géog? ord? du Roi / Avec Privilege 1748.

br: 167 x 203 mm.
pl: 175 x 247 mm.

Engravers:

RF: c.1:10,500,000

Scales: 42 mm. = 80 Lieues d'une heure:

Comments:

Book:

AP: 8
AU:
NAP:

Location:

68	**Sweden and Norway**	1756	RV

LES ROYAUMES / DE SUEDE ET DE NORWEGE, / divisés / en leurs Provinces ou / Gouvernemens, /Par le S. Robert de Vaugondy / Geog. ord. du Roi, de S.M. Pol. Duc de Lorraine et de Bar, / et de l'Académie Rle. des Sciences et Belles Lettres de Nancy / Avec Privilege / 1756

Inset: *[Iceland]*

br: 477 x 572 mm.;
pl: 502 x 592 mm.
Inset: 141 x 196 mm.

Engravers: (cartouche) E. haussard fecit

RF: c.1:1,385,000

Scales: 113 mm. = 240 Mille pas géométriques de 60 au degré; 48 Lieues de Suede de 12 au degré; 42 Lieues communes de Norwege de 10 2/3 au degré; 80 Lieues d'une heure de Chemin ou de Marine de 20 au degré.

Comments: on later pulls, cartouche very faint

Book:

AP:
AU: 20
NAP:

Location:

69	**Norway**	1747	R

Royaume de Norvege subdivisé en ses principaux Gouvernements...[tiré des mémoires de G. Sanson]

Paris: ches l'auteur

525 x 570 mm.

Engravers:

RF: c.1:

Scales:

Comments: Julien Cat. (1763), II: 6

Book:

AP:
AU:
NAP:

Location: BN: Ge.BB.565-II(3)

70	**Norway & Denmark**	1762	RV

NORWEGE / Par le S. Robert de Vaugondy / Géog? ordinaire du Roy &c. / Avec Privilege 1762

Inset: *ROYAUME DE DANEMARCK*

A Paris chés l'Auteur Quai de l'Horloge près le Pont neuf

br: 242 X 221 mm.

Engravers: (map) Gravé par E. Dussy
(cartouche) Arrivet inv. & sculp.

RF: c.1:7,000,000

Scales: 50 mm. = 180 mille pas géometriques de 60 au degré; 60 Lieues d'une heure de 20 au degré

Inset: 34 mm. = Echelle de 20 lieues

Inset c. 1:3,200,000

Comments:
State 1: 1762

State 2: [1778]: privilege, date and imprint removed; Fortin publication

Book:

AP:
AU:
NAP: 11

Location:

71	**Denmark**	1748	R

Royaume de / DANEMARCK / Par le Sr. Robert, Géog? du Roi / avec Privilege / 1748

br: 160 x 180 mm.
pl: 166 x 250 mm.

Engravers:

RF: c.1:3,450,000

Scales: 32 mm. = 20 Lieues d'une heure

Comments: latitude number, top right, written as 68 instead of 58

Book:

AP: 9
AU:
NAP:

Location:

72 **Denmark** 1750 R

Royaume de / DANEMARCK, / qui comprend le Nort-Jutland divisé / en ses quatre Dioceses, / le Sud-Jutland divisé en ses deux Duchés / de Sleswick et de Holstein, / et les Isles de Fionie, Selande, Laland, &c. / Par le Sr. Robert Geog. ordin. ru Roi / Avec Privilege 1750

br: 487 x 571 mm.
pl: 500 x 585 mm.

Engravers:

RF: c.1:l,180,000

Scales: 92 mm. = 60 Mille pas geometriques; 50 Lieues communes de Danemarck de 15 au degré

Comments:
State 1: 1750

State 2: [1768]; lettering changes from italic to Roman and vice versa; some spelling alterations, e.g., Malmuye to Malmo, Goteborg to Gotheborg

Book:

AP:
AU: 19
NAP:

Location:

73 **Russia** 1753 RV

Carte Generale de / L'EMPIRE DES RUSSES / en Europe et en Asie, / dressée d'après les cartes de l'Atlas Russien / Par le Sr. Robert de Vaugondy / Géographe ord? du Roy / Avec Privilege. / 1753

Paris: chés l'Autheur sur le Quay de l'horloge / du Palais proche le Pont neuf.

2 sheets: br: 460 x 1195 mm. assembled [CUL]

Engravers: (map) Delahaye l'ainé Sculps. et Scripsit
(cartouche) Gobin Sculp.

RF: c.1:7,500,000

Scales: 14 mm. = 60 Mille pas géometriques de 60 au Degré; 104 Woerstes de Russie de 104 au Degré; 90 Woerstes fixés par Pierre le grande à 90 au Degré; 12 Milles Suédois de 12 au Degré

Comments: [top right]: "Explication / de quelques denominations Russes / employées dans cette Carte..." [20 terms follow] "Termes usites dans les / confins de la Tatarie..." [10 terms follow]
between Asia and N. America: "Terre vue en 1723" (75°N. lat.); "Terre vue par Mr. Tchirikow en 1741" (53°N. lat, 187°E. long.); "I. de Beerings" (54°N, 182°E); "Terre vue par Jean de Gama" (45°N, 175°E); "Terre de la Compagnie" (45°N, 168°E, using dotted line)

State 1: 1753

State 2: [after 1786]: 'ord? du Roy' removed from title; imprint added: 'A Paris / Chez le Cn. Delamarche Geog./ rue du Foin St. Jacques, au / au [sic] Collège de Mtre Gervais

Julien Cat. (1763) no. 816: price 2# 10s.

Book:

AP:
AU:
NAP:

Location: CUL: Maps aa.3.75.1; BN: Ge. D. 7047 [Delamarche]; WLCL: Julien, Theatre du Monde I (103)

74 **Russia, European** 1748 R

RUSSIE BLANCHE / ou MOSCOVIE / Par le Sr. Robert Géog? ordin? du Roi / avec Privilege 1748.

br: 163 x 145 mm.
pl: 167 x 166 mm.

Engravers:

RF: c.1:16,000,000

74 **Russia, European** (continued)

Scales: 35 mm = 100 Lieues d'une heure

Comments:

Book:

AP: 10
AU:
NAP:

Location:

75 **Russia, European** 1749 RV

Partie Occidentale de / la RUSSIE POLONOISE / où se trouvent / la Russie Noire, la Haute / Volhynie, et la Haute Podolie / Par le Sr. Robert de Vaugondy / Fils de Mr. Robert Geog du Roi / Avec Privilege 1749

br: 165 x 238 mm.
pl: 179 x 242 mm.

Engravers:

RF: c.1:2,600,000

Scales: 42 mm. = 20 Lieues d'une heure

Comments:

Book:

AP: 121
AU:
NAP:

Location:

76 **Russia, European** 1749 RV

Partie Orientale / de la RUSSIE POLONOISE / où se trouvent / la Basse Volhynie et la / Basse Podolie / Par le Sr. Robert de Vaugondy / Fils de Mr. Robert Geog. du Roi / Avec Privilege 1749

br: 163 x 193 mm.
pl: 178 x 203 mm.

Engravers:

RF: c.1:3,350,000

Scales: 32 mm = 20 Lieues d'une heure

Comments: in Petite Tatarie: "Isles où est le trésor de l'armee des Cosaques"

Book:

AP: 122
AU:
NAP:

Location:

77 **Russia (European)** 1750 RV

RUSSIE / EUROPEENNE / Par le Sr. Robert de Vaugondy / Avec Privilége / 1750.

br: 177 x 153 mm.

Engravers:

RF: c.1:15,000,000

Scales: 30 mm. = 416 Voerstes de 104 au degre

Comments: "Pag. 361" outside border, top right cardinal points: Septentrionale and Midi outside border; Occidental and Orientale between lines

Book: Méthode Abrégée et facile pour apprendre la Géographie

AP:
AU:
NAP:

[l'Abbé Le François] Paris: Libraires Associés, 1770

Location: WLCL: C2 1770 Le

78 **Russia, European** 1752 RV

Partie Méridionale / de la / RUSSIE EUROPÉENNE / ou sont distinguées exactement / toutes les Provinces, / d'après le détail de l'Atlas Russien; / Par le Sr. Robert de Vaugondy fils / Géographe ordinaire de Roi / Avec Privilege / 1752

br: 481 x 593 mm.
pl: 498 x 635 mm.

Engravers:

RF: c.1:3,400,000

Scales: 98 mm.= 180 Mille Pas Géometriques de 60 au Degré; 312 Woerstes de Russie de 104 au Degré; 270 Woerstes fixés par Pierre le Grand a 90 au Degré

Comments:

Book:

AP:
AU: 22
NAP:

Location:

79 **Russia, European** 1753 RV

Partie / Septentrionale / de la / RUSSIE EUROPÉENNE / où sont distinguées exactement / toutes les Provinces, / d'après le détail de / l'Atlas Russien, / Par le Sr. Robert de Vaugondy fils / Géog. ordin? du Roi. / Avec Privilege. / 1753

[Paris]:

br: 480 x 594 mm.
pl: 497 x 636 mm.

Engravers:

RF: c.1:3,350,000

Scales: 98 mm. = 180 Mille Pas geometriques de 60 au degré; 312 Woerstes de Russie de 104 au degré

Comments: 'Moskwa' labeled in the space between neat line and the scale line; its symbol lies outside neat line
Au "Préface," 22b: gives criticism of Russian Atlas of 1745, esp. for coordinates of cities and rivers: e.g. Kiovie, Akerman, Asow, course of the Dnieper, extent of Black Sea; and Cape of Crimea

Book:

AP:
AU: 21
NAP:

Location:

80 **Russia, European** 1762 RV

Partie Meridionale / de la / RUSSIE EUROPEENNE, / TARTARIE RUSSIENNE, / PETITE TARTARIE. / Par le Sr. Robert de Vaugondy. / 1762.

A Paris, ches l'Auteur Quai de l'Horloge près le Pont neuf

br: 242 x 298 mm.

Engravers: (map) Gravé par E. Dussy
(cartouche) Arrivet inv. & sculp.

RF: c.1:7,500,000

Scales: 61 mm. = 240 Mille pas géometriques de 120 au degré; 416 Milles de Russie de 208 au degré

Comments:
State 1: 1762

State 2: 1778: date changed; imprint removed; Fortin publication

Book:

AP:
AU:
NAP: 12

Location:

81 **Russia, European** 1762 RV

Partie Septentrionale / de la / RUSSIE EUROPÉENE / Par le Sr. Robert de Vaugondy / Géog? Ordinaire du Roi &c.

A Paris chés l'Auteur Quai de l'Horloge près le Pont neuf

br: 245 x 298 mm.

Engravers: (map) Gravé par E. Dussy
(cartouche) Arrivet inv. & sculp.

RF: c.1:

Scales: 69 mm. = 220 Milles pas géometriques de 60 au degré; 100 Milles de Russie de 104 au degré

Comments:
State 1: 1762

State 2: [1778]: date and imprint removed; Fortin publication

Book:

AP:
AU:
NAP: 12

Location:

82 **France** 1749 RV

La France / divisée / par Gouvernemens généraux. / Par le Sr. Robert de Vaugondy / fils de Mr. Robert Geogr. ord. ru Roi / avec Privilege 1749.

br: 162 x 179 mm.
pl: 181 x 242 mm.

Engravers:

RF: c.1:6,250,000

Scales: 35 mm. = 40 Lieues d'une heure

Comments:

Book:

AP: 12
AU:
NAP:

Location:

83 **France** 1750 RV

LA FRANCE / divisée / Par Gouvernements généraux / Par le Sr. Robert de Vaugondy. / Avec Privilége 1750.

br: 150 x 152 mm.

Engravers:

RF: c.1:7,500,000

Scales: 30 mm. = 40 lieues d'une heure

Comments: "Pag. 61" outside border, top right cardinal points outside border

Book: Méthode Abrégée et facile pour apprendre la Géographie

AP:
AU:
NAP:

[l'Abbé Le François] Paris: Libraires Associés, 1770

Location: WLCL: C2 1770 Le

84 **France** 1750 R

LE ROYAUME DE / FRANCE / divisé suivant les Gouvernemens / généraux./ Par le Sr. Robert Géographe ord. du roi/ Avec Privilege / 1750.

[Paris]:

br: 481 x 598 mm.
pl: 496 x 639 mm.

84 France (continued)

Engravers:

RF: c.1:2,350,000

Scales: 94 mm. = 120 Mille pas geometriques ou Milles d'Italie; 50 Lieues communes de France de 25 au Degré; 40 Lieues d'une heure de Chemin de 20 au Degré

Comments: Meridian of Paris (20° E long.) shown

Book:

AP:
AU: 23
NAP:

Location:

85 France [1762] RV

ROYAUME DE / FRANCE / Par le S. Robert de Vaugondy / Géog? ordinaire du Roy &c.

none

br: 243 x 284 mm.

Engravers: (map) Gravé par E. Dussy
(cartouche) Arrivet inv. & sculp.

RF: c.1:4,800,000

Scales: 48 mm. = 120 Mille pas geometriques de 20 au degré;
39 mm. = 40 Lieues de France de 25 au degré

Comments:
State 1: [1762]: imprint?

State 2: [1778]: imprint removed? Fortin publication

Book:

AP:
AU:
NAP: 14

Location:

86 France, roads 1758 R

Carte / du Royaume de FRANCE / où sont tracées exactement / Les Routes des Postes, / Par le Sr. Robert / Géographe ordinaire du Roy. / Avec Privilege 1758.

br: 474 x 517 mm.
pl: 498 x 532 mm.

Engravers: (cartouche) M. Cne haussard sculp.

RF: c.1:2,400,000

Scales: 92 mm. = 120 Mille pas geometriques de 60 au degré; 40 Lieues de 20 au degré ou d'une heure; 50 Lieues de 25 au degré

Comments:

Book:

AP:
AU: 108
NAP:

Location:

87 France, historical 1738 R

GALLIARUM DESCRIPTIO / ex Sansonum Tabulis editis et manuscriptis / excerpta / ope animadversionum / Domini Martini Bocquet Benedictine / et Dissertationum Domini LeBoeuf / Autissiodor. Canonici emendata / ab Aegidio Roberto Geographo Regis ordinario //

DESCRIPTION DES GAULES / Tirée des Cartes imprimées et manuscrites / des Srs. Sanson, / Corrigée sur les Remarques / de Dom Bouquet Benedicitin. / et sur les Dissertations de Mr LeBoeuf / Chanoine d'Auxerre, / Par le Sr. G. Robert. Geographe ordinaire du Roy.

Paris: rue St. Jacques à l'image St. Paul / avec privilege 1738

87 France, historical (continued)

br: 495 x 673 mm.

Engravers:

RF: c.1:2,250,000

Scales: 97 mm. =
150 Vetera Milaria Romana;
94 Veteres Leucae Gallicae;
40 Leucae Majores Gallicae quorum 20 graduum efficiunt

Comments: Table (top right): Division of Gaul during different emperors: Cesar, Augustus, Otho, Diocletian, Constantine, Honorius (two divisions).
BL: 5 copies, coloured for five of emperors
Robert does maps for Bouquet's Receuil des Historiens de la France (RV. AU Preface, p. 21)
Julien Cat. (1763) no. 1302: price: 1#;
Lelong: Il y a sur cette carte des rémarques importantes dans les Mémoires de l'Ac.Roy. des Belles-lettres, tom. XIV, p. 162, où M. Freret parle au long de diverses anciennes routes de la Gaule & des mésures qu'on y employoit. (Other ref. cited)

Book:

AP:
AU:
NAP:

Location: WLCL:AF (37); BN: Ge DD.2987 B (9786); Ge C. 1984; BL: K.2.105 (1-5); Julien Cat. (1763): 1302; LeLong, BH: 36

88 France, historical 1740 R

FRANCIAE STATUS / sub regibus Primae Stirpis, / ex Observationibus D. Mart. Bouquet Benedictini / atque / ex Dissertationibus D. LeBeuf Canonici Autissiod. / Excerptus / ab Aegidio Robert, Geographo Regis Ordinario / cum Privilegio 1740. //

ETAT DE LA FRANCE / sous Les Rois de la Premiere Race / tiré des Observations de D. Bouquet Benedictin, / et des Dissertations de Mr. LeBeuf Chanoine d'Auxerre. / Par le Sr. G. Robert, Géographe Ord. du Roi. / [continue in Imprint]

A Paris, / Chéz l'Auteur, sur le Quai de l'horloge Palais, / Proche la Rue de Harlay, / Avec Privil? du Roi / 1740.

br: 500 x 666 mm.

Engravers:

RF: c.1:2,300,000

Scales: 96 mm. = 150 Vetera Millaria Romana; 94 Veteres Leucae Gallicae; 40 Leucae Majores Gallicae quarum 20 gradum efficiunt

Comments: to accompany Bouquet's Receuil des Historiens de la France
Julien Cat. (1763) no. 1401: price 1#

Book:

AP:
AU:
NAP:

Location: WLCL: AF (49); BN: Ge.D.6719; Ge.DD.2987 B (364); Julien Cat.(1763): no. 1401; Robert Cat. (1743)

89 France, historical 1742 R

CARTE GENERALE / DE LA MONARCHIE / DES GOTS [sic] / TANT DANS LES GAULES QU'EN ESPAGNE / Pour servir à la lecture / de l'histoire d'Espagne / de Jean de Ferreras / Dressée par le Sr. Robert / Geog. ord. du Roi / 1742

br: 407 x 533 mm.
pl: c. 424 x 538 mm.

Engravers:

RF: c.1: 3,700,000

Scales: 60 mm. = 150 Milles Anciens; 1200 Stades

Comments: shows France south of Orleans, all of Spain, Balearic Islands, Corsica and Sardinia; Italian peninsula to Rome; part of North Africa

Julien Cat. gives as title: 'l'Espagne sous les Visigots,' n.d., price: 1#5s.

Book: Histoire Générale d'Espagne

AP:
AU:
NAP:

Jean de Ferreras (trad. d'Hermilly) Paris: [Ganeau] 1751

Location: Vatican: R.G. Storia III. 115 (1 & 2);BN: 4° Oa. 80; BL: 178.c.7-16; L-D (1745), p. 247; Julien Catalogue (1763): 14000

90 France, historical 1742 R

Le Royaume de France / et ses Païs circonvoisins, / jusqu'à l'étendue de / l'Ancienne Gaule, / par les Srs Sanson...revue et assujetti aux Observations astronomiques par le Sr. Robert...

Paris: chés le Sr. Robert, Quay de l'Horloge du Palais, près la rue du Harlai

br: 500 x 682 mm.
pl: 510 x 690 mm.

Engravers: (map) Delahaye sculpsit

RF: c.1:2,220,000

Scales: 48 mm. = 25 Lieues communes de France; 60 Mille pas geometriques ou Milles d'Italie; 20 Lieues d'une heure du chemin

Comments: Table to right: 'Le Royaume de France peut être consideré en diverses sortes de gouvernements et peut se diviser...pour l'Administration de la Religion en 18 Provinces ecclesiastiques ou archevêchés et cent onze Evêchés... On peut encore donner une division pour le Départment des Quatre Secretaires d'Etat, Pour l'intendance du Commerce dans l'intérieur du Royaume, Pour les Quatre Cours des Monnoyes' [Table possibly by RV, as first work?]
These divisions are shown on the same map by different color outlines: BN: Ge.BB.565-VII (19) = Eveches; ibid. (28) = Cour des monnaies; ibid. (29) = Intendance du Commerce BL: maps.183.d.1 = l'Administration des Armes
L-D: 'Cette Carte est de même grandeur & de même plan que les Cartes de l'ancienne Gaule & de la première Race de nos Rois, donnée au Public par le même Auteur'

Book:

AP:
AU:
NAP:

Location: BN: Ge.BB. 565-VII (19), (28), (29); Ge. C.6197 (listed as ed. Sanson); BL: Maps 183.d.1 (7); L-D (1745), p. 246.

91 France, historical 1743 R

IMPERIUM CAROLI MAGNI / Occidentis Imperatoris / Complectens universam Galliam, in Hispania / quidquid à Pyrenaeis jugis occurrit ad Iberum et / Rubricatum usque fluvios; cum Balearibus insulis; / in Italiâ quidquid Langobardicae ditionis fuerat, cum / adjacentibus Istriâ, Liburniâ, Dalmatiâ, Insulâque / Corsica; Rhaetiam quoque, Vindeliciam, Noricum, / et quidquid Danubium et Alpes interjacet; atque / ultra Danubiam partem non modicam pannoniarum / et Germaniam non tantum cis Albim, set etiam / ulterius ad Oderam usque flumen et ad mare / Balticum. // Hanc tabulam conatibus Geographicis discribebat / AEg. Robert Regis Christianisss. Geographus. /

Parisiis / Apud auctorem in ripâ Sequanae ab / Horlogio Palatii dicta / Cum privilegio Regis / 1743

br: 475 x 648 mm.

Engravers:

RF: c.1:5,500,000

Scales: 60 mm. = 225 vetera Millaria Romana; 140 Veteres Leucae Galliae; 60 Majores Leucae Galliae

Comments: RV: 'Mon père chargé de composer les cartes du magnifique recuil des Historiens de France en composa aussi une qui devint supérieure à la précédent par les lumières des scavans pour l'ouvrage desquels

elle avoit été faite; j'ai cru ne pouvoir mieux faire que d'en donner une copie dans cet atlas' [AU 'Préface,' p. 21]

Julien (1763): price 1#
Lelong: 'Ce n'est qu'une copie de la précédente (Hondius/Bertius 1620) où l'on ne voit point les bornes de cet Empire.'

Book: [Historiens de France]

AP:
AU:
NAP:

[Dom Bouquet]

Location: WLCL: AF (54); BN: Ge DD.2987 B (10362); Ge C. 4889; Julien cat. (1763): no. 1404; Lelong, BH: p.411; Robert cat.(1743)

92 France, historical 1750 R

GALLIA ANTIQUA / in Provincias et Populos divisa, / Geographicis Sansonum Conatibus / et recentioribus Astronomicis Observationibus accommodata. / ab Aeg. Robert Geographo regis ordinario / Cum Privilegio Regis / 1750.

br: 480 x 545 mm.
pl: 501 x 632 mm.

Engravers:

RF: c.1:2,300,000

Scales: 94 mm. = 150 Millaria Romana quorum 75 gradui uni respondent; 100 Veteres Leuca Gallicae quorum 50 in uno gradu requiruntur; 40 Leucae Majores Gallicae quorum 20 gradum efficiunt

Comments:
State 2: [1768?]; Belgian/Netherlands coastline straightened

Book:

AP:
AU: 6
NAP:

Location:

93 France, historical 1752 RV

IMPERIUM / CAROLI MAGNI / occidentis Imperatoris / complectens / Galliam, Marcam Hispanicam, Baleares insulas / Corsicam, Italiam Vulturnum usque et Ausidum fluv? / Istriam, Liburniam, Dalmatiam, / Rhaetiam, Vindeliciam, Noricum, / et quidquid Germaniae a Rheno ad Oderam / usque et Mare Balticum / protenditur / Autore D. Robert de Vaugondy Geographo / Regis Ordinario / Cum Privilegio. 1752.

br: 475 x 545 mm.
pl: 494 x 572 mm.

Engravers: (map) [Delahaye]
(cartouche) haussard Sculp.

RF: c.1:5,000,000

Scales: 87 mm. = 300 Millaria Romana, quorum 75 gradum efficiunt; 200 Leucae Gallicae quorum gradum 50 efficiunt; 80 Leucae Francicae, vel unius horae quarum 20 gradui respondent

Comments:
State 1: 1752

State 2: [1768]: italics added and place name changes; Vectis I. [Isle of Wight] labelled

This map was the subject of controversy in the lawsuit brought by the Delahayes against A. Boudet concerning the italics. (BN: Ge FF. 13374) . See Chapter Three of text.

Book:

AP:
AU: 12
NAP:

Location:

94	**France, historical**	[1762]	RV

LA GAULE / Par le S. Robert de Vaugondy / Géographe ord? du Roi,/ de S. Maj? Pol? Duc de Lorr? et de Bar / et de la Société Royale de Nanci. /

[A Paris ches l'Auteur Quai de l'Horloge près le pont neuf]

br: 248 x 282 mm.

Engravers: (map) Gravé par E. Dussy
(cartouche) Arrivet inv. & Sculp.

RF: c.1:4,700,000

Scales: 47 mm. = 150 Milles Romains de 75 au degré; 38 mm. = 80 Lieues Gauloises de 50 au degré

Comments:
State 1: [1762]

State 2: [1778]: date and imprint removed? "et Censeur Royal" added to title; Fortin publication

Book:

AP:
AU:
NAP: 52

Location:

95	**France, ecclesaistical**	1748	R

FRANCE / ECCLESIASTIQUE / Par le S. Robert Géog. Ord. du Roi / avec Privilege / 1748

br: 160 X 176 mm.
pl: 175 x 185 mm.

Engravers: (map)

RF: c.1:6,250,000

Scales: 35 mm.= 40 Lieues d'une heure

Comments:

Book:

AP: 11
AU:
NAP:

Location:

96	**Picardy, Artois**	1748	R

Gouvernemens / Generaux de / PICARDIE ET ARTOIS, / du Boulonois, de Flandre, / et de Dunquerque. / Par le Sr. Robert. Geog. du Roy / avec Privilege. / 1748.

br: 160 x 168 mm.
pl: 165 x 250 mm.

Engravers:

RF: c.1:7,700,000

Scales: 35 mm. = 10 Lieues d'une heure

Comments:

Book:

AP: 13
AU:
NAP:

Location:

97	**Picardy, Artois**	1753	R

Gouvernement Général de / PICARDIE ET ARTOIS, / qui comprend / Le Comté d'Artois, Le Ponthieu, l'Amienois, / le Santerre, le Vermandois et le Thierrache. / Gouvernemens Généraux / du BOULENOIS / et de la FLANDRE FRANCOISE. / Par le Sr. Robert / Géographe ordinaire du Roi / Avec Privilege / 1753.

br: 481 x 590 mm.
pl: 500 x 632 mm.

97	**Picardy, Artois**(continued)

Engravers:

RF: c.1:398,000

Scales: 92 mm. = 20 Mille pas geometriques de 60 au degré; 6 2/3 Grandes lieues de France et d'une heure de 20 au degré; 8 1/3 Lieues communes de Picardie et d'Artois de 25 au degré; 7 1/3 Lieues communes de Flandre de 23 au degré; 8 Lieues de Haynaut de 24 au degré

Comments: Note Douvres [Dover] and bit of coast of England shown outside border, top left

Book:

AP:
AU: 24
NAP:

Location:

98	**Picardy, Artois**	[1759]	RV

[Carte de Picardie, Artois, Boulenois, Flandre]

Engravers:

RF: c.1:

Scales:

Comments: not seen in any location; known only by contemporary catalogue entries; cited in Mémoires de Trévoux, Nov. 1759, p. 2877, as one of eight new maps; price: 1# 5s.
Julien: title as "Picardie et Artois, Flandre Francois, Hainault Francois, et Cambresis, avec Table Méthodique" Price 1#5s. [cf. maps of Brittany and Normandy with Tables, Cat. 102, 106]

Book:

AP:
AU:
NAP:

Location: RV: Notice Raisonnee.(1774); Catalogue (1777);
Julien Catalogue (1763): no. 246

99	**Picardy, Artois, Flandres**	1762	RV

FLANDRE FRANCOISE, / PICARDIE ET ARTOIS, / ISLE DE FRANCE, ORLEANOIS, / CHAMPAGNE, LORRAINE, / ET ALSACE./ Par le S. Robert de Vaugondy Géog? ord? du Roi. / de S.M. Polon? Duc de Lorraine et de Bar. / et de la Société royale de Nanci / 1762

A Paris ches l'Auteur quai de l'Horloge près le pont neuf

br: 241 x 302 mm.

Engravers: (map) Gravé par E. Dussy
(cartouche) Arrivet inv. & sculp.

RF: c.1:1,775,000

Scales: 62 mm. = 60 Mille pas géométriques de 60 au degré; 25 Lieues communes de France de 25 au degré

Comments:
State 1: 1762

State 2: [1778]: date and imprint removed; "et Censeur Royal" added to title; Fortin publication

Book:

AP:
AU:
NAP: 15

Location:

100 **Normandy** 1748 R

Gouvernement / General / de NORMANDIE / Par le Sr. Robert Geog? ordin? du Roi / 1748

br: 160 x 197 mm.
pl: 165 x 207 mm.

Engravers: (map) Gravé par Guill? Delahaye

RF: c.1:1,380,000

Scales: 40 mm. = 10 Lieues d'une heure de chemin

Comments: long./lat. graticule visible

Book:

AP: 14
AU:
NAP:

Location:

---

101 **Normandy** 1751 R

Gouvernement Général de / NORMANDIE / divisé en ses sept Bailliages de / Coutances, Caen, Caux, Rouen, Evreux, / Gisors, et Alençon, / Par le Sr. Robert Géographe ordinaire du Roy / 1751 / Avec Privilege

Inset: *Supplément pour les Isles / GRENEZEY [sic] et JERSEY / appartenantes aux Anglois /*

br: 482 x 580 mm.;
pl: 499 x 595 mm.
Inset: 140 x 116 mm.

Engravers:

RF: c.1:550,000

Scales: 97 mm. = 30 Mille Pas Géometriques de 60 au Degré; 12 Lieues de Normandie de 25 au Degré; 10 Grandes Lieues de France et d'une heure de 20 au Degré

Comments: Paris and its symbol are between the border and plate mark, right edge; from AU Preface, p. 24a: problem of how to mark divisions: decided to do by bailliages based on the Sanson manuscripts

Book:

AP:
AU: 25
NAP:

Location:

---

102 **Normandy** 1758 RV

Carte / du Gouvernement de / NORMANDIE / par le S. Robert de Vaugondy Géog. ord. du Roi / de S.M.Polonoise Duc de Lorraine et de Bar / et Associé de l'Académie Royale des Sciences et / Belles Lettres de Nancy / Avec Privilege / 1758

Inset: *[Channel Islands]*

Paris / Chès l'Auteur, Quay de l'horloge du Palais / près le pont neuf.

br: 478 x 720 mm.
pl: 480 x 725

Engravers:

RF: c.1:468,000

Scales: 79 mm.= 20 mille pas géometriques de 60 au degré; 8 1/3 Lieues de Normandie de 25 au degré; 6 2/3 Grandes lieues de France de 20 au degré

Comments: Table of geographical divisions to the right of map; cited in Mémoires de Trévoux, Nov. 1759, p. 2877, as one of eight new maps; price 1#5s. Julien Catalogue (1763): 150; price 1#5s.

Book:

AP:
AU:
NAP:

Location: BN: Ge. DD. 2987 B (690); WLCL: Julien, Theatre du Monde (19)

---

103 **Normandy, Brittany, Touraine** 1762 RV

NORMANDIE, / BRETAGNE, / MAINE, ANJOU, / TOURAINE, / ET SAUMUROIS / Par le S. Robert de Vaugondy / Géog ordinaire du Roy &c. / 1762

A Paris ches l'Auteur quai de l'Horloge près le pont neuf

br: 243 x 302 mm.

Engravers: (map) Gravé par E. Dussy
(cartouche) Arrivet inv. & sculp.

RF: c.1:1,775,000

Scales: 52 mm. = 50 Mille pas géometriques de 60 au deg?; 62 mm. = 25 Lieues de France de 25 au degré

Comments:
State 1: 1762

State 2: [1778]: date and imprint removed; Fortin publication

Book:

AP:
AU:
NAP: 16

Location:

---

104 **Brittany** 1748 R

Gouvernement / General / de BRETAGNE / Par le Sr. Robert Geographe / ordin? du Roi. / 1748

br: 162 x 200 mm.
pl: 169 x 208 mm.

Engravers: (map) Gravé par Guill? Delahaye

RF: c.1:1,590,000

Scales: 35 mm. = 10 Lieues d'une heure de chemin

Comments: long./lat. graticule visible

Book:

AP: 15
AU:
NAP:

Location:

---

105 **Brittany** 1751 R

Gouvernement / General de / BRETAGNE / divisé en ses Dioceses ou Receptes / par le Sr. Robert Géog? ordin? / du Roy. / Avec Privilege / 1751

br: 480 x 595 mm.
pl: 500 x 637 mm.

Engravers:

RF: c.1:500,000

Scales: 109 mm. = 10 Grandes Lieues de France et d'une heure de 20 au degré; 12 Lieues communes de France de 25 au degré; 11 Lieues de Bretagne de 22 au degré

Comments:

Book:

AP:
AU: 26
NAP:

Location:

106	**Brittany**	1758	RV

Carte du gouvernment de BRETAGNE par le Sr. Robert de Vaugondy, Géog. ord. du Roi, de S.M. Pol. Duc de Lorraine & Bar, Associé de l'Académie Royale des Sciences et Belles Lettres de Nancy / avec privilege / 1758

Inset: *Passage de l'Iroise*

Paris, Ches l'Auteur, Quay de l'horloge, pres le Pont Neuf

br: 480 x 705 mm.

Engravers:

RF: c.1:460,000

Scales: 120 mm.= 10 Grandes Lieues de France; 12 Lieues communes de France; 11 Lieues de Bretagne

Comments: Table of geographical divisions to the right of map

State 1: 1758

State 2: 1768: the following changes: Inset: Guidelmede to Guitalmezeau (48°30' lat., 13° long.) Inset: place names added to circles which were blank on the 1758 state: Kemuon, Plourin, Bresles, P. de Renan, Lanioure; Lanfeutre; Harmour et S. Paul to Lampoul Plouarzel; Lambart to Ploudar. On the map: Gobelic to Baverie; Bel isle to Belle Isle ; Belain to Blain; Gouarec to Gouezec. On the table: Gouarec to Gouezec; new date to title.

State 3: 1778: Fortin imprint

RV claimed map was copied by Daumont (Inst. Geog., 1766, p. xxviii)

Book:

AP:
AU:
NAP:

Location: BN: Ge. DD. 2987 B (455) [1758]; Ge. DD. 2987 B; (266) [1768]; Ge.DD.2987 B (456) [1768]; WLCL: W-2-D, Julien, Theatre du Monde, no. 24.; N-2-C, Delisle-Buache atlas [I778]

107	**Maine and Perche**	1748	R

Gouvernemens / Generaux / du Maine et Perche, / d'Anjou, du Saumurois, / et de Touraine / Par le Sr. Robert Geog. du Roi. 1748.

br: 160 x 163 mm.
pl: 163 x 173 mm.

Engravers: (map) Gravé par Guill. Delahaye

RF: c.1:1,500,000

Scales: 37 mm. = 10 Lieues d'une heure de Chemin

Comments:

Book:

AP: 16
AU:
NAP:

Location:

108	**Maine, Perche, Anjou, Touraine**	1753	R

Gouvernemens / Généraux / du MAINE ET PERCHE, / DE L'ANJOU / DE LA TOURAINE / ET DU SAUMUROIS. / Par le Sr. Robert Géographe / Ordinaire du Roi. / Avec Privilege / 1753

br: 478 x 457 mm.
pl: 497 x 488 mm.

Engravers: (map) Gravé par De la Haye l'Ainé

RF: c.1:l:480,000

Scales: 154 mm. = 40 Mille pas géometriques de 60 au Degré; 16 Lieues de Maine, et du Perche de 24 au Degré; 18 5/6 Lieues de Touraine, de 28 au Degré; 13 1/3 Lieues d'une heure, de 20 au Degré

108 Maine, Perche, Anjou, Touraine(continued)

Comments: Chartres symbol (top rt.) just outside degree line

Book:

AP:
AU: 27
NAP:

Location:

109	**Tours, Diocese**	1754	RV

Carte / de la / Touraine / Dressée par M. Robert de Vaugondy / & Dediée / A Mgr. de Fleury / Archeveque de Tours / par S.T.H.S.F.L. / 1754

br: 172 x 154 mm.

Engravers:

RF: c.1:

Scales:

Comments:

Book:

AP:
AU:
NAP:

Location: BN: Mss. Coll. Tour et Anjou 14, fol. 17-18

110	**Tours, Diocese**	1762	RV

Carte / du Diocèse de Tours / Par M. Robert de Vaugondy. / Dediée / à Mgr. de Fleuri / Archevêque de Tours / par S.T.H.S.F.L. / 1762

br: 215 x 210 mm.

Engravers:

RF: c.1:

Scales:

Comments:

Book:

AP:
AU:
NAP:

Location: BN: Mss. Coll. Tour et Anjou 14, fol. 17-18

111	**Orleanois**	1748	R

Gouvernement General / d'ORLÉANOIS / Par le Sr. Robert Géographe / Ordinaire du Roy. 1748.

br: 163 x 182 mm.
pl: 170 x 187 mm.

Engravers: (map) Lattré Sculp.

RF: c.1:1,285,000

Scales: 43 mm. = 10 Lieues d'une heure.

Comments: 'avec privilege du Roy' outside cartouche line

Book:

AP: 17
AU:
NAP:

Location:

112	**Orleanois, Beauce, Gatinois**	1753	RV

Gouvernement Général / d'ORLÉANOIS, / où se trouvent / l'Orleanois Propre, le Blaisois, le Gatinois, / et LA BEAUCE, / qui comprend / Le Vendomois, le Dunois, et le Pays Chartrain. / Par le Sr. Robert de Vaugondy fils,/ Geographe ordin? du Roi / Avec Privilege / 1753

br: 480 x 590 mm.
pl: 496 x 633 mm.

Engravers: (cartouche) Moreau inv. et fecit

RF: c.1:408,000

Scales: 147 mm. = 30 Mille pas Geometriques de 60 au degré; 17 Lieues de Beauce et de Gatinois de 34 au degré; 14 Lieues de Sologne de 29 au degré; 10 Lieues d'une heure de 20 au degré

Comments:

Book:

AP:
AU: 28
NAP:

Location:

113	**Isle de France**	1754	R

Gouvernement Général / de / L'ISLE DE FRANCE / divisé par Pays. / Par le Sr. Robert Geographe ord. / du Roi. / Avec Privilege. 1754

br: 480 x 523 mm.
pl: 500 x 540 mm.

Engravers: (cartouche) haussard fecit

RF: c.1:397,000

Scales: 115 mm. = 25 Mille pas géometriques de 60 au degré; 6 3/4 Grandes Lieues de France, ou d'une heure de chemin; 10 Lieues moyennes de 25 au degré; 12 Lieues du Parisis de 28 au degré

Comments: meridian of Paris visible at 20° long. with 'l'Observatoire' marked south of Paris

Book:

AP:
AU: 29
NAP:

Location:

114	**Isle de France, Champagne,**	1748	R

Gouvernemens / Generaux, / de l'ISLE DE FRANCE, / de PARIS, et de la / CHAMPAGNE ET BRIE / Par le Sr. Robert, / Géographe ordinaire du Roy. / Avec Privilege 1748

br: 162 x 198 mm.
pl: 165 x 252 mm.

Engravers:

RF: c.1:1,875,000

Scales: 30 mm = 10 Lieues d'une heure

Comments:

Book:

AP: 18
AU:
NAP:

Location:

115	**Paris, environs**	1748	R

Banlieue / et Environs de / PARIS / Par le Sr. Robert Geog ord? du Roi / avec Privilége 1748.

br: 161 x 198 mm.
pl: 180 x 207 mm.

Engravers:

RF: c.1:520,000

Scales: 39 mm. = 1 1/2 Lieues d'une heure.

Comments:

Book:

AP: 19
AU:
NAP:

Location:

116	**Paris, environs**	1753	R

ENVIRONS / DE PARIS / Par le Sr. Robert / Geographe ordinaire / du Roi / Avec Privilege / 1753

br: 488 x 592 mm.
pl: 497 x 630 mm.

Engravers:

RF: c.1:176,500

Scales: 102 mm. = 10 Mille Pas Géometriques de 60 au Degré; 4 1/6 Lieues communes de France de 25 au Degré; 3 1/3 Grandes Lieues de France de 20 au Degré; 4 2/3 Lieues Parisiennes de 28 1/2 au Degré; 9510 Toises de Paris de 57060 au Degré

Comments:

Book:

AP:
AU: 30
NAP:

Location:

117	**Paris, Environs**	1761	RV

Environs de PARIS / Divisé par Pays. / dans lesquels l'on trouve / l'Etendue du Diocèse divisé en Archidiaconés et Doyennés, / d'après la description du Diocèse publiée / par l'abbé LeBeuf, de l'Académie Royale / des Inscriptions et B. Lettres. / Par le S. Robert de Vaugondy, / Géog? ord? du Roi, de S. Majeste Pol? Duc de Lorr? et de Bar, et / de l'Academie Royale des Sciences et B. Lettres de Nancy / Avec Privilege 1761

[Paris]

br: 484 x 590 mm.
pl: 498 x 607 mm.

Engravers: (map) De-la-Haye Sculps.
(cartouche) Arrivet inv. & sculp.

RF: c.1:182,000

Scales: 102 mm.= 10 Mille pas geometriques de 60 au degré; 4 1/6 Lieues communes de France de 25 au degré; 3 1/3 Grandes Lieues de France de 20 au degré; 4 2/3 Lieues Parisiennes de 28 1/2 au degré; 9510 Toises de Paris de 57060 au degré

Comments: Avertissement describes boundaries of the diocese of Paris and of the three Archdioceses: Le Gland, de Josas, de Brie. Coloring of the names corresponds to that of the archdiocese; red dots are used to outline the diocese of Paris

[State 1: 1760: as per Julien Cat. (1763) dated 1760]

State 2: 1761: dated in title

State 3: 1768; no changes noted except date

117 **Paris, Environs**(continued)

Book:

AP:
AU:
NAP:

Location: BN: Ge. D. 15348 [1761];Ge. DD. 2987 B (276 bis) [1768]; BL: 16090 (77) [1768]; Bod: (E)C21:29(12) [1768]; Julien Cat. (1763): no. 135

118 **Paris, Environs** 1761 RV

Plan Général de la Ville et des Faubourgs de Paris / par le Sr. Robert de Vaugondy Géog? ord? du Roi, de S.M. Polon? Duc de Lorr? et de Bar, et de l'Acad. Rle. de Nancy

Paris: Chès l'Auteur, Quay de l'Horloge du Palais, près le P. Neuf.

br: 188 x 210 mm.

Engravers: (map) De-la-Haye sculpsit

RF: c.1:

Scales: [10 mm. = 160 Toises]; Echelle de 320 Toises; 1/6 de Lieue de 28 1/2 au Deg.

Comments:

Book: Les Promenades des Environs de Paris

AP:
AU:
NAP:

Robert de Vaugondy (Paris: RV and Boudet, 1761)

Location: BN:Imprimées 8° Z le Senne 11.815

119 **Paris** 1760 RV

Plan de la Ville et des Faubourgs de PARIS / Divisé en ses vingt Quartiers / par le Sr. Robert de Vaugondy Géog? Ord? du Roi, du S.M.Pol? Duc de Lorr? et de Bar, de la Société royale de Nancy //

A Paris / Chès l'Auteur, Quay de l'horloge du Palais / près le Pont-neuf 1760

br: 547 x 830 mm.
pl: 582 x 861 mm

Engravers: (map) De-la-Haye Sculps.

RF: c.1:9400

Scales: 18 mm. = 100 toises de 500 Toises à 13T 5 piéds 1 Pouc. pour une seconde de Grand Cercle; 29 mm. = 1/16 d'une Lieue Parisienne Quart d'une Lieue Parisienne de 28 1/2 au degré

Comments: 'Table alphabetique des Rues' left and right margins

State 1: 1760

State 2: 1766

State 3: 1771: changes many streets and sites, e.g. the Colosseum (Le Colisée) near present l'Etoile, constructed in 1769-1771 by le Camus de Mezières as a "pleasure palace" for dances and balls; closed in 1779 and demolished in 1780 (Hillairet, Dict. Hist. des Rues, I, p.369)

State 4: 1772: shows the Nouvelle Halle au Blé and new streets around it; Palais Bourbon redrawn; Avertissement tells that RV will make a plan of Paris with division by parishes if anyone desires; In title: 'du feu Roi de Pologne' replaces 'S.M.Pol?'; 'Censeur Royal' added (early 1773?)

State 5: 1776: Bib. de la Ville de Paris, not seen

State 6+: 1797+: Delamarche editions

Julien Cat. (1763): price 3#; "grand aigle"

Book:

AP:
AU:
NAP:

Location: BN:Ge.DD.2987 B (814); Ge.C.3681 [both 1760]; Ge.C.6963 [1766]; Ge.DD.2987, B.no.816 [1771]; Ge.FF.11878 [1818, Delamarche]; Bod: (E)C21:50 Paris (23) [1766];(E)C21:50 Paris (2) [1772]; BL: 16110 (174) [1797, Delamarche]

120 **Paris, Bourgs de la Reine** 1761 RV

Partie des Environs de Paris qui renferme ceux du BOURG DE LA REINE

Paris: chés le Sr. Robert de Vaugondy Quay de l'Horloge 1761

br: 190 x 214 mm.

Engravers: (map) Arrivet Sculp.

RF: c.1:

Scales: 13 mm. = 1/4 d'une lieue parisienne; 500 [de] Echelle de deux mille toises

Comments:

Book: Les Promenades des Environs de Paris

AP:
AU:
NAP:

Robert de Vaugondy (Paris: RV and Boudet, 1761)

Location: BN:Imprimées 8° Z le Senne 11.815

121 **Paris, Ecole Militaire** [1760] RV

Hotel Royal de l'ECOLE MILITAIRE pour servir de supplément au Plan de Paris / par le Sr. Robert de Vaugondy

[Paris]

br: 190 x 212 mm.

Engravers: (map) DelaHaye sculps.

RF: c.1:

Scales: 29 mm. = 150 toises

Comments:

Book: Tablettes Parisiennes

AP:
AU:
NAP:

Robert de Vaugondy (Paris: 1760)

Location: BL: 16195 (3); BN: Ge.FF.4423 [1760]

122 **Paris, St. Cloud** 1761 RV

Partie des Environs de Paris qui renferme ceux de ST. CLOUD

Paris: chés le Sr. Robert de Vaugondy Quay de l'Horloge 1761

br: 190 x 215 mm.

Engravers: (map) Arrivet Sculp.

RF: c.1:

Scales: 13 mm. = 1/4 d'une lieue parisienne; 500 toises

Comments:

Book: Les Promenades des Environs de Paris

AP:
AU:
NAP:

Robert de Vaugondy (Paris: RV and Boudet, 1761)

Location: BN:Imprimées 8° Z le Senne 11.815

123 **Paris, St. Denys** 1761 RV

Partie des Environs de Paris qui renferme ceux de ST. DENYS.

Paris: ches le Sr. Robert de Vaugondy Quay de l'Horloge 1761

br: 190 x 213 mm.

Engravers: (map) Arrivet sculp.

RF: c.1:

123 **Paris, St. Denys** (continued)

Scales: 13 mm. = 1/4 d'une lieue parisienne; 500 toises

Comments:

Book: Les Promenades des Environs de Paris

AP:
AU:
NAP:

Robert de Vaugondy (Paris: RV and Boudet, 1761)

Location: BN:Imprimées 8° Z le Senne 11.815

124 **Paris, Vincennes** 1761 RV

Partie des Environs de Paris qui renferme ceux de VINCENNES

Paris: chés le Sr. Robert de Vaugondy Quay de l'Horloge 1761

br: 190 x 212 mm.

Engravers: (map) Arrivet Sculpt.

RF: c.1:

Scales: 13 mm. = 1/4 d'une lieue parisienne; 500 toises

Comments:

Book: Les Promenades des Environs de Paris

AP:
AU:
NAP:

Robert de Vaugondy (Paris: RV and Boudet, 1761)

Location: BN:Imprimées 8° Z le Senne 11.815

125 **Champagne, north** 1752 R

Partie Septentrionale / du Gouvernem? General de / CHAMPAGNE / où se trouvent / le Retelois, le Rhemois, la Champagne propre, / Partie de la Brie et du Pertois, &c. / Par le Sr. Robert Géog? Ord. du Roy / Avec Privilege 1752

br: 488 x 601 mm.
pl: 508 x 623 mm.

Engravers: (cartouche) E. haussard fecit

RF: c.1:313,000

Scales: 57 mm. = 10 Mille pas geomet. de 60 au degré; 2 Lieues de Champagne de 25 au degré

Comments: From AU 'Préface,' p. 24b-25c: map was criticised for looking like the 2-sheet map by Jaillot; RV replied that it is only because Jaillot's is a 'servile copy' of Sanson's maps
BN examples undated — Delamarche?

Book:

AP:
AU: 31
NAP:

Location: BN: Ge.DD. 1467 (38); Ge.CC.1520 (45): undated — Delamarche publications? ['du Roy / Avec Privilege 1752' removed?]

126 **Champagne, south** 1752 R

Partie Méridionale / du Gouvernemt. Général de / CHAMPAGNE / qui comprend / la Champagne propre, le Vallage, / le Bassigny, le Senonois, et une partie / de la Brie et du Perthois. / Par le Sr. Robert / Geographe ordinaire du Roi / Avec Privilege 1752

br: 478 x 642 mm.
pl: 494 x 664 mm.

Engravers: (cartouche) E. haussard sculp.

RF: c.1:319,000

Scales: 85 mm. = 15 Mille pas geometriques de 60 au degré; 5 Grandes Lieues de France de 20 au degré; 6 1/4 Lieues de Champagne de 25 au degré

126 **Champagne, south** (continued)

Comments: 'Occident' (cardinal point) printed outside left border
BN examples undated — Delamarche states?

Book:

AP:
AU: 32
NAP:

Location: BN: Ge.DD.1467 (39), Ge.CC.1520 (46): not dated — Delamarche?

127 **Alsace** 1754 R

L'ALSACE / divisée en / Haute et Basse / et / le Sundgau. / Par le Sr. Robert. / Geographe ordinaire du Roy. / Avec Privilege / 1754.

br: 482 x 702 mm.
pl: 500 x 710 mm.

Engravers: (cartouche) E. haussard sculp.

RF: c.1:209,500

Scales: 86 mm. = 10 Mille pas geometriques de 60 au degré; 3 1/3 Grandes Lieues de France de 20 au degré; 2 1/2 Mille d'Allemagne de 15 au degré

Comments: two maps side by side on one plate accommodate the north-south configuration of Alsace which follows the course of the Rhine

Book:

AP:
AU: 34
NAP:

Location:

128 **Lorraine and Alsace** 1748 R

LORRAINE / et ALSACE / Par le Sr. Robert Géog? ord. du Roi / avec Privilege / 1748.

br: 160 x 187 mm.
pl: 165 x 252 mm.

Engravers:

RF: c.1:1,750,000

Scales: 32 mm. = 10 Lieues d'une heure

Comments:

Book:

AP: 20
AU:
NAP:

Location:

129 **Lorraine and Bar** 1756 RV

Carte / de la LORRAINE ET DU BARROIS / dans laquelle se trouvent / la Généralité de Metz et autres Enclaves; / Dressée pour la lecture du Mémoire de M. Durival l'ainé / et Dediée / au Roy de Pologne / Duc de Lorraine et de Bar, /Par Le Sr. Robert de Vaugondy / Géog? ord? du Roi, et de Sa Majesté Polonse et Associé de l'Acad. Rle. des Sciences / et des Belles-Lettres de Nancy / 1756 / Avec Privilege

br: 482 x 590 mm.
pl: 494 x 609 mm.

Engravers: (cartouche) M. C. haussard

RF: c.1:396,000

Scales: 91 mm. = 20 Mille Pas Géometriques de 60 au Degré; 6 2/3 Grandes Lieues de France de 20 au Degré; 5 Lieues communes d'Allemagne de 15 au Degré; 7 2/3 Lieues de Lorraine de 2660 T. ou de 22 1/5 au Degré; 7 3/5 Lieues de Barrois de 2500 T. ou de 22 4/5 au Degré

Comments: Avertissement (top left) re: colour key; Inset (bot. rt.): geographical table

State 1: 1756: title as ill. pp. 114-115

129 **Lorraine and Bar** (continued)

State 2: 1756: "et de sa Majesté Polonse et Associé" to title

From AU 'Préface:' the divisions of bailliages made according to edict of June 1751 by King of Poland (p. 25a)

Book:

AP:
AU: 33
NAP:

Location:

130 **Lorraine, Verdun, Meuse R** 1731 R

VERODUNI / Evesché de Verdun: / ou sont les/ Comté et Ballge. de Verdun: / Le Barrois Ducal, ou / Balliage de St. Mihel; et / Terres Adjacentes, Scav. / Comté de Clermont,/ Marqsat. d'Hatton Chastel, / Balliage d'Aspremont, / Seignrie. de Iametz, etc. / Par N. Sanson Geogr. ordre. du Roy / Avec Privilege pour Vingt Ans / A Paris/ 1656/
COURS DE LA MEUSE / où se trouvent / Commercy, St. Mihel,/ Verdun, Stenay, Mouson./ A Paris/ Ches le Sr. Robert Geographe ord. du Roi / quai de l'horloge / 1731

A Paris / Chès le Sr. Robert Geographe ord. du Roi / quai de l'horloge / 1731

br: 431 x 489 mm.
pl: 442 x 497 mm.

Engravers:

RF: c.1:250,000

Scales: 8 mm. = 1 Mille pas geometriques; 19 mm. = 1 Lieues communes de France

Comments: Bodleian copy gives author's name (l'abbé Roussel) in mss. on flyleaf; adds that book was published by l'abbé Jean le Beuf.
'3e pour la Meuse' outside border, bott. left; suggests 2 more maps; one may be at BN: Ge D. 10003

Book: HISTOIRE ECCLESIASTIQUE ET CIVILE DE VERDUN/AVEC LE POUILLE; LA CARTE DU DIOCESE ET LE PLAN DE LA VILLE/

AP:
AU:
NAP:

[L'Abbé Roussel] (Paris: Simon, 1745)

Location: Bod: 3 delta 1145

131 **Lorraine, Verdun** [1745] R

Plan / de la Ville / de Verdun

[Book: Paris: Pierre-Guillaume Simon, Imprimeur / de Clergé de France, rue de la Harpe, à l'Hercule]

br: 337 x 457 mm.
pl: 361 x 459 mm.

Engravers: (map) P.L. Charpentier

RF: c.1:

Scales: 25 mm. = 100 Toises de France

Comments:

Book: HISTOIRE ECCLESIASTIQUE ET CIVILE DE VERDUN AVEC LE POUILLE; LA CARTE DU DIOCESE ET LE PLAN DE LA VILLE./

AP:
AU:
NAP:

[l'abbé Roussel] (Paris: Simon, 1745)

Location: Bod: 3 delta 1145

132 **Poitou, Aunis, Saintonge, Angoumois** 1748 R

Gouvernemens Generaux / de POITOU de SAINTONGE, / et ANGOUMOIS, et d'AUNIS / Par le Sr. Robert Géog. Ord. ru Roi. / avec Privilege. 1748.

br: 162 x 200 mm.
pl: 169 x 203 mm.

Engravers: (map) Lattré Sculp.

RF: c.1:1,585,000

Scales: 35 mm. = 10 Lieues d'une heure.

Comments:

Book:

AP: 21
AU:
NAP:

Location:

133 **Poitou, Aunis, Saintonge, Angoumois** 1753 RV

Gouvernemens Généraux / du POITOU, DU PAYS D'AUNIS, / ET DE / SAINTONGE-ANGOUMOIS, / Par le Sr. Robert de Vaugondy fils / Geographe ordinaire du Roi. / Avec Privilege. / 1753.

br: 482 x 588 mm.
pl: 496 x 630 mm.

Engravers: (cartouche) h [sic] elisabeth haussard fecit

RF: c.1:476,000

Scales: 114 mm. = 30 Mille pas géometriques de 60 au degré; 10 Grandes Lieues de France ou d'une heure de 20 au degré; 12 1/2 Lieues moyennes de France, de 25 au degré

Comments:

Book:

AP:
AU: 35
NAP:

Location:

134 **Poitou, Aunis, Saintonge, Angoumois** 1759 RV

Carte des / COTES DE FRANCE / depuis les Confins de la Bretagne, / jusques à l'embouchure de la Gironde, / et qui comprend / les Gouvernemens Généraux de / POITOU,d'AUNIS, / et de SAINTONGE-ANGOUMOIS. / Par le Sr. Robert de Vaugondy / Géographe ordin? du Roi, de S.M.Pol? Duc de Lorraine et de Bar, / et Associé de l'Académie royale des Sçien. et Belles-Lettres / de Nancy. / Avec Privilege / 1759

Paris: Ches l'Auteur Quay de l'horloge du Palais / près le Pont Neuf

br: 684 x 521 mm.
pl: 696 x 534 mm.

Engravers: (cartouche) Arrivet inv. & Sculpsit

RF: c.1:460,000

Scales: 20 mm = 5 Mille pas Geometriques de 60 au degré; 12 mm. = 1 Lieue d'une heure de Chemins de 20 au degré; 10 mm. = 1 Lieue commune de France de 25 au degré; 10 mm. = 1 Lieue de Poitou de 24 au degré

Comments: Table along right hand side of map: Les Gouvernemens Généraux que Cette Carte renferme [etc.]
dotted lines connecting towns: roads? (cf. Brittany, 1758)
Map announced in the Mémoires de Trévoux, Novembre, 1759, p. 2877, as one of eight new maps; price 1# 5s.
Julien Catalogue (1763): 185; price 1# 5s.

Book:

AP:
AU:
NAP:

Location: Bod: (E)C21: 6(3)

135 **Poitou, Aunis, Saintonge, Angoumois** 1762 RV

POITOU, AUNIS, / SAINTONGE ET ANGOUMOIS / LA MARCHE, LIMOSIN, / ET AUVERGNE. / Par le S. Robert de Vaugondy Géog? ord? du Roi. / de S. Maj. Polon? Duc de Lorr? et de Bar. / et de la Société royale de Nancy / 1762

A Paris ches l'Auteur quai de l'Horloge prè le pont neuf

br: 242 x 287 mm.

Engravers: (map) Gravé par E. Dussy
(cartouche) Arrivet inv. & sculp.

RF: c.1:2,400,000

Scales: 83 mm. = 80 Mille pas geometriques de 60 au dégré; 87 mm. = 35 Lieues communes de France de 25 au dégré

Comments:

State 1: 1762

State 2: [1778]: date and imprint removed; "et Censeur Royal" added to title; Fortin publication

Book:

AP:
AU:
NAP: 18

Location:

136 **Marche, Limousin** 1748 R

Gouvernemens Generaux / de la MARCHE / et du LIMOSIN avec les Pays de Combraille / et de Franc Aleu / Par le Sr. Robert Géog. ord. du Roy. / avec Privilege. 1748.

br: 161 x 160 mm.
pl: 170 x 166 mm.

Engravers:

RF: c.1:1,150,000

Scales: 48 mm. = 10 Lieues d'une heure de chemin

Comments:

Book:

AP: 22
AU:
NAP:

Location:

137 **Marche, Limousin, Auvergne** 1753 RV

Gouvernemens / Généraux / de la MARCHE, DU LIMOSIN, / ET DE L'AUVERGNE. / Par le Sr. Robert de Vaugondy fils. / Geographe ordin? du Roi./ Avec Privilege / 1753.

br: 480 x 590 mm.;
pl: 490 x 619 mm.

Engravers: (cartouche) haussard

RF: c.1:438,000

Scales: 83 mm. = 20 Mille pas géométriques de 60 au degré; 8 1/3 Lieues communes de France de 25 au degré; 6 2/3 Grandes Lieues de France ou d'une heure de 20 au degré

Comments: "haussard" etched in much firmer hand than usual

Book:

AP:
AU: 36
NAP:

Location:

138 **Berry** 1748 R

Gouvernemens Generaux / du BERRY, / du BOURBONNOIS, / et du NIVERNOIS. / Par le Sr. Robert Géog. ord. du Roi. / avec Privilege. 1748.

br: 161 x 172 mm.
pl: 166 x 262 mm.

Engravers: (map) Lattré Sc.

RF: c.1:1,400,000

Scales: 39 mm. = 10 Lieues d'une heure de chemin.

Comments:

Book:

AP: 23
AU:
NAP:

Location:

139 **Berry, Nivernois, Bourbonnois** 1753 R

Gouvernements Généraux / du BERRY, DU NIVERNOIS, / ET DU BOURBONOIS./ Par le Sr. Robert / Géographe ordinaire du Roi. / Avec Privilege. / 1753.

br: 480 x 545 mm.;
pl: 493 x 567 mm.

Engravers: (cartouche) Gravé par E. haussard

RF: c.1:459,000

Scales: 166 mm. = 42 Mille pas Geometriques de 60 au degré; 14 Grandes Lieues de France et d'une heure de chemin de 20 au degré; 20 1/7 Lieues de Berry de 28 4/5 au degré; 16 1/12 Lieues de Bourbonois de 23 au degré

Comments: meridian of Paris marked (20° E)
From AU 'Préface,'p. 25b: 'one might think ours a copy of Jaillot's Généralité de Bourge, 1707, if one didn't know that his is a servile reduction of the Diocèse de Bourges of Sanson, 1678'

Book:

AP:
AU: 37
NAP:

Location:

140 **Berry, Nivernois, Bourbonnois** [1759] RV

[Carte de Berri, Nivernois, Bourbonnois...]

Engravers:

RF: c.1:

Scales:

Comments: Not seen in any location ; known only by RV catalogue entries cited in Mémoires de Trévoux, Nov. 1759, p. 2877, as one of eight new maps; price: 1# 5s.
Julien Catalogue (1763): 176; price 1# 5s.

Book:

AP:
AU:
NAP:

Location: RV: Tablettes Parisiennes (1760); Catalogue (1777): shown as Sanson map, but see Comments

141	**Berry to Burgundy**	1762	RV

BERRI, NIVERNOIS, BOURBONOIS, / LYONOIS, BOURGOGNE, BRESSE, / ET FRANCHE-COMTÉ. / Par le S. Robert de Vaugondy Géog? ord? du Roi. / de S. Maj? Polon? Duc de Lorr? et de Bar. / et de la Société royale de Nanci. / 1762.

A Paris ches l'Auteur quai de l'Horloge près le pont neuf

br: 243 x 270 mm.

Engravers: (map) Gravé par E. Dussy
(cartouche) Arrivet inv. & sculp.

RF: c.1:1,775,000

Scales: 61 mm. = 60 Mille pas géometriques de 60 au degré; 25 Lieues communes de France de 25 au degré

Comments:
State 1: 1762

State 2: [1778]: date and imprint removed; "et Censeur Royal" added to title; Fortin publication

Book:

AP:
AU:
NAP: 17

Location:

142	**Burgundy, north**	1752	R

Partie / Septentrionale du Gouvernement Général / DE BOURGOGNE / ou se trouvent / L'Auxerrois, les Bailliages de Semur / en Auxois, de Chatillon-sur-Seyne, / ou de la Montagne, de Bar-sur-Seyne, / d'Autun, de Dijon, partie de celui de / Challon-sur-Saone et du Charolois. / Par le Sr. Robert Geographe / ordinaire du Roi / Avec Privilege 1752.

br: 481 x 508 mm.
pl: 496 x 530 mm.

Engravers: (map) Gravé par Delahaye l'ainé

RF: c.1:346,000

Scales: 105 mm.= 20 Mille pas géometriques de 60 au Degré; 6 2/3 Grandes Lieues de France et d'une heure de 20 au Degré; 8 1/3 Lieues communes de France de 25 au Degré; 7 1/6 Lieues de Bourgogne de 21 1/2 au Degré; 5 Lieues de Lyon de 15 au Degré

Comments:

Book:

AP:
AU: 38
NAP:

Location:

143	**Burgundy, south**	1752	R

Partie Meridionale / du Gouvernement Genl. de / BOURGOGNE / ou se trouvent / les Bailliages d'Autun, et de Challon, le Mâconois, / La Bresse et le Bugey / divisés en leurs Mandemens. // Gouvernement General du / LYONOIS / divisé en / Lyonois, Forez et Beaujolois. / Par le Sr. Robert Géographe / ord. du Roy. 1752 / Avec Privilege

br: 485 x 550 mm.
pl: 495 x 582 mm.

Engravers: (cartouche) E. haussard sculp.

RF: c.1:345,000

Scales: 105 mm. = 20 Mille Pas géometriques de 60 au degré; 6 2/3 Grandes Lieues de France ou d'une heure de 20 au degré; 8 1/3 Lieues communes de France de 25 au degré; 7 1/6 Lieues de Bourgogne de 21 1/2 au degré; 5 Lieues de Lyonois de 15 au degré

Comments: Geneva and section of Lake G. in margin between border (right) and plate mark

143	**Burgundy, south** (continued)

Book:

AP:
AU: 39
NAP:

Location:

144	**Burgundy, Franche-Comté**	1748	R

Gouvernemens Generaux / de BOURGOGNE, et,/de FRANCHE-COMTÉ / Par le Sr. Robert, Géog? / Ordin? du Roy. / Avec Privilege

br: 156 x 185 mm.
pl: 162 x 195 mm.

Engravers: (map) Lattré Sculp.

RF: c.1:1,870,000

Scales: 60 mm. = 20 Lieues d'une heure de chemin.

Comments:

Book:

AP: 24
AU:
NAP:

Location:

145	**Burgundy, Franche-Comté**	1749	R

Partie Méridionale / du COMTÉ DE BOURGOGNE / OU FRANCHE-COMTÉ, / où sont les Bailliages de / Orgelet, Lons-Le-Saunier, / Poligny, Arbois, Salins, Pontarlier, / Les Terres de St. Claude, / et partie des Bailliages de / Dole et Dornans. / Par le Sr. Robert, Geog? ordin? du Roy. / avec Privilege. 1749.

br: 470 x 568 mm.
pl: 498 x 633 mm.

Engravers:

RF: c.1:237,500

Scales: 115 mm. = 15 Mille Pas Géometriques, de 60 au Degré; 5 Grandes Lieues de France, ou d'une heure de 20 au Deg.; 5 3/8 Lieues de Bourgogne, de 21 1/2 au Deg.

Comments: Undated examples are perhaps Delamarche issues, with dates erased?

Book:

AP:
AU: 41
NAP:

Location: BN: Ge.DD.1467 (48); Ge.CC.1520 (48): not dated

146	**Burgundy, Franche-Comté**	1749	R

Partie Septentrionale / du COMTÉ DE BOURGOGNE / OU FRANCHE-COMTÉ. / où sont les Bailliages de / Vesoul, Gray, Besancon, Baume, / le Comté de Montbelliard, / et partie des Bailliages de / Dole, Quingey, et Ornans, / Par le Sr. Robert, Géographe ordinaire du Roy. / Avec Privilege 1749.

br: 467 x 562 mm.
pl: 495 x 589 mm.

Engravers: (cartouche) St.1: Gobin fecit

RF: c.1:238.500

Scales: 77 mm. = 10 Mille pas géométriques de 60 au degré; 3 1/3 Grandes Lieues de France de 20 au degré; 3 1/2 Lieues de Bourgogne de 21 1/2 au degré

Comments:
State 1: 1749: cartouche signed 'Gobin fecit'

State 2: 1749: cartouche signed 'E. haussard sculp.'; design itself unchanged

Undated examples from Delamarche re-issue, with dates erased?

146 **Burgundy, Franche-Comté** (continued)

Book:

AP:
AU: 40
NAP:

Location: BN: Ge.DD.1467 (47); Ge.CC.1520 (47): examples not dated

147 **Lyon, Auvergne** 1748 R

Gouvernemens Generaux / du LYONNOIS et / de l'AUVERGNE / Par le Sr. Robert Géog. ord. du Roy / avec Privilege. 1748.

br: 162 x 183 mm.
pl: 171 x 189 mm.

Engravers: (map) Lattré Sculp.

RF: c.1:1,350,000

Scales: 41 mm. = 10 Lieues d'une heure

Comments:

Book:

AP: 25
AU:
NAP:

Location:

148 **Guienne, east** 1753 R

Partie Orientale / du Gouvernement Général / de la / GUIENNE / où se trouvent / le Quercy, et le Rouergue. / Par le Sr. Robert Geographe / ordinaire du Roi. / Avec Privilege. / 1753.

br: 482 x 513 mm.
pl: 501 x 538 mm.

Engravers:

RF: c.1:416,000

Scales: 131 mm. = 30 Mille pas geometriques de 60 au Degré; 10 Grandes Lieues de France et d'une heure de 20 au Degré; 9 1/2 Lieues de Guienne et Gascogne de 19 au Degré

Comments:

Book:

AP:
AU: 43
NAP:

Location:

149 **Guienne, north** 1752 R

Partie Septentrionale / du Gouvernement Général / de la GUIENNE / ou se trouvent / le Bourdelois, le Perigord, / l'Agenois, le Bazadois, / et une partie / du Condomois et de la Lomagne. / Par le Sr. Robert Geographe / ordinaire du Roi. / Avec Privilege 1752.

br: 477 x 586 mm.
pl: 495 x 630 mm.

Engravers: (map) Gravé par De la Haye l'ainé

RF: c.1:414,000

Scales: 85 mm. = 20 Mille pas geometriques de 60 au degré; 6 2/3 Grandes Lieues de France de 20 au degré; 6 1/3 Lieues de Guyenne et Gascogne de 19 au degré

Comments:

Book:

AP:
AU: 42
NAP:

Location:

150 **Guienne, south; Basse Navarre** 1753 R

Partie Méridile. / du Gouvernement de / GUIENNE / où se trouvent / le Condomois, la Chalosse, le Pays / de Soule, le Labour, l'Armagnac, les Landes, / le Cominge, le Bigorre, le Conserans, etc. // Gouvernement de / BASSE NAVARRE ET BEARN. / Par le Sr. Robert Géog. ord. du Roy / avec Privilége 1753.

br: 482 x 613 mm.
pl: 492 x 664 mm.

Engravers:

RF: c.1:419,000

Scales: 85 mm. = 20 Mille Pas Géométriques de 60 au degré; 6 2/3 Grandes Lieues de France de 20 au degré; 6 1/3 Lieues de Guienne et Gascogne de 19 au degré

Comments
State 1: 1753: shield in cartouche is empty

State 2: 1753: shield filled with lion

Book:

AP:
AU: 44
NAP:

Location:

151 **Guienne, Gascony, Basse Navarre** 1748 R

Gouvernemens Generaux / de la GUIENNE PROPRE, / et de la BASSE NAVARRE / unie au Bearn // GUIENNE et GASCOGNE / divisées par pays / Par le Sr. Robert Géographe ord. du Roi /

br: 162 x 200 mm.
pl: 171 x 240 mm.

Engravers: (map) Lattré Sculp.

RF: c.1:2,300,000

Scales: 48 mm. = 20 Lieues d'une heure de chemin.

Comments: 'avec Privilege 1748' outside border, bot. right.

Book:

AP: 26
AU:
NAP:

Location:

152 **Guienne, Gascony, Basse Navarre** 1762 RV

GUIENNE ET GASCOGNE, / BASSE NAVARRE / et BEARN. / Par le S. Robert de Vaugondy Géog? ord. du Roi. / de S. Maj? Polon? Duc de Lorr? et de Bar. / et de la Société royale de Nanci. / 1762.

A Paris ches l'Auteur quai de l'Horloge près le pont neuf

br: 242 x 264 mm.

Engravers: (map) Gravé par E. Dussy
(cartouche) Arrivet inv. & sculp.

RF: c.1:1,775,000

Scales: 63 mm. = 60 Mille pas géométriques de 60 au degré; 25 Lieues communes de France de 25 au degré

Comments:
State 1: 1762

State 2: [1778]: date and imprint removed; "et Censeur Royal" added to title; Fortin publication

Book:

AP:
AU:
NAP: 19

Location:

153 **Foix and Roussillon** 1748 R

Gouvernemens / Généraux / de FOIX et de ROUSSILLON, / et les pays Circonvoisins / Par le Sr. Robert Géog? ord? du Roi. / avec Privilege 1748.

br: 163 x 198 mm.
pl: 166 x 250 mm.

Engravers:

RF: c.1:960,000

Scales: 48 mm. = 8 Lieues d'une heure

Comments:

Book:

AP: 27
AU:
NAP:

Location:

154 **Languedoc** 1748 R

Gouvernement / General / du LANGUEDOC. / Par Le S. Robert Geographe / Ordinaire du Roi / 1748

br: 160 x 197 mm.
pl: 168 x 205 mm.

Engravers: (map) Gravé par Guill? Delahaye

RF: c.1:1,950,000

Scales: 55 mm.= 20 Lieues d'une heure de Chemin

Comments:

Book:

AP: 28
AU:
NAP:

Location:

155 **Languedoc** 1752 R

Gouvernement Général du / LANGUEDOC / divisé par Diocèses, / avec les pays voisins de la Catalogne, / qui comprennent les / Gouvernemens Généraux / de FOIX et / de ROUSSILLON. / Par le Sr. Robert Géog? ordin? / du Roi. / Avec Privilege. 1752.

br: 480 x 490 mm.
pl: 498 x 506 mm.

Engravers: (cartouche) Moreau inv. et sculp.

RF: c.1:715,000

Scales: 153 mm. = 60 Mille Pas Géométriques, de 60 au Degré; 20 Grandes Lieues de France, de 20 au Degré; 25 Lieues moyennes de France, de 25 au Degré; 19 Lieues de Languedoc, de 19 au Degré

Comments:

Book:

AP:
AU: 45
NAP:

Location:

156 **Languedoc** 1759 RV

Gouvernemens généraux de LANGUEDOC, / de FOIX et de ROUSSILLON, / par le Sr. Robert de Vaugondy, Geographe ord. du roi, de S.M. Polon? Duc de Lorraine et de Bar / et Associé de l'Académie Royale des Sciences et Belles-Lettres de Nancy / Avec Privilège 1759

Inset: *Supplément pour les Cevennes où se trouvt. les Diocèses du Puy, de Mende, et de Viviers*

Paris, chès l'auteur, Quay de l'Horloge près le Pont Neuf

br: 745 x 530 mm.

156 **Languedoc** (continued)

Engravers: (cartouche) Arrivet inv. & sculp.

RF: c.1:470,000

Scales: 118 mm. = 30 mille pas géometriques de 60 au degré; 10 grandes lieues de France de 20 au degré; 9 lieues de Languedoc de 18 au degré

Comments: Table of geographical divisions to the right of map
Cited in Mémoires de Trévoux, Nov. 1759, p. 2877, as one of eight new maps; price 1# 5s.
Julien Catalogue (1763): 201; price 1# 5s.

Book:

AP:
AU:
NAP:

Location: BN: Ge. DD. 2987 B (651)

157 **Languedoc, Provence** 1762 RV

LANGUEDOC, / DAUPHINÉ, / PROVENCE. / Par le S. Robert de Vaugondy / Géog? ordinaire du Roy &c / 1762.

A Paris ches l'Auteur quai de l'Horloge près le pont neuf

br: 242 x 283 mm.

Engravers: (map) Gravé par E. Dussy
(cartouche) Arrivet inv. & sculp.

RF: c.1:1,775,000

Scales: 52 mm. = 50 Mille pas géometriques de 60 au deg?; 63 mm. = 25 Lieues de France de 25 au degré

Comments
State 1: 1762

State 2: [1778]: date and imprint removed; Fortin publication

Book:

AP:
AU:
NAP: 20

Location:

158 **Dauphiné** 1748 R

Gouvernement / General / du DAUPHINÉ / Par le Sr. Robert Geographe / ordinaire du Roi. / 1748

br: 162 x 198 mm.
pl: 172 x 206 mm.

Engravers: (map) Gravé par Guill. Delahaye

RF: c.1:1,000,000

Scales: 45 mm.= 10 Lieues d'une heure de chemin.

Comments:

Book:

AP: 29
AU:
NAP:

Location:

159 **Dauphiné** 1751 R

Gouvernement Général / du DAUPHINÉ. / divisé par Bailliages, / Par le Sr. Robert. / Géographe ordin? du Roy. / Avec Privilege / 1751.

br: 486 x 513 mm.
pl: 501 x 539 mm.

Engravers:

RF: c.1:394,000

159 **Dauphiné** (continued)

Scales: 94 mm. = 20 Mille pas geometriques de 60 au degré; 6 1/2 Grandes Lieues de France de 20 au degré; 6 1/4 Lieues de Dauphiné de 19 au degré

Comments:

Book:

AP:
AU: 46
NAP:

Location:

160 **Provence** 1748 R

Gouvernement / General / de PROVENCE. / Par le Sr. Robert Géog? ordin? du Roi / 1748.

br: 160 x 198 mm.
pl: 170 x 208 mm.

Engravers: (map) Gravé par Guill? Delahaye.

RF: c.1:1,440,000

Scales: 39 mm. = 10 Lieues d'une heure de Chemin

Comments:

Book:

AP: 30
AU:
NAP:

Location:

161 **Provence** 1754 RV

Gouvernement Général de / PROVENCE, / divisé en ses Vigueries, / avec les pays circonvoisins, / le Comtat Vénaissin, la Principauté / d'Orange, / et les Terres adjacentes, / par le Sr. Robert de Vaugondy fils / Géographe ord? du Roy. / Avec Privilège 1754.

br: 477 x 535 mm.
pl: 493 x 561 mm.

Engravers:

RF: c.1:473,000

Scales: 114 mm. = 30 Mille pas géometriques de 60 au Degré; 12 1/2 Lieues communes de France de 25 au Degré; 9 1/2 Lieues de Provence de 19 au Degré; 10 Lieues d'une heure, de 20 au Degré

Comments:

Book:

AP:
AU: 47
NAP:

Location:

162 **Corsica** 1748 RV

Carte de l'Isle de / CORSE / Par le Sr. Robert de Vaugondy / fils de Mr. Robert Géog? ord? / du Roi / Avec Privilege 1748

br: 163 x 202 mm.
pl: 166 x 249 mm.

Engravers:

RF: c.1:1,150,000

Scales: 38 mm. = 4 Lieues d'une heure

Comments:

162 **Corsica** (continued)

Book:

AP: 153
AU:
NAP:

Location:

163 **Corsica** 1756 RV

Carte Nouvelle / de l'ISLE DE CORSE / dressée d'après d'une grande carte manuscrite / levée sur les lieux / par ordre de M. le Maréchal de Maillebois, / par le Sr. Robert de Vaugondy / Géographe ord. du Roi et de S.M.Pol. Duc de Lorraine et de Bar, / de l'Académie royale des Sciences et Belles Lettres de Nancy / 1756

A Paris: Chès le Sr. Robert Géog. ord du Roy / Quai de l'Horloge du Palais / avec privilege

br: 590 x 450 mm.
pl: 611 x 485 mm.

Engravers: (cartouche) Arrivet inv. & sculp. [State 2]

RF: c.1:326,755

Scales: 13 mm. = 1 Lieue de 2739 pas géometriques; 27 mm. = 5 Milles de Corse de 833 toises 2 pied chacun

Comments
State 1: 1756

State 2: 1768: cartouche: Arrivet Inv. & sculp.

State 3: 1769: tipped into AU, BN: Ge.DD.4520 (54)

another copy: A Lausanne / Chez Marc Chapuis
'Legende des camps et Attaques' for letters marking different camps and battles
island divided into jurisdicitons
cf. Carte de L'Isle de Corse in Atlas Portatif et Militaire [Cat.162]
Julien catalogue (1763):no. 725; no price given; described as 'rare'
See Berthelot & Ceccaldi, Les Cartes de la Corse de Ptolémée au XIX siècle, pp. 179-180.
1768 was year of cession of Corsica to France

Book:

AP:
AU:
NAP:

Location: BN: Ge.DD.4520 (54); Bodleian: E C 21:22(9) [Buache collection]; BL: K.85.69 (1769); R. Douwma Prints: (Chapuis imprint, 1769); WLCL: Julien, Theatre du Monde II (96)

164 **Belgium** 1750 RV

PAYS BAS / CATHOLIQUES / Par le Sr. Robert de Vaugondi / Avec Privilége 1750.

br: 148 x 154 mm.

Engravers:

RF: c.1:2,350,000

Scales: 37 mm. = 45 mille pas géometriques; 10 grandes lieues de France

Comments: "Pag. 179" outside border cardinal points between border and neat line

Book: Méthode Abrégée et facile pour apprendre la Géographie

AP:
AU:
NAP:

[l'Abbé Le François] Paris: Libraires Associés, 1770

Location: WLCL: C2 1770 Le

165 **Belgium, Luxemburg** 1748 R

Carte / des PAYS BAS / CATHOLIQUES / Par le Sr. Robert, Géog? ord? du Roi / Avec Privilege 1748.

br: 161 x 196 mm.
pl: 166 x 250 mm.

Engravers:

RF: c.1:2,300,000

Scales: 48 mm. = 20 Lieues d'une heure

Comments:

Book:

AP: 31
AU:
NAP:

Location:

166 **Belgium, Luxemburg** 1751 R

PAYS-BAS / CATHOLIQUES / où sont distinguées / les limites de la France qui comprennent / le Comté d'Artois, / partie des Comtés / de Flandre et de Haynaut, / et du Duché de Luxembourg. / Par Le Sr. Robert / Geographe ordinaire du Roy. / Avec Privilege. / 1751.

br: 482 x 556 mm.
pl: 495 x 563 mm.

Engravers:

RF: c.1:630,000

Scales: 116 mm. = 40 Mille pas geometriques de 60. au Degré; 15 Lieues de Hollande et Luxembourg de 22 au degré; 16 Lieues de Flandres de 24 au degré; 19 Lieues d'Artois de 28 au degré; 13 1/4 Lieues de Brabant de 20 au degré; 10 Milles communes d'Allemagne de 15 au degré; 17 Lieues communes de France de 25 au degré; 13 1/4 Grandes Lieues de France et d'une heure de 20 au degré

Comments:

Book:

AP:
AU: 48
NAP:

Location:

167 **Belgium, Netherlands, Westphalia** 1762 RV

PAYS-BAS, / PROVINCES-UNIES, / WESTPHALIE./ Par le S. Robert de Vaugondy Géog? / ord? du Roi, de S.M. Polon? Duc de Lorr. et de Bar / et de la Société royale de Nanci. 1762.

A Paris ches l'Auteur quai de l'Horloge près le pont neuf

br: 243 x 270 mm.

Engravers: (map) Gravé par E. Dussy
(cartouche) Arrivet inv. & Sculp.

RF: c.1:2,300,000

Scales: 50 mm. = 60 Mille pas géomet? de 60 au degré; 20 Lieues de Hollande de 20 au degré

Comments:
State 1: 1762

State 2: 1778: date changed in title; imprint removed; Fortin publication

Book:

AP:
AU:
NAP: 22

Location:

168 **Flanders** 1744 R

Theatre de la Guerre / dans les PAYS-BAS / où se trouvent / les comtés de Flandre, de Haynaut, d'Artois, / De Cambresis de Namur, le Duché de Brabant &c. / dans lequel on a marqué les Campemens des Armées du Roy / Com[m]andées par Mr. Le Maréchal de Luxembourg suivant les / Mémoires de Mr. Vaultier, Commissaire Ordre. de l'Artillerie. / Dressé par le Sr. Moullart-Sanson, / Revue corigé et Augmenté des Compagnes du Roy / Par le Sr. Robert Géog. Ordre. de Sa Majesté

A Paris / Chez l'Auteur Quai de l'Horloge du Palais / 1744 / Avec Privilege du Roy.

br: 432 x 697 mm.
pl: 444 x 713 mm.

Engravers:

RF: c.1:455,000

Scales: 61 mm. = 15 Milles Pas Géom. ou Milles Pas d'Italie; 5 Lieues d'une heure de chemin

Comments: under scales: 'Remarques' with 4 symbols for camps and battles. Julien Cat. (1763) citation; price 1#

Book:

AP:
AU:
NAP:

Location: WLCL: Julien, Theatre du Monde I (52)
Julien cat. (1763): no. 271

169 **Flanders** 1748 R

Carte Generale / du Comté de / FLANDRE. / Par le Sr. Robert, / Géographe ordinaire du Roy. / Avec Privilege 1748.

br: 161 x 198 mm.
pl: 166 x 250 mm.

Engravers:

RF: c.1:840,000

Scales: 45 mm. = 7 Lieues d'une heure.

Comments: Finder map for the following nine large scale maps of Flanders in the AP [Cat. 171 - 179]

Book:

AP: 32
AU:
NAP:

Location:

170 **Flanders** 1752 RV

COMTÉ DE FLANDRE / où se trouvent distingués / les Bailliages de Berg St. Winox, de Furnes, de / Bourbourg, de Bailleul, et de Douay, / les Chatellenies de Cassel, de Lille, d'Ypres, de / Courtray, d'Oudenarde, le Tournesis, le Franc-de-Bru= / ges, le Burgraviat de Gand, le Pays de Waes, / La Seigneurie de Dendermonde, / Le Comte d'Alost. / Par le Sr. Robert de Vaugondy fils / Géographe ordinaire du Roi. / Avec Privilege. / 1752

br: 480 x 621 mm.
pl: 501 x 629 mm.

Engravers:

RF: c.1:258,000

Scales: 139 mm. = 20 Mille pas geometriques de 60 au degré.; 7 2/3 Lieues communes de Flandres de 23 au degré.; 5 Lieues communes d'Allemagne de 15 au degré.; 8 2/3 Lieues communes de France de 25 au degré.; 6 2/3 Lieues d'une heure de chemin de 20 au degré.

Comments:

Book:

AP:
AU: 49
NAP:

Location:

171 **Flanders 1** 1748 R

I. Carte du Comté de Flandre ou sont les envs. de l'ECLUSE, d'ARDENBURG, de CADSANT, de BLA[N]CKENBG.etc. / Par le Sr. Robert Géog? ordin? du Roi avec Priv? 1748

br: 163 x 188 mm.
pl: 176 x 248 mm.

Engravers: (map) Gravé par Guil. Delahaye

RF: c.1:250,000

Scales: 45 mm. = 2 Lieues d'une heure

Comments:

Book:

AP: 33
AU:
NAP:

Location:

172 **Flanders 2** 1748 R

II.Carte de la Flandre òu sont les environs d'AXEL, d'HULST dans la Flandre Hollandse. / de MIDDELBURG, GOES, et TOLEN en Zélande, de BERG-OP-ZOOM; et de LILLO en Brabant. / Par le Sr. Robert Geog? du Roy Avec Privilége 1748

br: 158 x 217 mm.
pl: 171 x 229 mm.

Engravers:

RF: c.1:250,000

Scales: 45 mm. = 2 Lieues d'une heure

Comments: at 22° N.: "en 1377 le 12 Nov. 19 Villages ont été onondés en cet endroit"

Book:

AP: 34
AU:
NAP:

Location:

173 **Flanders 3** 1748 R

III. Carte de la Flandre où sont les Environs de GRAVELINE, de DUNKERQUE,/de BERG-ST.VINOX, de FURNES, et de NIEUPORT./ Par le Sr. Robert Géog? du Roi avec Privilege 1748.

br: 153 x 218 mm.
pl: 160 x 232 mm.

Engravers:

RF: c.1:250,000

Scales: 45 mm. = 2 Lieues d'une heure

Comments:

Book:

AP: 35
AU:
NAP:

Location:

174 **Flanders 4** 1748 R

IV. Carte du Comté de Flandre où sont les environs de BRUGES, d'OSTENDE, de DIXMUDE, / Par le Sr. Robert Géog? ord? du Roi Avec Privilege 1748.

br: 162 x 213 mm.
pl: 170 x 273 mm.

Engravers:

RF: c.1:250,000

174 **Flanders 4** (continued)

Scales: 45 mm.= 2 Lieues d'une heure

Comments:

Book:

AP: 36
AU:
NAP:

Location:

175 **Flanders 5** 1748 R

[V.] Carte / de la Flandre / où sont les environs de / GAND, du SAS DE GAND / de DENDERMO[N]DE, d'ALOST /et d'ANVERS . /Par le Sr. Robert Géog? du Roi / avec Privilege 1748.

br: 166 x 237 mm.
pl: 170 x 252 mm.

Engravers:

RF: c.1:250,000

Scales: 35 mm. = 1 Lieues d'une heure

Comments: V top rt., outside border

Book:

AP: 37
AU:
NAP:

Location:

176 **Flanders 6** 1748 R

VI. Carte du Cté. de Flandres où sont les environs de St. OMER, de CASSEL, d'AIRE,/de St. VENANT. / Par le Sr. Robert, Géographe ordinaire du Roi Avec Priv 1748.

br: 165 x 231 mm.
pl: 171 x 246 mm.

Engravers:

RF: c.1:250,000

Scales: 45 mm. = 2 Lieues d'une heure

Comments: battle marked west of Cassel: crossed swords, 1328, 1677.

Book:

AP: 38
AU:
NAP:

Location:

177 **Flanders 7** 1748 R

VII. Carte de la Flandre où sont les environs d'YPRES, COURTRAY, MENIN ARMENTIERS / LILLE et TOURNAY / Par le Sr. Robert Géographe ordinaire du Roi. avec Privilege 1748.

br: 160 x 203 mm.
pl: 170 x 210 mm.

Engravers:

RF: c.1:250,000

Scales: 45 mm. = 2 Lieues d'une heure

Comments:

Book:

AP: 39
AU:
NAP:

Location:

178	**Flanders 8**	1748	R

VIII. Carte de Flandre ou sont les environs d'OUDENARDE, NINOVE, / GERARDMONT. / Par le Sr. Robert Géog. ordin. du Roi avec Privilege 1748.

br: 161 x 191 mm.
pl: 165 x 207 mm.

Engravers:

RF: c.1:250,000

Scales: 45 mm. = 2 Lieues d'une heure.

Comments:

Book:

AP: 40
AU:
NAP:

Location:

179	**Flanders 9**	1748	R

IX. Carte du Comté de Flandre ŏu sont les environs de la BASSÉE / de DOUAY, de ST. AMAND &C. / Par le Sr. Robert Géographe ordin? du Roi. Avec Priv. 1748.

br: 163 x 206 mm.
pl: 179 x 222 mm.

Engravers:

RF: c.1:250,000

Scales: 45 mm. = 2 Lieues d'une heure

Comments: Battle marked at Fontenoy [33° N., 2° 17' E.]: "11 may 1745"

Book:

AP: 41
AU:
NAP:

Location:

180	**Brabant**	1748	R

Carte Generale / du Duché / de BRABANT / Par le Sr. Robert, Géog? du Roi / Avec Privilege 1748

br: 163 x 177 mm.
pl: 178 x 247 mm.

Engravers:

RF: c.1:910,000

Scales: 31 mm. = 5 Lieues d'une heure

Comments: finder map for the following ten maps of Brabant in the AP [Cat. 181-190]

Book:

AP: 42
AU:
NAP:

Location:

181	**Brabant 1**	1748	R

I. Carte du Brabant où sont les environs de BERG-OP-ZOOM, de TOLEN dans la Zélande,/et de VILLEMSTAT dans la Hollande. / Par le Sr. Robert Géog. ord. du Roi avec Privilége 1748.

br: 160 x 194 mm.
pl: 180 x 209 mm.

Engravers:

RF: c.1:235,000

Scales: 47 mm. = 2 Lieues d'une heure

Comments:

181	**Brabant 1** (continued)

Book:

AP: 43
AU:
NAP:

Location:

182	**Brabant 2**	1748	R

II. Carte du Brabant où sont les environs de BRÉDA, et de BOSLEDUC. / Par le Sr. Robert Géog? ord? du Roi. avec Privilége 1748.

br: 160 x 197 mm.
pl: 170 x 208 mm.

Engravers:

RF: c.1:235,000

Scales: 47 mm. = 2 Liueues d'une heure

Comments:

Book:

AP: 44
AU:
NAP:

Location:

183	**Brabant 3**	1748	R

III. Carte du Brabant ou se trouve[n]t le MAESLAND, la Seigneurie de RAVENSTEIN, / la Terre de CUYCK &c. Par le Sr. Robert Géog? ord? du Roi. avec Privilége 1748.

br: 162 x 198 mm.
pl: 172 x 208 mm.

Engravers:

RF: c.1:235,000

Scales: 47 mm. = 2 Lieues d'une heure

Comments:

Book:

AP: 45
AU:
NAP:

Location:

184	**Brabant 4**	1748	R

IV. Carte du Brabant / où se trouve[n]t la Mairie de SANTHO / VEN, partie du Marquisat de BERG / OP-ZOOM, et la Terre de RYEN. / Par le Sr. Robert Geog. du Roi. / avec Privilége 1748.

br: 162 x 196 mm.
pl: 172 x 208 mm.

Engravers: (map) Gravé par / Guil. Delahaye

RF: c.1:235,000

Scales: 35 mm. = 1 1/2 Lieues d'une heure

Comments:

Book:

AP: 46
AU:
NAP:

Location:

185	**Brabant 5**	1748	R

V. Carte du Brabant où se trouve[n]t les Mair[ie]s. d'HERENTALS, de TURNHOUT, part[i]e/ de Celle de BOSLEDUC et du Cte. d'HOCHSTRATEN. Par le Sr. Robert Geog? du Roi Avec Privil? 1748.

br: 160 x 198 mm.
pl: 175 x 207 mm.

Engravers:

RF: c.1:235,000

Scales: 47 mm. = 2 Lieues d'une heure de Chemin

Comments:

Book:

AP: 47
AU:
NAP:

Location:

186	**Brabant 6**	1748	R

VI. Carte du Brabant ou se trouv[en]t les envir[on]s. d'EYNDHOVEN dans la Mairie de BOSLEDUC./ Par le Sr. Robert Geographe ord. du Roi Avec Privilege 1748.

br: 162 x 198 mm.
pl: 170 x 207 mm.

Engravers:

RF: c.1:235,000

Scales: 47 mm. = 2 Lieues d'une heure de Chemin

Comments:

Book:

AP: 48
AU:
NAP:

Location:

187	**Brabant 7**	1748	R

VII. Carte du Brabant / ou sont les environs / de BRUXELLES, de MALINES / et LIERE. / Par le Sr. Robert Géog? du Roi / avec Privilege 1748.

br: 161 x 207 mm.
pl: 170 x 217 mm.

Engravers: (map) Gravé par / Guil. Delahaye.

RF: c.1:235,000

Scales: 45 mm. = 2 Lieues d'une heure

Comments:

Book:

AP: 49
AU:
NAP:

Location:

188	**Brabant 8**	1748	R

VIII. Carte du Brabant / ou sont les envs. de / LOUVAIN, d'ASCOT, de DIEST / de LEEUWE &c./ Par le Sr. Robert Géog. du Roi / avec Privilege 1748.

br: 164 x 207 mm.
pl: 172 x 238 mm.

Engravers:

RF: c.1:235,000

Scales: 34 mm. = 1 1/2 Lieues d'une heure

Comments:

188 Brabant 8 (continued)

Book:

AP: 50
AU:
NAP:

Location:

189	**Brabant 9**	1748	R

[IX.] Carte du Brabant / ou sont les Mairies de / ROO, de NIVELLE, de HULPEN, et de / GENEPPE / Par le Sr. Robert Géog. ord. ru Roi / avec privilege 1748.

br: 164 x 198 mm.
pl: 166 x 208 mm.

Engravers:

RF: c.1:230,000

Scales: 48 mm. = 2 Lieues d'une heure

Comments: IX top left, outside border

Book:

AP: 51
AU:
NAP:

Location:

190	**Brabant 10**	1748	R

[X.] Carte du Brabant / où sont les environs / de TILLEMONT, de JUDOIGNE, / de GEMBLOURS &c. / Par le Sr. Robert Geog? du Roi. / avec Privilége 1748.

br: 160 x 190 mm.
pl: 172 x 203 mm.

Engravers: (map) Gravé par J.B.H. Delahaye

RF: c.1:240,000

Scales: 47 mm. = 2 Lieues d'une heure

Comments: X top left, outside border

Book:

AP: 52
AU:
NAP:

Location:

191	**Brabant, Namurois, north**	1752	RV

Partie Septentrionale du / DUCHÉ DE BRABANT / où se trouvent / les Mairies de Bosleduc de Turnhout, de / Santhoven, d'Herentals, les Marquisats de / Berg-op-Zoom, et du St. Empire, la Baronie de Bréda / partie de la Seigneurie de Malines, et de / l'Evêché de Lyége &c. / Par le Sr. Robert De Vaugondy fils / Géographe ordinaire du Roi. / Avec Privilége. / 1752.

br: 480 x 588 mm.
pl: 496 x 628 mm.

Engravers:

RF: c.1:220,000

Scales: 98 mm. = 12 Mille Pas Geometriques de 60 au Degré; 4 Lieues de Brabant de 20 au Degré; 3 Lieues communes d'Allemagne de 15 au Degré

Comments:

Book:

AP:
AU: 50
NAP:

Location:

192 **Brabant, Namurois, south** 1752 RV

Partie Meridion. du / DUCHÉ DE BRABANT / où se trouvent / le Quartier de Bruxelles et de Louvain / la Seigneurie de Malines; et une / partie de l'Evechê de Lyege, / le COMTÉ DE NAMUR / avec les Confins / du Haynaut et du Luxembourg / Par le Sr. Robert de Vaugondy Geog? / ordin? du Roi. / Avec Privilége. / 1752.

br: 483 x 590 mm.
pl: 501 x 637 mm.

Engravers: (cartouche) haussard S.

RF: c.1:217,500

Scales: 125 mm. = 15 Mille Pas Geometriques de 60 au Degré; 5 Lieues de Brabant de 20 au Degré; 3 3/4 Lieues communes d'Allemagne de 15 au Degré

Comments:

Book:

AP:
AU: 51
NAP:

Location:

193 **Hainault, Cambresis** 1754 R

Comtés / de HAINAUT, / et de CAMBRESIS, / avec les confins / du Comté de Flandre, du Duché / de Brabant, du Comté de Namur, / et une partie / de la Seigneurie de Liége. / Par le Sr. Robert Géographe ordinaire / du Roi. / Avec Privilege. / 1754.

br: 484 x 555 mm.
pl: 499 x 585 mm.

Engravers: (cartouche) E. haussard Sculpcit

RF: c.1:240,000

Scales: 112 mm. = 15 Milles pas géometriques de 60 au degré; 3 3/4 Milles d'Allemagne de 15 au degré; 5 Grandes Lieues de France ou d'une heure de 20 au degré; 6 Lieues de Haynaut de 24 au degré

Comments: From AU 'Préface,' p.27a: "We have not left out the small details of Friex' map: censes, mills, and other little spots interesting to the military; sometimes a shortage of space stopped the name but not the mark."

Book:

AP:
AU: 53
NAP:

Location:

194 **Hainault, Namur, Cambresis** 1748 R

Carte Générale des / Comtés / de HAYNAUT, de NAMUR, et / de CAMBRESIS. / Par le Sr. Robert Geographe ordin? du Roi. / Avec Privilége 1748.

br: 163 x 200 mm.
pl: 176 x 248 mm.

Engravers: (map) Gravé par Guil? Delahaye

RF: c.1:815,000

Scales: 40 mm. = 6 Lieues d'une heure

Comments: Finder map for following ten large scale maps of Hainault in AP [Cat. 195-204]

Book:

AP: 53
AU:
NAP:

Location:

195 **Hainault 1** 1748 R

I. Carte du Comté du Haynaut où sont les environs d'AHT, de LESSINES, / et Frontieres du Comté de Flandre où se trouve TOURNAY. / Par le Sr. Robert Géographe ordinaire du Roi. Avec Privilege 1748.

br: 163 x 156 mm.
pl: 182 x 171 mm.

Engravers:

RF: c.1:240,000

Scales: 35 mm. = 1 Lieues d'une heure

Comments:

Book:

AP: 54
AU:
NAP:

Location:

196 **Hainault 2** 1748 R

II. Carte du Cté. de Haynaut où sont les environs de HALLE / D'ENGHIEN, de BRAINE LE COMTE / Par le Sr. Robert Géog? odi. du Roi. Avec Priv? 1748

br: 159 x 154 mm.
pl: 175 x 169 mm.

Engravers:

RF: c.1:240,000

Scales: 35 mm. = 1 Lieues d'une heure

Comments: lower right hand corner: portion of engraver's signature?

Book:

AP: 55
AU:
NAP:

Location:

197 **Hainault 3** 1748 R

III. Carte du Cté de Haynaut et des Frontieres du Cté. de Flandre / ôu sont DOUAY ST. AMAND, VALENCIENNES et BOUCHAIN / Par le Sr. Robert Géographe ord? du Roi. Avec Privilege 1748.

br: 167 x 150 mm.
pl: 176 x 165 mm.

Engravers:

RF: c.1:250,000

Scales: 45 mm. = 2 Lieues d'une heure

Comments:

Book:

AP: 56
AU:
NAP:

Location:

198 **Hainault 4** 1748 R

IV. Carte du Cté. de Haynaut ou sont les environs de MONS, de MAUBEUGE / de CONDÉ du QUENOY de BINCHE &c. Par le Sr. Robert Géog? ord? du Roi Avec Priv? 1748

br: 168 x 183 mm.
pl: 176 x 198 mm.

Engravers:

RF: c.1:250,000

Scales: 35 mm. = 1 Lieues d'une heure

198 **Hainault 4** (continued)

Comments: 30' lat. engraved as 35 (right)

Book:

AP: 57
AU:
NAP:

Location:

199 **Hainault 5** 1748 R

V. Carte des Ctes. de Haynaut et de Namur, et partie du Lyegeois, ou se trouvent FONTAINE l'EVÊQUE, / CHARLEROY, CHASTELLET, FLORENNES, WALCOURT. / Par le Sr. Robert Géog? ord. du Roi. Avec Priv? 1748

br: 166 x 175 mm.
pl: 175 x 246 mm.

Engravers:

RF: c.1:245,000

Scales: 45 mm. = 2 Lieues d'une heure

Comments:

Book:

AP: 58
AU:
NAP:

Location:

200 **[Hainault] 6** 1748 R

VI. Carte du Cté. de Namur ou sont les environs de NAMUR; et partie du LYEGEOIS / ou se trouve Huy——— Par le Sr. Robert Géog? ord? du Roi. Avec Privilege 1748.

br: 165 x 199 mm.
pl: 175 x 246 mm.

Engravers:

RF: c.1:240,000

Scales: 34 mm.= 1 Lieues d'une heure

Comments:

Book:

AP: 59
AU:
NAP:

Location:

201 **Hainault 7** 1748 R

VII. Carte du Cté de Haynaut ou sont les environs de CAMBRAI, / Par le Sr. Robert Géog? du Roi avec Privilege 1748

br: 158 x 170 mm.
pl: 178 x 180 mm.

Engravers: (map) Gravé par Guill. Delahaye

RF: c.1:250,000

Scales: 34 mm. = 1 Lieues d'une heure

Comments:

Book:

AP: 60
AU:
NAP:

Location:

202 **Hainault 8** 1748 R

VIII. Carte du Cté. de Haynaut où sont les environs de LANDRECHIES, / d'AYMERIES, d'AVESNE, &c. Par le Sr. Robert Géog? ord? du Roi. Avec Privilege 1748.

br: 157 x 183 mm.
pl: 174 x 245 mm.

Engravers:

RF: c.1:250,000

Scales: 42 mm. = [2] Lieues d'une heure

Comments:

Book:

AP: 61
AU:
NAP:

Location:

203 **Hainault 9** 1748 R

IX. Carte du Cté. de Haynaut où sont les environs de PHILIPPEVILLE, / MARIENBOURG, ROCROIX &c. Par le Sr. Robert Géog? du Roi avec Priv? 1748.

br: 156 x 173 mm.
pl: 175 x 247 mm.

Engravers: (map) Gravé par / Guil. Delahaye

RF: c.1:245,000

Scales: 45 mm. = 2 Lieues d'une heure

Comments: no lat. minutes numbered, though marked

Book:

AP: 62
AU:
NAP:

Location:

204 **[Hainault] 10** 1748 R

X. Carte du Comté du Namur où se trouvent les environs de DINANT. / Par le Sr. Robert Géographe ordinaire du Roi. avec Privilege 1748.

br: 163 x 199 mm.
pl: 177 x 206 mm.

Engravers:

RF: c.1:240,000

Scales: 45 mm. = 2 Lieues d'une heure

Comments:

Book:

AP: 63
AU:
NAP:

Location:

205 **Liege, Limbourg** 1748 R

Carte Générale / de la Seigneurie / de LYEGE; / et du Duché / de LIMBOURG. / Par le Sr. Robert Géog? du Roi. / avec Privilege 1748.

br: 159 x 159 mm.
pl: 166 x 249 mm.

Engravers:

RF: c.1:670,000

Scales:

Comments: Finder map for following six large scale maps from AP [Cat. 206-211]

205 **Liege, Limbourg** (continued)

Book:

AP: 64
AU:
NAP:

Location:

206 **Liege/Limbourg 1** 1748 R

I. Carte de l'Etat de Lyege ou se trouvent la TERRE de LOEN, et la CAMPINE / LYEGEOISE. —- Par le Sr. Robert Géographe ordinaire du Roi avec Privilege 1748.

br: 163 x 198 mm.
pl: 171 x 209 mm.

Engravers:

RF: c.1:240,000

Scales: 46 mm. = 2 Lieues d'une heure

Comments:

Book:

AP: 65
AU:
NAP:

Location:

207 **Liege/Limbourg 2** 1748 R

II. Carte de L'Etat de Lyege ou se trouvent le Cté. de HORN et les Environs / de MAESEYCK, et partie du D[uc]hé. de LIMBOURG. Par le Sr. Robert Géog? du Roi. avec Priv 1748.

br: 162 x 196 mm.
pl: 170 x 207 mm.

Engravers:

RF: c.1:235,000

Scales: 47 mm. = 2 Lieues d'une heure

Comments:

Book:

AP: 66
AU:
NAP:

Location:

208 **Liege/Limbourg 2** 1748 R

III. Carte de l'Etat de Liege où sont les environs de MAESTRICHT, de TONGRES, de / ST. TRON, d'HASSELT &c. Par le Sr. Robert Géog ordin. du Roi. avec Privilege 1748.

br: 164 x 208 mm.
pl: 172 x 229 mm.

Engravers:

RF: c.1:240,000

Scales: 46 mm. = 2 Lieues d'une heure

Comments:

Book:

AP: 67
AU:
NAP:

Location:

209 **Liege/Limbourg 4** 1748 R

IV. Carte du Duché de Limbourg où sont les environs de FAUQUEMONT, de DALEM, / de ROLDUC, d'AIX la Chapelle &c. Par le Sr. Robert Géographe ord? du Roi. avec Privilége 1748.

br: 162 x 211 mm.
pl: 169 x 223 mm.

Engravers:

RF: c.1:235,000

Scales: 46 mm. = 2 Lieues d'une heure

Comments: curious symbol, center of map, at Strithagen

Book:

AP: 68
AU:
NAP:

Location:

210 **Liege/Limbourg 5** 1748 R

V. Carte de l'Etat de Lyege ou se trouvent LYEGE, HANNUYE, HUY, FRAYPONT &c. / Par le Sr. Robert Géog. ordin. du Roi avec Privilege 1748.

br: 161 x 211 mm.
pl: 165 x 220 mm.

Engravers:

RF: c.1:235,000

Scales: 47 mm. = 2 Lieues d'une heure

Comments:

Book:

AP: 69
AU:
NAP:

Location:

211 **Liege/Limbourg 6** 1748 R

VI. Carte du Duché de Limbourg où sont les environs de LIMBOURG. / Par le Sr. Robert, Géographe ordinaire du Roi avec Privilege 1748.

br: 161 x 211 mm.
pl: 173 x 230 mm.

Engravers:

RF: c.1:235,000

Scales: 47 mm. = 2 Lieues d'une heure

Comments:

Book:

AP: 70
AU:
NAP:

Location:

212 **Liege, Limbourg** 1754 RV

La Principauté / de LIÉGE / et / le Duché / de LIMBOURG./ Par le Sr. Robert de Vaugondy / Géographe ordinaire du Roy. / Avec Privilege 1754.

br: 481 x 453 mm.
pl: 506 x 471 mm.

Engravers:

RF: c.1:240,000

Scales: 77 mm. = 10 Milles pas géometriques de 60 au degré; 3 1/3 Lieues communes de Pays-bas de 22 au degré

Comments:

212 **Liege, Limbourg** (continued)

Book:

AP:
AU: 54
NAP:

Location:

213 **Luxemburg** 1748 R

Carte Générale / du Duché / de LUXEMBOURG / Par le Sr. Robert, Géog? du Roi / avec Privilege / 1748

br: 162 x 180 mm.
pl: 166 x 253 mm.

Engravers:

RF: c.1:910,000

Scales: 38 mm. = 6 Lieues d'une heure

Comments: finder map for following eleven large scale maps from the AP [Cat. 214-224]

Book:

AP: 71
AU:
NAP:

Location:

214 **Luxemburg 1** 1748 R

I. Carte du Dché de Luxembourg ou sont les Frontieres du NAMUROIS / et du LYEGEOIS. / Par le Sr. Robert Géog? ordin? du Roi avec Privilege 1748.

br: 160 x 165 mm.
pl: 166 x 256 mm.

Engravers:

RF: c.1:250,000

Scales: 34 mm. = 1 Lieues d'une heure

Comments:

Book:

AP: 72
AU:
NAP:

Location:

215 **Luxemburg 2** 1748 R

II. Carte du Dé. de Luxembourg où se trouvt. la Mairie de la MARCHE, le Cté. de DURBUY. / et Partie de la Seignrie. de LYEGE &C. Par le Sr. Robert Geog? du Roi avec Privilége 1748.

br: 161 x 210 mm.
pl: 181 x 257 mm.

Engravers:

RF: c.1:250,000

Scales: 45 mm. = 2 Lieues d'une heure

Comments:

Book:

AP: 73
AU:
NAP:

Location:

216 **Luxemburg 3** 1748 R

III. Carte du Dhé. de Luxembourg où sont les Ctés. de ST. VEIT, de REYFERSCHEID, les Baronies / de WILDENBORG, CRONENBURG &c. Par le Sr. Robert Géog? ordin? du Roi. avec Privilege 1748.

br: 164 x 205 mm.
pl: 180 x 254 mm.

Engravers:

RF: c.1:235,000

Scales: 36 mm. = 1 Lieues d'une heure

Comments:

Book:

AP: 74
AU:
NAP:

Location:

217 **Luxemburg 4** 1748 R

IV. Carte du Duché de Luxembourg où sont les Seignries. de LOMPRÉ / MIRWART et VILLANCE, Partie du Duché de BOUILLON et les Confins du LYEGEOIS / Par le Sr. Robert Géographe ordinaire du Roi. Avec Privilege 1748.

br: 165 x 190 mm.
pl: 176 x 204 mm,

Engravers:

RF: c.1:240,000

Scales: 46 mm. = 2 Lieues d'une heure

Comments:

Book:

AP: 75
AU:
NAP:

Location:

218 **Luxemburg 5** 1748 R

V. Carte du Dhé. de Luxembourg où sont les Ctés. de LA ROCHE, et Prevoté de BASTOGNE, et l'Abbaye / et Seigneurie de S. HUBERT dans le Lyegeois. Par le Sr. Robert Géog? ordin? du Roy. Avec Privilege 1748.

br: 160 x 210 mm.
pl: 165 x 250 mm.

Engravers:

RF: c.1:240,000

Scales: 35 mm. = 1 Lieues d'une heur [sic]

Comments:

Book:

AP: 76
AU:
NAP:

Location:

219 **Luxemburg 6** 1748 R

VI. Carte du Dché. de Luxembourg ôu sont les Prevôté de BIDBURG, Comté de VIANDEN / et Seigneurie de PRUYM. Par le Sr. Robert Géographe ordinaire du Roi Avec Privilege 1748.

br: 162 x 205 mm.
pl: 167 x 219 mm.

Engravers:

RF: c.1:240,000

219 **Luxemburg 6** (continued)

Scales: 46 mm. = 1 Lieues d'une heure

Comments:

Book:

AP: 77
AU:
NAP:

Location:

220 **Luxemburg 7** 1748 R

VII. Carte des Frontieres du Luxembourg où sont les environs de BOUILLON, de CHARLEVILLE / de MEZIERES, de SEDAN, d'YVOIX &c. Par le Sr. Robert Géographe ordinaire du Roi. Avec Privilege 1748.

br: 163 x 206 mm.
pl: 178 x 216 mm.

Engravers:

RF: c.1:250,000

Scales: 34 mm. = 1 Lieues d'une heure

Comments:

Book:

AP: 78
AU:
NAP:

Location:

221 **Luxemburg 8** 1748 R

VIII. Carte du Dché. de Luxembourg où sont les Prevotés de NEUFCHATEL, de CHINY, / et les Environs d'ARLON. Par le Sr. Robert Géographe du Roi Avec Privilege 1748.

br: 161 x 211 mm.
pl: 215 x 227 mm.

Engravers:

RF: c.1:240,000

Scales: 46 mm. = 2 Lieues d'une heure

Comments:

Book:

AP: 79
AU:
NAP:

Location:

222 **Luxemburg 9** 1748 R

IX. Carte du Dché. du Luxembourg ou sont les environs de LUXEMBG. GREVEN MACHEREN DIEKIRCH &c. / Par le Sr. Robert Géog? ordin? du Roi avec Privilege 1748.

br: 159 x 210 mm.
pl: 165 x 261 mm.

Engravers:

RF: c.1:240,000

Scales: 34 mm. = 1 Lieues d'une heure

Comments:

Book:

AP: 80
AU:
NAP:

Location:

223 **Luxemburg 10** 1748 R

X. Carte du Dché. de Luxembourg òu se trouvent, MOUSON, MONTMEDY, STENAY, JAMETZ &c. / Par le Sr. Robert Géog? ordin? du Roi avec Privilege 1748.

br: 161 x 222 mm.
pl: 169 x 238 mm.

Engravers:

RF: c.1:240,000

Scales: 46 mm. = 2 Lieues d'une heuere

Comments:

Book:

AP: 81
AU:
NAP:

Location:

224 **Luxemburg 11** 1748 R

XI. Carte du Dché. de Luxembourg et des Confins de la LORRAINE où sont les environs de THIONVILLE. / Par le Sr. Robert Géographe ordinaire du Roy Avec Privilege 1748.

br: 161 x 222 mm.
pl: 166 x 251 mm.

Engravers:

RF: c.1:240,000

Scales: 35 mm. = 1 Lieues d'une heure

Comments:

Book:

AP: 82
AU:
NAP:

Location:

225 **Luxemburg** 1753 RV

Carte du Duche de / LUXEMBOURG, / dans laquelle on trouve / la partie Meridionale / & Limitrophe / de la Seigneurie / de Lyege. / Par le Sr. Robert de Vaugondy / Géographe Ord. du Roy. / Avec privilege 1753.

br: 482 x 606 mm.
pl: 510 x 623 mm.

Engravers: (cartouche) E. haussard sculp.

RF: c.1:275,500

Scales: 97 mm. = 15 Mille pas géometriques de 60 au degré; 3 3/4 Milles d'Allemagne de 15 au degré; 5 Lieues de Luxembourg de 20 au degré; 6 1/4 Lieues communes de France de 25 au degré

Comments:

Book:

AP:
AU: 52
NAP:

Location:

226 **Netherlands** 1748 R

Carte / des Provinces-Unies / des PAYS-BAS. / Par le S. Robert Géog? ordin? du Roi. avec Privilege 1748.

br: 162 x 196 mm.
pl: 166 x 255 mm.

Engravers:

RF: c.1:2,300,000Scales:

Scales: 44 mm. = 18 Lieues d'une heure

226 **Netherlands** (continued)

Comments:

Book:

AP: 83
AU:
NAP:

Location:

227 **Netherlands** 1750 RV

PROVINCES UNIES / DES PAYS BAS / Par le Sr. Robert de Vaugondy. / avec Privilége du Roi. 1750.

br: 148 x 153 mm.

Engravers:

RF: c.1:2,250,000

Scales: 46 mm. = 20 lieues d'une heure

Comments: "Pag. 188" outside border, top right cardinal points between border and neat line

Book: Méthode Abrégée et facile pour apprendre la Géographie

AP:
AU:
NAP:

[l'Abbé Le François] Paris: Libraires Associés, 1770

Location: WLCL: C2 1770 Le

228 **Netherlands** 1753 R

LES PROVINCES-UNIES / DES PAYS-BAS / qui Comprennent / le Duché de Gueldre, les Comtés de / Hollande, de Zelande, de Zutphen, / les Seigneuries d'Utrecht, d'Ouest-Frise, / d'Ower-Issel, et de Groningue. / Par Le Sr. Robert / Geographe ordinaire du Roi. / Avec Privilege. 1753.

br: 481 x 588 mm.
pl: 494 x 612 mm.

Engravers: (cartouche) haussard

RF: c.1:555,000

Scales: 66 mm. = 20 Mille pas géometriques de 60 au degré; 6 3/4 Lieues d'Hollande de 20 au degré; 5 Lieues communes d'Allemagne de 15 au deg.; 8 1/4 Lieues communes de France de 25 au deg.

Comments:

Book:

AP:
AU: 55
NAP:

Location:

229 **Zeeland** 1748 R

Carte du Comté / de ZÉLANDE / Par le Sr. Robert Géog? ordin? du Roi / avec Privilege 1748.

br: 161 x 235 mm.
pl: 171 x 257 mm.

Engravers: (map) Gravé par / Guill? Delahaye.

RF: c.1:240,000

Scales: 46 mm. = 2 Lieues d'une heure

Comments:

Book:

AP: 84
AU:
NAP:

Location:

230 **Holland** 1748 R

Carte Generale des IX / de la HOLLANDE; / avec / la Seig. d'Utrecht, Partie /de la Gueldre, &c. / Par le Sr. Robert. Géog? du Roi / Avec Privilege 1748.

br: 168 x 212 mm.
pl: 176 x 248 mm.

Engravers:

RF: c.1:865,000

Scales: 35 mm. = 5 Lieues d'une heure

Comments: Finder map for the following nine large scale maps from the AP [Cat. 231-239]

Book:

AP: 85
AU:
NAP:

Location:

231 **Holland 1** 1748 R

I. Carte du Comté de Hollande où se trouve l'ISLE DE TEXEL. / Par le Sr. Robert Géog. ord. du Roi Avec Privilége 1748.

br: 163 x 212 mm.
pl: 171 x 239 mm.

Engravers: (map) Gravé par / Guil. Delahaye

RF: c.1:290,000

Scales: 36 mm. = 2 Lieues d'une heure de chemin

Comments:

Book:

AP: 86
AU:
NAP:

Location:

232 **Holland 2** 1748 R

[II] Partie du Comté / de Hollande / ou sont les environs / d'ALCMAAR, de MEDENBLICK, / d'ENCKHUYSEN, d'EDAM &c. / Par le Sr. Robert Géog. ord. du Roi. / Avec Privilége 1748.

br: 158 x 212 mm.
pl: 176 x 246 mm.

Engravers: (map) Gravé par Guill. / Delahaye

RF: c.1:290,000

Scales: 37 mm. = 2 Lieues d'une heure

Comments: II top left, outside border

Book:

AP: 87
AU:
NAP:

Location:

233 **Holland 3** 1748 R

Carte de la Hollande / où se trouve une partie de la VELUWE, / et de l'OVER-YSSEL. / Par le Robert Géog? ord?/ du Roi / Avec Privilege 1748.

br: 158 x 209 mm.
pl: 171 x 230 mm.

Engravers: (map) Gravé par Guill. Delahaye

RF: c.1:290,000

Scales: 37 mm. = 2 Lieues d'une heure

Comments:

233 **Holland 3** (continued)

Book:

AP: 88
AU:
NAP:

Location:

234 **Holland 4** 1748 R

[IV] Partie du Comté / de Hollande, / où sont les environs / de HAARLEM, LEYDEN, LA HAYE, / DELFT GONDA [sic] &c. / Par le Sr. Robert Géographe ord. du Roi / avec Privilege 1748.

br: 157 x 211 mm.
pl: 171 x 230 mm.

Engravers: (map) Gravé par / Guill. Delahaye

RF: c.1:290,000

Scales: 37 mm. = 2 Lieues d'une heure

Comments: IV top left, outside border

Book:

AP: 89
AU:
NAP:

Location:

235 **Holland 5** 1748 R

V. Carte de la Hollande et de la Seigneurie, d'UTRECHT, où sont les environs d'AMSTELDAM / de MUYDEN, NAARDEN, WOERDEN, et d'UTRECHT Par le Sr. Robert Géog? du Roi Avec Priv? 1748.

br: 165 x 203 mm.
pl: 180 x 253 mm.

Engravers:

RF: c.1:290,000

Scales: 35 mm. = 2 Lieues d'une heure

Comments:

Book:

AP: 90
AU:
NAP:

Location:

236 **Holland 6** 1748 R

[VI] Partie du Duche /de GUELDRE / où se trouve / la VELUWE. / Par le Sr. Robert Géog? du Roi / avec Privilege 1748.

br: 156 x 182 mm.
pl: 168 x 203 mm.

Engravers:

RF: c.1:290,000

Scales: 35 mm. = 2 Lieues d'une heure

Comments: VI top left, outside border

Book:

AP: 91
AU:
NAP:

Location:

237 **Holland 7** 1748 R

VII. Carte de la Hollande ou sont les envs. de ROTTERDAM, d'HELLEVOETSLUYS, de GOERÉE, / de VILLEMSTAT &c. Par le Sr. Robert Géographe ordin. du Roi. Avec Privilege 1748.

br: 162 x 203 mm.
pl: 170 x 265 mm.

Engravers: (map) Gravé par Delahaye l'ainé

RF: c.1:290,000

Scales: 36 mm. = 2 Lieues d'une heure

Comments:

Book:

AP: 92
AU:
NAP:

Location:

238 **Holland 8** 1748 R

VIII. Carte de la Hollande ôu son [sic] les environs de SCHOONHOVE, de UYANE, de GORCUM, / d'HEUSDEN, de GEERTRUYDENBERG, Partie d'UTRECHT, et de la GUELDRE. / Par le Sr. Robert, Géographe ordinaire du Roi. Avec Privilege 1748.

br: 166 x 203 mm.
pl: 175 x 246 mm.

Engravers:

RF: c.1:290,000

Scales: 37 mm. = 2 Lieues d'une heure

Comments:

Book:

AP: 93
AU:
NAP:

Location:

239 **Holland 9** 1748 R

IX. Carte de la Hollande ôu sont les environs de NIMEGUE, de GRAVE, / d'ARNHEM dans le Dché. de GUELDRE Par le Sr. Robert Géog? du Roi Avec Privilege 1748.

br: 165 x 180 mm.
pl: 172 x 201 mm.

Engravers: (map) Gravé par Delahaye l'ainé

RF: c.1:290,000

Scales: 27 mm. = 1 Lieues d'une heure

Comments:

Book:

AP: 94
AU:
NAP:

Location:

240 **Germany** 1745 R

L'Empire d'Allemagne / divisé en ses dix Cercles, / ces Cercles en leurs principaux Etats; / et les Etats, Principautés et Souverainetés / qui passent ou qui ont passé jusqu'à présent / sous le nom d'Allemagne, / Dressé sur les Cartes des Srs. Sanson. Geog. du Roi / et assujettie aux observations astronomiques / Par le Sr. Robert Géog. ord du Sa Majesté

A Paris Chés l'Auteur Quay de l'horloge du Palais./ Avec Privilege du Roy. 1745

br: 497 x 714 mm.
pl: 502 x 720 mm.

240 **Germany** (continued)

Engravers: (map) G. N. Delahaye
(cartouche) Gobin Sc.

RF: c.1:2,200,000

Scales: 49 mm= 60 Mille Pas Geometriques; 15 Lieues communes d'Allemagne; 12 Grandes Lieues d'Allemagne; 25 Lieues communes de France; 20 Lieues d'une heure de Chemin

Comments: Geographical and Political table on right
Remarks explaining political and religious abbreviations, top left.
Julien Cat. (1763): price, 1#

Book:

AP:
AU:
NAP:

Location: BN: Ge.C.7443; Ge.C.7449; BL: Maps.183.d.1 (9); WLCL: Julien, Theatre du Monde II (3); Julien Cat. (1763): no. 413

241 **Germany** 1748 R

L'Empire / d'ALLEMAGNE / divisé par Cercles / Par le Sr. Robert Géographe ord. du Roi / avec Privilege / 1748.

br: 159 x 195 mm.
pl: 166 x 250 mm.

Engravers:

RF: c.1:7,200,000

Scales: 46 mm. = 60 Lieues d'une heure; 45 Grandes Lieues d'Allemagne

Comments:
State 1: 1748: "95" top right, outside border

State 2: 1769: date looks altered from 1748

Book:

AP: 95
AU:
NAP:

Location: BN: Ge.F.1364 (1769)

242 **Germany** 1750 R

ALLEMAGNE / divisée par Cercles / Par le Sr. Robert Géogrape. du Roi. / 1750

br: 150 x 155 mm.

Engravers:

RF: c.1:7,000,000

Scales:

Comments: Robert geographe du roi [i.e. Gilles]
"Pag. 277" outside border, top right; Cardinal points: Septentrional & Midy outside border; Orient & Occident between bdr & nt ln.

Book: Méthode Abrégée et facile pour apprendre la Géographie

AP:
AU:
NAP:

[l'Abbé Le François] Paris:Libraires Associés, 1770

Location: WLCL: C2 1770 Le

243 **Germany** 1756 RV

Carte de / L'EMPIRE D'ALLEMAGNE / divisé en ses dix cercles, / dessinée d'après les Cartes Nationales / assujetie aux observations astronomiques. / Par le Sr. Robert de Vaugondy / Géogr? ord? du Roi, de l'Académie / Royale des Sciences et Belles Lettres de Nancy. / Avec Privilége 1756.

br: 480 x 557 mm.
pl: 495 x 571 mm.

243 **Germany** (continued)

Engravers: (cartouche) rd Sculp. [=haussard]

RF: c.1:2,250,000

Scales: 63 mm.= 80 Mille pas géométriques de 60 au degré; 20 Lieues communes d'Allemagne de 15 au degré; 13 Grandes lieues de Saxe de 10 au degré; 15 Lieues de Hesse de 11 au degré; 16 Lieues de Bohême de 12 au degré; 11 Lieues d'Autriche de 8 au degré; 13 1/2 Lieues moyennes de Suisse de 10 au degré; 32 Lieues du Palatinat de 24 au degré; 30 Lieues de Hollande de 22 au degré; 32 Lieues de Flandre de 24 au degré; 33 1/2 Lieues communes de France de 25 au degré; 26 2/3 Lieues d'une heure de Chemin de 20 au degré

Comments:

Book:

AP:
AU: 56
NAP:

Location:

244 **Germany** 1762 RV

Empire / d'ALLEMAGNE / Par le S. Robert de Vaugondy / Géog? ordinaire du Roy &c. / 1762

A Paris ches l'Auteur quai de l'Horloge près le pont neuf

br: 242 x 283 mm.

Engravers: (map) Gravé par E. Dussy
(cartouche) Arrivet inv. & sculp.

RF: c.1:4,800,000

Scales: 47 mm. = 120 Mille pas géometriques de 60 au degré
40 mm. = 25 Milles d'Allemagne de 15 au degré

Comments:
State 1: 1762

State 2: [1778]: date and imprint removed; Fortin publication

Book:

AP:
AU:
NAP: 21

Location:

245 **Germany, roads** 1757 R

Carte / de l'Empire / D'ALLEMAGNE / où sont marquées exactement / Les Routes des Postes / Par le Sr. Robert / Géographe ordinaire du Roi. / Avec Privilége 1757.

br: 484 x 552 mm.
pl: 498 x 572 mm.

Engravers: (cartouche) E. haussard sculp.

RF: c.1:2,250,000

Scales: 73 mm. = 90 Mille pas geometriques de 60 au degré; 25 Lieues d'Allemagne de 15 au degré; 30 Lieues d'une heure de 20 au degré

Comments:

Book:

AP:
AU: 106
NAP:

Location:

246	Germany, historical	1745	R

Carte / de la GERMANIE / sous les Empereurs / Carlovingiens et Saxons, / pour servir à l'Histoire d'Allemagne / du R.P.Barre Chanoine régulier / de la Congrégation de France, / dressée sous les yeux de l'auteur / par le Sr. Robert Geographe / ordinaire du Roi./ 1745.

[Paris]

br: 398 x 543 mm.
pl: 403 x 551 mm.

Engravers:

RF: c.1:5,000,000

Scales: 55 mm.= 225 Anciens Milles Romains; 140 Anciennes Lieues Gauloises; 60 Grandes Lieues Gauloises

Comments: In fact, a map of Europe

Book: Histoire générale de l'Allemagne

AP:
AU:
NAP:

le P. Joseph Barre (Paris, 1748)

Location: BN: Ge.D.11032; Ge. D. 11033

247	Germany, historical	1756	RV

GERMANIA / ANTIQUA / in quattuor magnos populos, / in minores et minimos distincta; / et REGIONES Danubium / inter et mare Adriaticum contentae /Autore Robert de Vaugondy Geog. / Regis ord? et regiae Scient? ac Litterarum / Academiae Nanceiensis socio. / Cum Privilegio / 1756.

br: 470 x 575 mm.
pl: 495 x 623 mm.

Engravers: (cartouche) E. haussard sculp.

RF: c.1:2,300,000

Scales: 73 mm. = 113 Millaria Romana quorum ad gradum 75; 900 Stadia quorum ad Gradum 600

Comments:
State 1: 1756

State 2: [1768], changes in lettering (Roman to italic)

Book:

AP:
AU: 7
NAP:

Location:

248	Germany, historical	1770	RV

GERMANIA / ex / Cornelio Tacito. / Auctor Robert de Vaugondy Geog? Regis / et Academiae Regiae Nanceianae / Socius. 1770

[Book: Paris: L.F. Delatour, rue St. Jacques; London: Paul Vaillant]

br: 237 x 317 mm.
pl: 260 x 330 mm.

Engravers: (map) Guill? De-la-Haye.

RF: c.1:

Scales: 47 mm. = 150 Milliaria Romana

Comments:

Book: C. Corneli TACITI OPERA recognovit, emendavit, supplementis explevit, NOTIS, DISSERTATIONIBUS TABULIS GEOGRAPHICIS ILLUSTRAVIT

AP:
AU:
NAP:

Gabriel Brotier (Paris and London, 1771)

Location: BL: 53.g.1-4

249	Rhine	1746	R

Le Cercle Electoral du RHIN, / les Duchés de Gueldre, de Limbourg, de Juliers, / de Luxembourg, de Berg, de Westphalie, de 2 Ponts, &c. / Les Comtés de la Marck, de Waldeck, de Witgetstein, &c./Les Landgraviats de Hesse-Cassel, de Darmstat, les Seign.ries. / de l'Evêché de Spire, des Abbayes de Fulde, d'Hirschfelt &c. / Partie des Evêchés de Munster, de Paderborn, / des Duchés de Cleves et de Brunswick, de la Thuringe, / de Franconie, de Lorraine, &c. Dressé sur les Cartes des Srs. Sanson, et rectifié sur des Cartes gravées / et manuscrites du pays. / Par le Sr. Robert Géographe ordinaire du Roi.

A Paris / chés l'Auteur Quay de l'horloge du Palais 1746 / Avec Privilége du Roi

br: 544 x 706 mm.
pl: 546 x 714 mm.

Engravers: (map) Gravé par G. N. Delahaye

RF: c.1:635,000

Scales: 88 mm.= 30 Mille Pas Geometriques; 10 Lieues d'une heure de chemin; 7 Lieues communes d'Allemagne

Comments: Table below title with divisions
Julien Cat. (1763): price, 1#

Book:

AP:
AU:
NAP:

Location: WLCL: Julien, Theatre du Monde II (50);
BN: Ge.DD.2987 B (4052)
Julien Cat. (1763): no. 582;
Lelong, BH: no. 1998

250	Rhine 1	1748	R

I. Carte du COURS DU RHIN ou sont les environs de WALTSHUT, / et de LAUFENBURG Par le Sr. Robert Géog? ord? du Roi Avec Privilege 1748.

br: 163 x 159 mm.
pl: 177 x 190 mm.

Engravers:

RF: c.1:240,000

Scales: 34 mm.= 1 Lieues d'une heure

Comments:

Book:

AP: 109
AU:
NAP:

Location:

251	Rhine 2	1748	R

II. COURS DU RHIN ou sont les environs de BASLE, et de RHINFELDEN. / Par le Sr. Robert Géog? ord? du Roi, Avec Privilege 1748.

br: 165 x 158 mm.
pl: 176 x 189 mm.

Engravers:

RF: c.1:250,000

Scales: 37 mm. = 1 Lieues d'une heure

Comments:

Book:

AP: 110
AU:
NAP:

Location:

252 **Rhine 3** 1748 R

III. COURS DU RHIN où sont les environs de BRISACH, NEWENBURG, FRIBOURG &c. / Par le Sr. Robert Geog? du Roy. Avec Privilége 1748.

br: 161 x 198 mm.
pl: 175 x 246 mm.

Engravers:

RF: c.1:240,000

Scales: 45 mm. = 2 Lieues d'une heure.

Comments: Outside border: map number engraved as 3, not 111.

Book:

AP: 111
AU:
NAP:

Location:

253 **Rhine 4** 1748 R

IV. COURS DU RHIN où sont les environs de RHINAUW, CAPPLE, SCHLETAT, COLMAR &c. / Par le Sr. Robert Géographe ordinaire du Roi. Avec Privilege 1748.

br: 162 x 198 mm.
pl: 175 x 218 mm.

Engravers:

RF: c.1:240,000

Scales: 45 mm. = 2 Lieues d'une heure

Comments:

Book:

AP: 112
AU:
NAP:

Location:

254 **Rhine 5** 1748 R

V. Carte du COURS DU RHIN où sont les Environs de STRASBOURG. / Par le Sr. Robert &c. Avec Privilege 1748.

br: 163 x 197 mm.
pl: 182 x 213 mm.

Engravers:

RF: c.1:240,000

Scales: 45 mm. = 2 Lieues d'une heure

Comments:

Book:

AP: 113
AU:
NAP:

Location:

255 **Rhine 6** 1748 R

VI. Carte du COURS DU RHIN où sont les environs du FORT, LOUIS, de LAUTERBURG / d'HAGUENAU Par le Sr. Robert Géographe ordinaire du Roi Avec Privilege 1748.

br: 166 x 206 mm.
pl: 174 x 221 mm.

Engravers:

RF: c.1:250,000

Scales: 34 mm. = 2 Lieues d'une heure

Comments:

255 **Rhine 6** (continued)

Book:

AP: 114
AU:
NAP:

Location:

256 **Rhine 7** 1748 R

VII. COURS DU RHIN ou sont les environs de PHILISBOURG, SPIRE LANDAW &c. / Par le Sr. Robert Géographe ord? du Roi Avec Privilege 1748.

br: 163 x 196 mm.
pl: 176 x 212 mm.

Engravers:

RF: c.1:240,000

Scales: 45 mm. = 2 Lieues d'une heure

Comments:

Book:

AP: 115
AU:
NAP:

Location:

257 **Westphalia** 1749 RV

Cercle / de WESTPHALIE, / divisé suivant ses / differentes Principautés / Par le Sr. Robert de Vaugondy / Fils de Mr. Robert Geogr. ord. du Roi / avec Privilege 1749.

br: 196 x 163 mm.
pl: 205 x 177 mm.

Engravers:

RF: c.1:2,32,400,000

Scales: 36 mm. = 15 Lieues d'une heure

Comments:

Book:

AP: 96
AU:
NAP:

Location:

258 **Westphalia** 1751 R

CERCLE DE / WESTPHALIE / divisé en tous ses / Etats et Souverainetés, / Par le Sr. Robert / Géographe ordinaire du Roy. / Avec Privilége. / 1751.

Inset: *Supplément, où se trouvent la Principauté D'OOST FRISE et le Comté D'OLDENBOURG*

br: 481 x 562 mm.
pl: 497 x 587 mm.
Inset [top left]: 156 x 193 mm.

Engravers:

RF: c.1:620,000

Scales: 117 mm. = 40 Mille pas geometriques de 60 au degré; 10 Lieues communes d'Allemagne de 15 au degré; 8 Grandes Lieues d'Allemagne de 12 au degré; 13 1/3 Lieues d'une heure de 20 au degré

Comments:
State 1: 1751: in cartouche, shield above title empty

State 2: 1751: shield filled with cow rampant

n.b.: cardinal points engraved outside border

258 **Westphalia**

Book:

AP:
AU: 58
NAP:

Location:

259 **Saxony** 1762 RV

Cercles de / HAUTE-SAXE / et de / BASSE-SAXE / Par le S. Robert de Vaugondy &c.

A Paris ches l'Auteur quai de l'Horloge près le pont neuf 1762

br: 243 x 295 mm.

Engravers: (map) Gravé par E. Dussy
(cartouche) Arrivet inv. & Sculp.

RF: c.1:2,300,000

Scales: 48 mm. = 60 Mille pas géometr? de 60 au degré; 15 Milles d'Allemagne de 15 au degré

Comments:
State 1: 1762

State 2: [1778]: imprint and date removed; Fortin publication

Book:

AP:
AU:
NAP: 23

Location:

260 **Saxony, lower** 1749 RV

Cercle de la / BASSE SAXE, / divisée par ses Principautés / Par le Sr. Robert de Vaugondy / Fils Mr. Robert Geogr. ord. du Roi / Avec Privilege / 1749.

br: 163 x 199 mm.
pl: 174 x 206 mm.

Engravers:

RF: c.1:2,300,000

Scales: 37 mm. = 15 Lieues d'une heure

Comments:

Book:

AP: 97
AU:
NAP:

Location:

261 **Saxony, lower** 1752 R

Cercle de / BASSE SAXE / où sont distingués / Les Etats de Brunswich, Les Duchés / de Holstein, de Mecklenbourg et des / Evêchés d'Hildesheim, et d'Halberstadt / Par le Sr. Robert, / Géographe ordinaire du Roy. / Avec Privilege. / 1752.

br: 484 x 542 mm.
pl: 496 x 562 mm.

Engravers:

RF: c.1:725,000

Scales: 74 mm. = 30 Mille Pas Géometriques de 60 au Degré; 8 Lieues communes d'Allemagne 15 au Degré; 6 Grandes Lieues d'Allemagne de 12 au Degré; 5 Grandes Lieues de Saxe de 10 au Degré

Comments:
State 1: 1752: shield in cartouche empty

State 2: 1752: shield filled in with lion rampant on field of hearts

261 **Saxony, lower** (continued)

Book:

AP:
AU: 59
NAP:

Location:

262 **Saxony, upper** 1749 RV

Partie Meridionale / du Cercle de la / HAUTE SAXE / divisée par ses Principautés / Par le Sr. Robert de Vaugondy / Fils de M. Robert Géogr. ord. du Roi / avec Privilege 1749.

br: 160 x 203 mm.
pl: 177 x 210 mm.

Engravers:

RF: c.1:2,300,000

Scales: 48 mm. = 20 Lieues d'une heure

Comments:

Book:

AP: 99
AU:
NAP:

Location:

263 **Saxony, upper** 1749 RV

Partie Septle. du Cercle / de la HAUTE SAXE / Divisée par / ses Principautés / Par le Sr. Robert de Vaugondy / Fils de Mr. Robert Géogr. ord. du Roi / Avec Privilege 1749.

br: 163 x 210 mm.
pl: 173 x 216 mm.

Engravers:

RF: c.1:2,300,000

Scales: 37 mm. = 15 Lieues d'une heure

Comments:

Book:

AP: 98
AU:
NAP:

Location:

264 **Saxony, upper** 1751 RV

Partie Septentrionale / du Cercle de HAUTE SAXE / qui contient / Le Duché de Poméranie / et Le Marquisat de Brandebourg, / Par le Sr. Robert de Vaugondy fils / de M. Robert Géographe ordinaire du Roy / Avec Privilege / 1751.

Inset: *Supplément / pour le MARQUISAT DE / BRANDEBOURG / où se trouve / La Vieille Marche et / Le Quartier de Pregnitz.*

br: 485 x 553 mm.
pl: 498 x 574 mm.
Inset [lower right]: 170 x 212 mm.

Engravers:

RF: c.1:750,000

Scales: 72 mm. = 30 Mille Pas Géometriques de 60 au degré; 7 1/2 Lieues communes d'Allemagne de 15 au degré; 6 Grandes Lieues d'Allemagne de 12 au degré; 5 Grandes Lieues de Saxe de 10 au degré "Ce Supplément est de même Echelle que la Carte."

Comments:

Book:

AP:
AU: 61
NAP:

Location:

265	**Saxony, upper**	1756	R

Partie Meridionale du / Cercle de HAUTE SAXE / òu sont / Le Duché de Saxe, Le Marquisat de / Misnie, Le Landgraviat de Thuringe, / et autres Etats et Principautés, / Par le Sr. Robert / Géographe ordinaire du Roi. / Avec Privil. 1756.

br: 478 x 540 mm.
pl: 494 x 552 mm.

Engravers: (cartouche) Cne.haussard [Catherine Haussard]

RF: c.1:600,000

Scales: 120 mm. = 40 Mille pas géometriques de 60 au degré; 10 Lieues communes d'Allemagne de 15 au degré; 8 Grandes Lieues d'Allemagne de 12 au degré; 6 2/3 Grandes Lieues de Saxe de 10 au degré

Comments: AU Preface, p. 27b: maps of Saxony listed as numbers 59 and 60

Book:

AP:
AU: 60
NAP:

Location:

266	**Circle of the Four Electors**	1749	RV

Cercle des Quatre / Electeurs par le Sr. Robert de Vaugondy / Fils de Mr. Robert Geog. ord. du Roi / Avec Privilege 1749.

br: 163 x 158 mm.
pl: 178 x 165 mm.

Engravers:

RF: c.1:2,100,000

Scales: 40 mm. = 15 Lieues d'une heure

Comments:

Book:

AP: 100
AU:
NAP:

Location:

267	**Circle of the Four Electors**	1753	RV

Carte des Cercles / DU HAUT ET DU BAS RHIN / ou se trouvent, dans le premier, / le Duché de Deux-Ponts, les Evêchés de Worms / et de Spire, les Etats de Nassau et de Hesse, / les Comtés de Hanaw, de Solms, d'Isenburg, / les Abbayes de Fulde et d'Hirchfeld &c., / et dans le second, / les Electorats de Mayence, de Trêves / de Cologne, et le Palatinat du Rhin, / avec leurs enclaves. / Par le Sr. Robert de Vaugondy Géog? / ord. du Roy 1753 / Avec Privilege

br: 476 x 506 mm.
pl: 493 x 545 mm.

Engravers: (cartouche) hau[ssard]

RF: c.1:635,000

Scales: 85 mm. = 30 Mille pas géometriques de 60 au Degré; 6 Grandes Lieues d'Allemagne de 12 au Degré; 7 1/2 Lieues moyennes d'Allemagne de 15 au Degré; 5 3/8 Lieues de Hesse de 11 1/4 au Degré; 12 Lieues de Palatinat de 24 au Degré

Comments: note scratch on plate. lower rt. corner, between border and plate mark.

The Psalgraben (Roman retrenchment) is shown at 24° 50' - 26° 40' long. and 50° 10' - 50° 20' lat.

State 1: 1753

State 2: [1768?]: many italic place names erased (e.g., Munster on Moselle R.); some changed (e.g., Hedigenboren to Walthuren, Electorat de Mayence); some re-spelled (e.g., Waldeck to Valdeck).

267	**Circle of the Four Electors** (continued)

Book:

AP:
AU: 62
NAP:

Location:

268	**Franconia**	1749	RV

Cercle / de FRANCONIE / divisé en ses princip[a]ux. Etats/ Par le Sr. Robert de Vaugondy / Fils de Mr. Robert Géogr. ord. du Roi / avec Privilege / 1749.

br: 162 x 152 mm.
pl: 175 x 161 mm.

Engravers:

RF: c.1:1,600,000

Scales: 35 mm. = 10 Lieues d'une heure

Comments:

Book:

AP: 101
AU:
NAP:

Location:

269	**Franconia**	1752	R

Cercle de FRANCONIE / qui comprend / les Evêchés de Wurtzbourg, / de Bamberg et d'Aichstet; / Les Marquisats de Culmbach / et d'Anspach; / et les Comtés d'Henneberg,/d'Hohenlohe, de Vertheim, / de Reineck, d'Erpach &c. / Par le Sr. Robert Geographe ord. du Roy. / Avec Privilege 1752.

br: 474 x 560 mm.
pl: 495 x 581 mm.

Engravers: (cartouche) haussard sculp.

RF: c.1:485,000

Scales: 75 mm. = 20 Mille pas géometriques de 60 au degré; 5 Lieues communes d'Allemagne de 15 au degré; 4 Grandes Lieues d'Allemagne de 12 au degré; 6 2/3 Lieues d'une heure, de 20 au degré

Comments: Cf. later states of Haut et Bas Rhin [Cat. 267] which overlaps geographically with Franconia.
This plate does not seem to have undergone similar changes.
Later pulls of plate, engraver signature only "sard sculp."

Book:

AP:
AU: 63
NAP:

Location:

270	**Franconia, Swabia**	1761	RV

HAUT ET BAS RHIN, / FRANCONIE, / SOUABE. / Par le S. Robert de Vaugondy &c.

A Paris ches l'Auteur quai de l'Horloge près le pont neuf 1761

br: 247 x 205 mm.

Engravers: (map) Gravé par E. Dussy

RF: c.1:2,300,000

Scales: 50 mm. = 60 Mille pas géometr? de 60 au degré; 15 Milles d'Allemagne de 15 au degré

Comments:

State 1: 1761

State 2: [1778]; imprint and date removed; Fortin publication

270 **Franconia, Swabia** (continued)

Book:

AP:
AU:
NAP: 24

Location:

271 **Swabia** 1749 RV

Cercle de SOUABE / où sont distingués / les Enclaves de la Maison / d'Autriche / Par le Sr. Robert de Vaugondy / Fils de Mr. Robert Géogr. ord. du Roi / Avec Privilege 1749.

br: 162 x 191 mm.
pl: 173 x 200 mm.

Engravers:

RF: c.1:1,450,000

Scales: 37 mm. = 10 Lieues d'une heure

Comments:

Book:

AP: 102
AU:
NAP:

Location:

272 **Swabia** 1751 R

CERCLE DE SOUABE, / dans lequel sont distingués / les principaux Etats, et les Enclaves / de la Maison d'Autriche. / Par le Sr. Robert, Geographe / ord. du Roy. / Avec Privilege / 1751.

br: 480 x 522 mm.
pl: 500 x 589 mm.

Engravers:

RF: c.1:475,000

Scales: 76 mm. = 20 Mille pas Geometriques fr 60 au degré; 5 Milles d'Allemagne de 15 au degré

Comments:
State 1: 1751

State 2: [1768]: changes in lettering from italics to Roman, vice versa, e.g., *Arben* to ARBON on Lac Constance

Book:

AP:
AU: 64
NAP:

Location:

273 **Bavaria** 1748 R

Cercle de BAVIERE / où se trouvent le / Palatinat de Baviere / et l'Archev. de Saltzbourg / Par le S. Robert Géog. ord. du Roi / avec Privilege 1748.

br: 160 x 166 mm.
pl: 176 x 175 mm.

Engravers:

RF: c.1:2,300,000

Scales: 35 mm. = 10 Lieues d'une heure

Comments: map number engraved 130 instead of 103.

Book:

AP: 103
AU:
NAP:

Location:

274 **Bavaria** 1751 R

LE CERCLE DE BAVIERE / qui comprend / le Haut Palatinat, les Duchés de / Haute et Basse Baviere et de Neuburg, / l'Archevéché de Salzburg / les Evéchés de Ratisbonne, de Freisingen, / de Passau et la Prevôté de Berchtolsgaden, /Par le Sr. Robert Géog. ord. du Roy / Avec Privilege 1751.

br: 477 x 442 mm.
pl: 514 x 499 mm.

Engravers:

RF: c.1:640,000

Scales: 84 mm. = 30 Mille Pas Géométriques de 60 au Degré; 10 Grandes Lieues de France de 20 au Degré; 7 1/2 Milles d'Allemagne de 15 au Degré; 6 1/2 Lieues de Baviere de 13 au Degré

Comments: bottom center: map extends through the border

State 2: [1768?]: many placenames changed from italic to Roman; other changes, e.g. "Erlbach" to "Mark Erlback" (49°30' N., 28°20' E)

Book:

AP:
AU: 65
NAP:

Location:

275 **Austria** 1748 R

L'Archiduché / d'AUTRICHE / et les Duchés de / Stirie, Carinthie et Carniole / Par Le Sr. Robert Géog. du Roi / avec Privilege 1748.

br: 161 x 195 mm.
pl: 178 x 204 mm.

Engravers:

RF: c.1:2,300,000

Scales: 37 mm. = 15 Lieues d'une heure

Comments:

Book:

AP: 104
AU:
NAP:

Location:

276 **Austria, Bavaria** 1762 RV

Cercles de / BAVIERE / et d'AUTRICHE / Par le S. Robert de Vaugondy &c. / 1762

A Paris ches l'Auteur quai de l'Horloge près le pont neuf

br: 242 x 285 mm.

Engravers: (map) Gravé par E. Dussy
(cartouche) Arrivet inv. & Sculp.

RF: c.1:2,300,000

Scales: 50 mm. = 60 Mille pas géométriques de 60 au degré; 15 Milles d'Allemagne de 15 au degré

Comments:
State 1: 1762

State 2: 1778: date changed in title; imprint removed; Fortin publication

Book:

AP:
AU:
NAP: 25

Location:

277 **Austria, north** 1752 RV

Partie Septentrionale / du / CERCLE D'AUTRICHE / qui comprend / l'Archiduché d'Autriche, / divisé en ses huit Quartiers; / et La Haute Partie du / Duché de Stirie. / Par le Sr. Robert de Vaugondy fils / Géographe Ord. du Roi. / avec Privilege / 1752.

br: 480 x 619 mm.
pl: 495 x 632 mm.

Engravers:

RF: c.1:477,000

Scales: 114 mm. = 30 Mille pas geometriques de 60 au Degré; 7 1/2 Lieues communes d'Allemagne de 15 au Degré; 6 Grandes Lieues d'Allemagne de 12 au Degré; 4 Lieues d'Autriche de 8 au Degré

Comments: 32° E, 47° 20'N: "Il y avoit autrfois dans cet endroit des mines d'Argent très riches qui furent remplies d'eau en l'année 1210, et où 1400 mineurs furent engloutis."
33° E, 47° 20'N: "M. Rettelstein qui a une caverne longue de 2 milles d'où l'on tire des Os de Dragons"
n.b.: no decorative frame around scale

Book:

AP:
AU: 66
NAP:

Location:

278 **Austria, south** 1752 RV

Partie Méridionale / du CERCLE D'AUTRICHE, / qui comprend / La Basse Partie du Duché de Stirie, / Le Duché de Carinthie, / divisé en haute et basse; / Le Duché de Carniole, / divisé en haute, basse, moyenne et intere. Carniole, / et L'Istrie Impériale / Par le Sr. Robert de Vaugondy fils / Geographe ordinaire du Roi. / Avec Privilege 1752.

br: 475 x 615 mm.
pl: 485 x 630 mm.

Engravers: (map) Gravé par Delahaye l'aîné

RF: c.1:480,000

Scales: 113 mm. = 30 Mille Pas Geometriques de 60 au Degré; 7 1/2 Lieues communes d'Allemagne de 15 au Degré; 5 Grandes Lieues d'Allemagne de 12 au Degré; 4 Lieues d'Autriche de 8 au Degré

Comments:
State 1: 1752

State 2: [1768?]: accented "See" e.g. Cirkhnitzer Sée, Gottschée, Weissen Sée, Milstatter Sée, but not all of them

Book:

AP:
AU: 67
NAP:

Location:

279 **Tyrol** 1748 R

Comte de TIROL / et / Evêché de Brixen. / Par le Sr. Robert Geog. du Roi / Avec Privilege 1748.

br: 162 x 211 mm.
pl: 178 x 222 mm.

Engravers:

RF: c.1:860,000

Scales: 31 mm. = 5 Lieues d'une heure.

Comments: map number engraved outside border 185; should be 105.

Book:

AP: 105
AU:
NAP:

Location:

280 **Tyrol** 1753 R

LE TYROL / sous le nom duquel l'on comprend / Le Comté de Tyrol, / les Comtés annèxes de / Bregentz, Feldkirck, Monfort,/Pludentz, &c. / et les Evêchés de / Trente, et de Brixen, / sous la Protection des Comtes de Tyrol. / Par le Sr. Robert Geographe / ordinaire du Roi. / Avec Privilege 1753.

br: 475 x 545 mm.
pl: 493 x 580 mm.

Engravers: (map) Gravé par Delahaye l'aîné
(cartouche) haussard

RF: c.1:476,000

Scales: 113 mm. = 30 Mille pas geometriques de 60 au Degré; 7 1/2 Lieues communes d'Allemagne de 15 au Degré; 6 Grandes Lieues d'Allemagne de 12 au Degré; 4 Lieues d'Autriche de 8 au Degré; 10 Lieues d'une heure de chemin de 20 au Degré

Comments: Map concerned in lawsuit with the Delahaye family of engravers and the publisher Boudet. See Chapter Three.

Book:

AP:
AU: 68
NAP:

Location:

281 **Switzerland** 1748 R

Les Suisses / les Alliés de Suisses / et leurs Sujets / Par le Sr. Robert, Géog? ord? du Roi. / avec Privilege 1748.

br: 162 x 206 mm.
pl: 166 x 253 mm.

Engravers:

RF: c.1:1,650,000

Scales: 33 mm. = 10 Lieues d'une heure

Comments: map number engraved outside border upside down, looks like 901, instead of 106.

Book:

AP: 106
AU:
NAP:

Location:

282 **Switzerland** 1756 RV

Carte de la / REPUBLIQUE DES SUISSES / où sont distingués / les 13 Cantons et leurs alliés, / les Sujets de ces Cantons et ceux / de leurs alliés. / Par le Sr. Robert de Vaugondy Géog. / ord. du Roi, de l'Académie Rle. des Sciences / et Belles-Lettres de Nancy / Avec Privilege 1756.

br: 480 x 602 mm.
pl: 494 x 627 mm.

Engravers: (cartouche) E. haussard sculp.

RF: c.1:540,000

Scales: 103 mm. = 30 Mille pas géométriques ou Milles d'Italie de 60 au degré; 7 1/2 Lieues communes d'Allemagne de 15 au degré; 4 Grandes Lieues de Suisse de 8 au degré; 5 Lieues moyennes de Suisse de 10 au degré; 6 Petites Lieues de Suisse de 12 au degré; 10 Lieues d'une heure de 20 au degré

Comments: Insets [bot. left; top right]: descriptions of Swiss cantons, by subject, religion, and legal jurisdiction

Book:

AP:
AU: 57
NAP:

Location:

283 **Switzerland** 1762 RV

LES SUISSES / ET / LES GRISONS, / avec leurs Alliés et leurs / Sujets. / Par le S. Robert de Vaugondy / Géographe ord? du Roi. &c / 1762

A Paris ches l'Auteur quai de l'Horloge près le pont neuf

br: 241 x 300 mm.

Engravers: (map) Gravé par E. Dussy
(cartouche) Arrivet inv. & Sculp.

RF: c.1:1,050,000

Scales: 52 mm. = 30 Mille pas Géometriques de 60 au degré; 5 Lieues moyennes de Suisse de 10 au degré

Comments: Color key, bottom right: Les Suisses, leurs Sujets, leurs Allies, les Sujets des Grisons

State 1: 1762

State 2: 1778: date changed in title; imprint removed; Fortin publication

Book:

AP:
AU:
NAP: 27

Location:

284 **Bohemia** 1748 R

Royaume de BOHÉME, / Marquisat de MORAVIE, / et SILÉSIE Autrichienne. / Par le Sr. Robert Geog? du Roi / avec Privilege 1748.

br: 159 x 232 mm.
pl: 176 x 240 mm.

Engravers:

RF: c.1:2,300,000

Scales: 37 mm. = 15 Lieues d'une heure

Comments:

Book:

AP: 107
AU:
NAP:

Location:

285 **Bohemia** 1751 R

LE ROYAUME DE BOHEME, / Le Duché de Silesie, / et les Marquisats / de Moravie et Lusace, / Dressés d'apres les cartes de Muller, / Par le Sr. Robert / Geographe ordin. du Roi. / Avec Privilege. / 1751.

br: 482 x 556 mm.
pl: 496 x 581 mm.

Engravers:

RF: c.1:900,000

Scales: 59 mm. = 30 Mille pas Geometriques de 60 au deg.; 7 1/2 Milles d'Allemagne de 15 au deg.; 8 7/12 Milles communes de Silesie de 17 1/6 au deg.; 6 Milles de Boheme et Gr. M. de Silesie de 12 au deg.

Comments:
State 1: 1751

State 2: [1768]; 51°N 35°E: "Wratislaw" changed to "Breslau"; Prin. de "Wratislaw" to "Breslau"
n.b. no frame around Echelle (top left)

Book:

AP:
AU: 69
NAP:

Location:

286 **Bohemia** 1761 RV

BOHEME SILÉSIE, / MORAVIE, / LUSACE. / Par le S. Robert de Vaugondy / Géog? ord? du Roi, de S.M. Pol? / Duc de Lorr? et de Bar &c.

A Paris ches l'Auteur quai de l'Horloge près le pont neuf 1761

br: 245 x 231 mm.

Engravers: (map) Gravé par E. Dussy
(cartouche) Arrivet inv. & Sculp.

RF: c.1:2,300,000

Scales: 41 mm. = 50 Mille pas géom? de 60 au degré; 34 mm. = 10 Milles d'Allemagne de 15 au deg?

Comments:
State 1: 1761

State 2: [1778]: imprint and date removed; Fortin publication

Book:

AP:
AU:
NAP: 26

Location:

287 **Hungary** 1749 RV

Royaume de / HONGRIE. / Par le Sr. Robert de Vaugondy / Fils de Mr. Robert Géog. du Roi / Avec Privilege / 1749.

br: 163 x 222 mm.
pl: 178 x 240 mm.

Engravers:

RF: c.1:3,800,000

Scales: 28 mm. = 12 Lieues de Hongrie de 12 au degrés; 20 Lieues d'une heure

Comments:

Book:

AP: 108
AU:
NAP:

Location:

288 **Hungary** 1751 R

Royaume / de HONGRIE, / Principauté de Transilvanie, / Sclavonie, Croatie, / et partie de la / Principauté de Valaquie, de la / Bosnie, de la Servie et de la / Bulgarie. / Par le Sr. Robert, Géographe ordinaire / du Roi. / Avec Privilege / 1751.

br: 476 x 548 mm.
pl: 497 x 572 mm.

Engravers:

RF: c.1:410,000

Scales: 100 mm. = 80 Mille Pas géometriques, de 60 au Degré; 13 1/4 Lieues de Hongrie, de 10 au Degré; 20 Lieues communes d'Allemagne, de 15 au Degré

Comments: n.b.: 46°N, 37°30' E: Fort Romain; 41°30'E, 45-46°N: Chemin de l'Empereur Trajan

Book:

AP:
AU: 70
NAP:

Location:

289 **Poland** 1748 R

Etats de la Couron[n]e de / POLOGNE / ou sont les Royaues. de Pologne, le / Grd. Duche de Lithuanie, la Russie / Noire, la Volhinie &c. / Par le Sr. Robert, Géographe ord. du Roi. / avec Privilege 1748.

br: 161 x 212 mm.
pl: 166 x 252 mm.

Engravers:

RF: c.1:6,700,000

Scales: 46 mm. = 60 Lieues d'une heure

Comments:

Book:

AP: 116
AU:
NAP:

Location:

290 **Poland** 1749 RV

Royaume de POLOGNE / divisé en / Haute et Basse Pologne, et / Subdivisé en Palatinats / Par le Sr. Robert de Vaugondy / Fils de Mr. Robert Geog. du Roi A.Pr. 1749.

br: 164 x 196 mm.
pl: 176 x 203 mm.

Engravers:

RF: c.1:3,250,000

Scales: 35 mm. = 20 Lieues d'une heure

Comments:

Book:

AP: 118
AU:
NAP:

Location:

291 **Poland** 1750 RV

POLOGNE / Par le Sr. Robert de / Vaugondy / Avec Privilege 1750.

br: 147 x 165 mm.

Engravers:

RF: c.1:7,500,000

Scales: 30 mm. = 40 lieues d'une heure

Comments: "Pag. 350" outside border, top right. Cardinal points: only Septentrional outside border

Book: Méthode Abrégée et facile pour apprendre la Géographie

AP:
AU:
NAP:

[l'Abbé Le François] Paris: Libraires Associés, 1770

Location: WLCL: C2 1770 Le

292 **Poland** 1752 RV

Le Royaume de / POLOGNE / divisé / en ses Duchés et Provinces, et / subdivisé en Palatinats. / Par le Sr. Robert de Vaugondy / Géographe ordinaire du Roi. / Avec Privilege 1752.

br: 480 x 540 mm.
pl: 497 x 560 mm.

Engravers:

RF: c.1:2,400,000

292 **Poland** (continued)

Scales: 90 mm. = 120 Mille pas géometriques, de 60 au degré; 30 Lieues d'Allemagne et Pologne, de 15 au degré; 50 Lieues communes de France, de 25 au degré; 40 Lieues d'une heure de 20 au degré

Comments: n.b.: cartouche shows the Pacta Conventa, the agreement between the king and nobility reached at the beginning of each reign, from the 16th century

State 2: title altered "...divisé / selon les Partages faits en 1772, 1793, et 1795 / entre la Russie, la Prusse et l'Autriche. /" Colour code added under the Echelle: Russie (yellow), Prusse (green), Autriche (red); cf. coat of arms at top of cartouche

Book:

AP:
AU: 71
NAP:

Location: BN: Ge.AF Pf 95 (31) [State 2]

293 **Poland** 1759 RV

[Carte de Pologne]

Engravers:

RF: c.1:

Scales:

Comments: not seen in any location; known only from catalogues
Julien Cat. (1763): described as having a Table Methodique, price 1# 5s

Book:

AP:
AU:
NAP:

Location: RV Catalogues: Tablettes Parisiennes (1760); Notice Raisonne (1774); Catalogue (1777); Julien Catalogue (1763): no. 836

294 **Poland** 1762 RV

Royaume de / POLOGNE / Par le S. Robert de Vaugondy / Géog? ordinaire du Roy &c. / 1762

A Paris ches l'Auteur quai de l'Horloge près le pont neuf

br: 246 x 286 mm.

Engravers: (map) Gravé par E. Dussy.
(cartouche) Arrivet inv. & Sculp.

RF: c.1:4,800,000

Scales: 71 mm. = 180 Mille pas géometriques de 60 au degré; 45 Lieues de Pologne de 15 au degré

Comments:
State 1: 1762

State 2: [1778]: date and imprint removed; Fortin publication

Book:

AP:
AU:
NAP: 28

Location:

295 **Prussia** 1748 R

Royaume de PRUSSE / et / Prusse Rle. ou Polonoise. / Par le Sr. Robert Geog. du Roi / avec Privilege 1748.

br: 161 x 197 mm.
pl: 178 x 206 mm.

Engravers:

RF: c.1:2,250,000

295 **Prussia** (continued)

Scales: 37 mm. = 15 Lieues d'une heure

Comments:

Book:

AP: 117
AU:
NAP:

Location:

296 **Prussia** 1751 R

LA PRUSSE / divisée en / Prusse Royale, et / Prusse Ducale laquelle a été / érigée en Royaume par l'Empereur / Leopold en faveur de Frederic I. Elect? / de Brandebourg couronné en Janvier 1701/ Par le Sr. Robert Géog. ord. du Roy. / Avec Privilege / 1751.

br: 486 x 616 mm.
pl: 500 x 633 mm.

Engravers: (map) Guill? Delahaye Sculpsit

RF: c.1:715,000

Scales: 78 mm. = 30 Mille Pas Geometriques de 60 au Degré; 7 1/2 Milles d'Allemagne de 15 au Degré; 10 Lieues Marines d'une heure et de Pologne de 20 au Degré; 8 Lieues de Prusse de 16 au Degré

Comments:

Book:

AP:
AU: 72
NAP:

Location:

297 **Lithuania** 1749 RV

Grand Duché / de LITHUANIE / divisé par Palatinats. / Par le Sr. Robert de Vaugondy / Fils de Mr. Robert Geog. du Roi / Avec Privilege / 1749.

br: 163 x 214 mm.
pl: 175 x 238 mm.

Engravers:

RF: c.1:3,800,000

Scales: 29 mm. = 20 Lieues d'une heure

Comments:

Book:

AP: 119
AU:
NAP:

Location:

298 **Finland** 1749 RV

Curlande, Livonie, Ingrie / et Partie Meridionale / de FINLANDE. / Par le Sr. Robert de Vaugondy fils de / Mr. Robert Geogr. ordin. du Roi. / Avec Privilege / 1749.

br: 163 x 187 mm.
pl: 176 x 195 mm.

Engravers:

RF: c.1:4,300,000

Scales: 38 mm. = 150 Lieues Marines et d'une heure.

Comments:

Book:

AP: 120
AU:
NAP:

Location:

299 **Iberian Peninsula** 1748 R

Royaumes / d'ESPAGNE, et de / PORTUGAL. / Par le Sr. Robert, Géog? ord? du Roi. / avec Privilége 1748.

br: 161 x 196 mm.
pl: 168 x 252 mm.

Engravers:

RF: c.1:6,700,000

Scales: 32 mm. = 40 Lieues d'une heure, ou Communes d'Espagne

Comments:

Book:

AP: 123
AU:
NAP:

Location:

300 **Iberian peninsula** 1750 RV

Royaumes / d'ESPAGNE et de / PORTUGAL. / Par le Sr. Robert de Vaugondy / avec Privilége 1750.

br: 149 x 151 mm.

Engravers:

RF: c.1:9,000,000

Scales: 25 mm. = 40 [unit not given]

Comments: "Pag. 251" outside border, top right. Cardinal points: Septentrion & Midy outside border; Orient & Occident between border & neat line

Book: Méthode Abrégée et facile pour apprendre la Géographie

AP:
AU:
NAP:

[l'Abbé Le François], Paris: Libraires Associés, 1770

Location: WLCL: C2 1170 Le

301 **Iberian Peninsula** 1750 R

Royaumes / d'ESPAGNE et de / PORTUGAL, / Par le Sr. Robert Geographe / ordinaire du Roy. / Avec Privilege. / 1750.

br: 488 x 588 mm.
pl: 501 x 630 mm.

Engravers: (map) Guill? Delahaye Sculpsit

RF: c.1:2,300,000

Scales: 72 mm. = 90 Mille pas Geometriques; 30 Lieues d'Espagne de 20 au deg.; 27 Lieues d'Aragon de 18 au deg.; 30 Lieues d'une heure ou de Marine de 20 au deg.

Comments:

Book:

AP:
AU: 73
NAP:

Location:

302 **Iberian Peninsula** 1762 RV

Royaumes / d'ESPAGNE, / et de / PORTUGAL / Par le S. Robert de Vaugondy / Géog? ordinaire du Roy &c. / 1762

A Paris ches l'Auteur quai de l'Horloge près le pont neuf

br: 243 x 286 mm.

Engravers: (map) Gravé par E. Dussy
(cartouche) Arrivet inv. & Sculp.

RF: c.1:4,800,000

302 **Iberian Peninsula** (continued)

Scales: 71 mm. = 180 Mille pas Géometriques de 60 au degré; 60 mm. = 50 Lieues d'Espagne de 20 au degré; 48 mm. = 36 Lieues de Portugal de 18 au degré

Comments:
State 1: 1762

State 2: [1778]: date and imprint removed; Fortin publication

Book:

AP:
AU:
NAP: 29

Location:

303 **Iberian Peninsula, roads** 1757 R

Carte / des Royaumes / d'ESPAGNE et de PORTUGAL / dans laquelle sont tracées / Les Routes des Postes / Par le Sr. Robert. / Geographe ordin. du Roi. / Avec Privilege. / 1757.

br: 480 x 575 mm.
pl: 498 x 595 mm.

Engravers: (cartouche) E.haussard Sculp.

RF: c.1:2,300,000

Scales: 71 mm. = 90 Mille pas geometriques de 60 au degré; 30 Lieues communes d'Espagne de 20 au degré

Comments:

Book:

AP:
AU: 105
NAP:

Location:

304 **Spain** 1747 R

L'ESPAGNE / divisée en tous ses / Royaumes et Principautés / suivant qu'ils sont compris sous / les Couronnes de / Castille, et d'Aragon, possedées / Par le Roi Catholique; / sous la Courne. de Portugal possedée / par le Roi de Portugal. / Tirée des Mémoires les plus recents. / Par le Sr. Sanson Geog? du Roi / Revüe, Par le Sr. Robert Géog. ord de Sa Majeste /

A Paris / Chés l'Auteur quay de l'horloge du Palais. / 1747

br: 455 x 672 mm.
pl: 459 x 676 mm.

Engravers:

RF: c.1:

Scales: 63 mm. 120 Mille pas geometriques ou Milles d'Italie; 33 Lieues communes d'Espagne; 50 Lieues communes de France; 40 Grandes Lieues de France ou d'une heure de chemin

Comments: Table to left

Book:

AP:
AU:
NAP:

Location: BL: Maps.183.d.1 (6); BN: Ge.C.3865; Julien Cat. (1763): nos. 13, 62

305 **Spain, historical** 1741 R

CARTE DE / L'ESPAGNE ANCIENNE / divisée / en Trois Grandes Parties / et ses differens Peuples / avec / leurs villes, leurs Rivieres etc / et son rapport avec / l'Espagne Moderne / Pour servir a la lecture de l'histoire de Jean de Ferreras / Dressée par le Sr. Robert / Geog. ord. du Roi / 1741

br: 397 x 504 mm.
pl: 405 x 515 mm.

Engravers:

305 **Spain, historical** (continued)

RF: c.1: 2,500,000

Scales: 80 mm. = 150 Milles Anciens; 1200 Stades

Comments: two place names for each location: Latin and Spanish
shows Iberian peninsula with Balearic Islands and part of North Africa to 'Julia Caesaria'
see Cat. [36] for notes on price

Book: Histoire Générale d'Espagne...

AP:
AU:
NAP:

Jean de Ferreras (trad. par d'Hermilly) Paris: [Ganeau], 1751

Location: Vatican: R.G. Storia III.115 (1&2); BN: 4° Oa. 80; BL: 178.c.7-16; L-D (1745) p. 245-247

306 **Spain, historical** 1750 RV

HISPANIA / ANTIQUA / in tres praecipuas partes, / scilicet / TARRACONENSEM, LUSITANIAM, / ET BAETICAM; / nec non in Populos divisa, / a N. Sanson Geog. Regis ordinario / Revisa et ad observationes astronomicas / redacta / accurante Robert de Vaugondy filio. / Cum Privilegio Regis. / 1750.

br: 475 x 570 mm.
pl: 497 x 625 mm.

Engravers: (map) Guill? Delahaye Sculp?

RF: c.1:2,300,000

Scales: 95 mm. = 150 Millaria Romana quorum 75 gradus unus continet.; 1200 Stadia Graeca quorum 600 gradum efficiunt.

Comments:
State 1: 1750

State 2: [1768]: lettering changes (italics to Roman and vice versa); "Maenaria" added to Balearic Islands; "Colubraria" labels nothing
n.b.: Scale has no frame

Book:

AP:
AU: 8
NAP:

Location:

307 **Castille, north** 1749 RV

Partie Septentrionale / des Etats de CASTILLE / où se trouve / le Royaume de Gallice / Par le Sr. Robert de Vaugondy / Fils de Mr. Robert Geogr. ord. du Roi. / avec Privilege 1749.

br: 163 x 193 mm.
pl: 173 x 200 mm.

Engravers:

RF: c.1:1,650,000

Scales: 33 mm. = 10 Lieues d'une heure

Comments:

Book:

AP: 124
AU:
NAP:

Location:

308 Castille, north 1749 RV

Partie Septle. des Etats de / CASTILLE où se trouvent les / Roymes. des Asturies, de Leon / de la Castille Vieille, et La / Biscaye / Par le Sr. Robert de Vaugondy fils de / Mr. Robert Géog. du Roi A. Pr. 1749.

br: 163 x 191 mm.
pl: 170 x 198 mm.

Engravers:

RF: c.1:2,600,000

Scales: 41 mm. = 20 Lieues d'une heure

Comments:

Book:

AP: 125
AU:
NAP:

Location:

309 Castille, north 1752 RV

Partie Septentrionale de / LA COURONNE DE CASTILLE / où se trouvent / les Royaumes de Castille Vieille / de Leon, de Gallice, des Asturies, / La Biscaye et La Navarre, / en Partie. / Par le Sr. Robert de Vaugondy fils / Géographe ordin? du Roi. / Avec Privilege 1752.

br: 480 x 595 mm.
pl: 497 x 635 mm.

Engravers:

RF: c.1:1,250,000

Scales: 87 mm. = 60 Mille pas géometriques de 60 au degré; 17 1/2 Grandes Lieues Espagnoles de 17 au degré; 20 Lieues moyennes ou d'une heure de 20 au degré; 26 1/2 Petites Lieues de 26 1/2 au degré; 25 Lieues communes de France de 25 au degré

Comments:

Book:

AP:
AU: 74
NAP:

Location:

310 Castille, south 1749 RV

Partie Meridion[a]le des Etats de / CASTILLE où se trouvent / l'Estremadure-de-Leon / et la Castille Nouvelle / Par le Sr. Robert de Vaugondy Fils de / Mr. Robert Geogr. du Roi / Avec Privilege 1749.

br: 164 x 232 mm.
pl: 175 x 238 mm.

Engravers:

RF: c.1:2,300,000

Scales: 48 mm. = 20 Lieues d'une heure.

Comments:

Book:

AP: 126
AU:
NAP:

Location:

311 Castille, south 1749 RV

Partie Meridionale des / ETATS DE CASTILLE / òu se trouvent / l'Andalousie et le / Royme. de Grenade / Par le Sr. Robert de Vaugondy fils de / Mr. Robert, Géog. du Roi / avec Privilege 1749.

br: 161 x 217 mm.
pl: 173 x 240 mm.

Engravers:

RF: c.1:2,700,000

Scales: 41 mm. = 20 Lieues d'une heure

Comments:

Book:

AP: 130
AU:
NAP:

Location:

312 Castille, south 1751 RV

Partie Meridionale / des ETATS DE CASTILLE, / où se trouvent / la Castille Nlle., l'Estremadure, /l'Andalousie, les Royaumes / de Grenade et de Murcie; / et Partie des Etats d'Aragon, / qui contient le Royaume de / Valence. / Par le Sr. Robert de Vaugondy fils / de Mr. Robert Géographe ordinaire / du Roy. / 1751.

br: 482 x 558 mm.
pl: 497 x 577 mm.

Engravers:

RF: c.1:1,320,000

Scales: 81 mm. = 60 Mille Pas Geometriques de 60 au Degré; 20 Lieues communes d'Espagne de 20 au Degré

Comments:

Book:

AP:
AU: 75
NAP:

Location:

313 Navarre 1749 RV

Royaume / de NAVARRE / divisé en six Mérindades. / Par le Sr. Robert de Vaugondy Fils de / Mr. Robert Géog? du Roi. / avec Privilege 1749.

br: 221 x 163 mm.
pl: 236 x 171 mm.

Engravers:

RF: c.1:840,000

Scales: 47 mm. = 7 Lieues d'une heure

Comments:

Book:

AP: 127
AU:
NAP:

Location:

314 Aragon 1752 RV

Etats de la Couronne / d'ARAGON, / ou se trouvent / les Royaumes d'Aragon et / de Navarre / la Principauté de Catalogne. / Par le Sr. Robert de Vaugondy fils / Geographe ordin? du Roi. / Avec Privilége / 1752.

br: 478 x 588 mm.
pl: 501 x 633 mm.

Engravers:

RF: c.1:795,000

Scales: 137 mm. = 60 Mille pas Geometriques de 60. au degré; 20 Lieues communes d'Espagne de 20. au degré; 18 Lieues d'Aragon de 18. au degré; 20 Grandes Lieues de France de 20 au degré

314 **Aragon** (continued)

Comments: [bot. ctr.]: "Nota Comme la dimension de cette Carte n'a pu permettre d'y representer le Royaume de Valence, il faut recourir à la Carte de la partie méridionale des Etats de Castille, où il se trouve en entier."

Book:

AP:
AU: 76
NAP:

Location:

315 **Aragon, north** 1749 RV

Partie Septentle. de la / Couronne d'ARAGON / où se trouvent / Le Royme. d'Aragon et La / Principauté de Catalogne / Par le Sr. Robert de Vaugondy / Fils de Mr. Robert Geog. du Roi Av. Pr. 1749.

br: 164 x 215 mm.
pl: 177 x 222 mm.

Engravers:

RF: c.1:2,150,000

Scales: 52 mm. = 20 lieues d'une heure.

Comments:

Book:

AP: 128
AU:
NAP:

Location:

316 **Aragon, south** 1749 RV

Partie Méridionale de la / Couronne d'ARAGON / ou se trouvent les / Rmes. de Valence, de Maiorque / et de Murcie dans les / Etats de Castille / Par le Sr. Robert de Vaugondy Fils de / Mr. Robert Geog? du Roi. av. Privil? / 1749.

br: 163 x 198 mm.
pl: 176 x 208 mm.

Engravers:

RF: c.1:3,000,000

Scales: 37 mm. = 20 Lieues d'une heure

Comments:

Book:

AP: 129
AU:
NAP:

Location:

317 **Portugal** [1758] RV

[Carte de Portugal]

two sheets

Engravers:

RF: c.1:

Scales:

Comments: not seen; known only from RV catalogue which says map is corrected by RV

Book:

AP:
AU:
NAP:

Location: RV: Catalogue (1777)

318 **Portugal, north** 1749 RV

Partie Septentrionale / du Royaume / de PORTUGAL / Par le Sr. Robert de Vaugondy fils de / Mr. Robert Geog. du Roi / Avec Privilege / 1749.

br: 161 x 197 mm.
pl: 172 x 204 mm.

Engravers:

RF: c.1:2,000,000

Scales: 40 mm. = 15 Lieues d'une heure

Comments:

Book:

AP: 131
AU:
NAP:

Location:

319 **Portugal, north** 1751 R

Partie Septentrionale / du Royaume de / PORTUGAL. / Par le Sr. Robert / Géographe ordinaire du Roy / Avec Privilege. / 1751.

br: 482 x 521 mm.
pl: 500 x 540 mm.

Engravers:

RF: c.1:675,000

Scales: 79 mm. = 30 Mille Pas Géometriques de 60 au Degré; 10 Lieues Marines et d'une heure de 20 au Degré; 9 Lieues communes de Portugal de 18 au Degré

Comments:

Book:

AP:
AU: 78
NAP:

Location:

320 **Portugal, south** 1749 RV

Partie Meridionale / du Royaume / de PORTUGAL / Par le S. Robert de Vaugondy Fils de / Mr. Robert Géog? du Roi / avec Privilege / 1749.

br: 163 x 200 mm.
pl: 174 x 207 mm.

Engravers:

RF: c.1:2,000,000

Scales: 40 mm. = 15 Lieues d'une heure

Comments:

Book:

AP: 132
AU:
NAP:

Location:

321 **Portugal, south** 1751 R

Partie Meridionale / du Royaume de / PORTUGAL. / Par le Sr. Robert. / Géographe ordinaire du Roy. / Avec Privilege. / 1751.

br: 480 x 522 mm.
pl: 500 x 530 mm.

Engravers:

RF: c.1:675,000

Scales: 107 mm. = 40 Mille Pas Géométriques de 60 au Degré; 13 3/1 [sic] Lieues Marines et d'une heure de 20 au Degré; 12 Lieues communes de Portugal de 18 au Degré

Comments:

Book:

AP:
AU: 77
NAP:

Location:

322 **Italy** 1743 R

L'ITALIE / et ses Isles Circonvoisines / Sicile, Sardaigne, Corse, &c. / exactement divisées en leurs / Etats, Royaumes, Republiques, &c. / Par le Sr. Sanson Géog. Ord. du Roy. / Revue et Assujettie aux Observations / astronomiques / Par le Sr. Robert. Géog. Ord. du Roy./

A Paris / Quai de l'Horloge du Palais / avec Privilege du Roy / 1743. / Delahaye le jeune

br: 515 x 683 mm.
pl: 522 x 692 mm.

Engravers: (map) Delahaye le jeune

RF: c.1:2,300,000

Scales: 97 mm. = 120 Mille pas Geometriques ou Milles communs d'Italie; 100 Milles de Piemont; 140 Milles de Venise; 150 Milles de Rome; 50 Lieues Communes de France; 40 Lieues d'une heure de Chemin et Lieues Marines

Comments: Table to right

Book:

AP:
AU:
NAP:

Location: BL: Maps.183.d.1 (8)

323 **Italy** 1748 R

L'ITALIE / et ses Isles circonvoisines, / Sicile, Sardaigne, Corse &c. / Par le Sr. Robert Géog? ord? du Roi. / Avec Privilege 1748.

br: 162 x 165 mm.
pl: 166 x 253 mm.

Engravers:

RF: c.1:7,700,000

Scales: 36 mm. = 120 Milles d'Italie

Comments:

Book:

AP: 133
AU:
NAP:

Location:

324 **Italy** 1750 RV

ITALIE. / par le Sr. Robert / de Vaugondy / avec Privilége 1750.

br: 150 x 153 mm.

Engravers:

RF: c.1:8,500,000

Scales: 25 mm. = 40 [unit not given]

Comments: "Pag. 209" outside border, top right no cardinal points

Book: Méthode Abrégée et facile pour apprendre la Géographie

AP:
AU:
NAP:

[l'Abbé Le François] Paris: Libraires Associés, 1770

Location: WLCL: C2 1170 Le

325 **Italy** 1750 R

L'ITALIE / qui comprend les Etats de Piémont, / les Duchés de Milan, de Parme, / de Modene, de Mantoue &c. / les Républiques de Genes, de Venise, / de Luques etc. / le Grand Duché de Toscane, l'Etat du / Saint-Siége, / le Royaume de Naples, et les Isles / de Corse, de Sardaigne, et de Sicile. / Par le Sr. Robert Géographe ord. du Roi. / Avec Privilége 1750.

br: 485 x 554 mm.
pl: 496 x 628 mm.

Engravers:

RF: c.1:2,500,000

Scales: 87 mm. = 120 Milles d'Italie.; 40 Lieues d'une heure.

Comments:

Book:

AP:
AU: 79
NAP:

Location:

326 **Italy** 1761 RV

L'ITALIE / Par le S. Robert de Vaugondy Géog? / ord? du Roi, de S.M. Polon? Duc de / Lorraine et de Bar, et de / la Société royale de Nancy / 1761

A Paris ches l'Auteur quai de l'Horloge près le pont neuf

br: 245 x 283 mm.

Engravers: (map) Gravé par E. Dussy
(cartouche) Arrivet inv. & Sculp.

RF: c.1:4,800,000

Scales: 36 mm. = 90 Milles d'Italie de 60 au deg?; 30 Lieues d'une heure

Comments:
State 1: 1761

State 2: [1778]: date and imprint removed; "et Censeur Royal" added to title; Fortin publication

Book:

AP:
AU:
NAP: 30

Location:

327 **Italy, Roads** 1756 R

Carte de / L'ITALIE / dans laquelle sont tracées / les Routes des Postes; / Par le Sr. Robert / Géographe ordinaire du Roy. / Avec Privilege. / 1756. /

br: 477 x 545 mm.
pl: 496 x 570 mm.

Engravers: (cartouche) Elisabeth haussard Sculp.

RF: c.1:2,500,000

Scales: 87 mm. = 120 Milles d'Italie de 60 au degré

Comments: scale has no frame

Book:

AP:
AU: 104
NAP:

Location:

328 **Italy, historical** 1743 R

ANTIQUORUM ITALIAE et ILLYRICI occidentalis /orientalis episcopatuu[m] Geographica Descriptio

Paris: Apud A. Robert

two sheets
br: 370 x 500 mm. each

Engravers: (map) Abraham de la plaetsen fecit

RF: c.1:3,500,000

Scales: 31 mm. = 60 Milliarum Italiae seu passuum geometricoru[m] mille.

Comments: engraver's name suggests this is a re-publication of a Sanson map

Julien Cat. (1763): price, 1#

Book:

AP:
AU:
NAP:

Location: BN: Ge.C.4203/4204 (dated 1743); WLCL: AF (95) [same title, publ. Tavernier, 1640]; Julien Cat. (1763): no. 671); Robert, Intro. à la Geog., (1743)

329 **Italy, historical** 1743 R

ITALIAE ANTIQUAE / ex variis auctoribus decerptae / N. Sanson. Regis Geographo, / altera editio / ad Observationes astronom. redacta / et Itinerariis Romanis adornata / accurante / Aeg. Robert Regis Geog./

Parisiis / Apud auctorem / in ripa Sequanae / vulgo quai de l'horloge dicta/ Cum privilegio Regis / 1743

br: 512 x 682 mm.
pl: 523 x 690 mm.

Engravers: (map) Delahaye Filius
(cartouche) Gobin Sculpsit

RF: c.1:2,250,000

Scales: 97 mm. = 150 vetera Milliara Romana; 1400 Stadia Graeca

Comments: based on N. Sanson, 2nd ed.; Roman roads shown; Table Méthodique to right
Julien Cat. (1763): price, 1# 5s.
n.b.: BL copy: frame around scales outlined but not inked

Book:

AP:
AU:
NAP:

Location: WLCL: AF (58); BL: Maps.183.d.1 (18); Julien Cat. (1763): no. 1309; Robert, Intro. à la Geog., (1743)

330 **Italy, historical** 1751 RV

ITALIA ANTIQUA / cum Insulis / Sicilia, Sardinia, / et Corsica, / in suas majores partes et populos divisa, / A Nicolao Sanson Christ. Regis Geographo / revisa et ad observationes astronomicas redacta / accurante Robert de Vaugondy filio / Cum privilegio Regis / 1751

br: 478 x 545 mm.
pl: 499 x 570 mm.

Engravers: (map) Guill? Delahaye Sc.

RF: c.1:2,500,000

Scales: 86 mm. = 150 Millaria Romana quorum 75 gradum efficiunt; 1200 Stadia Graeca quorum 600 gradui respondent

Comments
State 1: 1751

State 2: [1768]: lettering and place name changes (italics to Roman and vice versa): e.g., Capua, Neapolis, Salernum are italics in State 1, Roman in State 2.

330 **Italy, historical** (continued)

Book:

AP:
AU: 9
NAP:

Location:

331 **Italy, historical** 1770 RV

ITALIA / ex / Cornelio Tacito. / Auctor Robert de Vaugondy Geog? Regis / et / Academiae regiae Nanceianae Socius. / 1770

[Book: Paris: L.F.Delatour, rue St. Jacques; London: Paul Vaillant]

br: 226 x 240 mm.
pl: 242 x 275 mm.

Engravers: (map) Guill. De-la-Haye

RF: c.1:

Scales: Millaria Romana

Comments: outside border, top rt.: "Tom.III, Pag. 5"

Book: C. Cornelii TACITI OPERA recognovit, emendavit, supplementis explevit, Notis, Dissertationibus Tabulis Geographicis Illustravit

AP:
AU:
NAP:

Gabriel Brotier (Paris, 1771)

Location: BL: 53.g.1-4

332 **Savoy** 1748 R

Duché de / SAVOYE / Par le Sr. Robert Géog? ord? du Roi / Avec Privilege / 1748.

br: 162 x 188 mm.
pl: 166 x 254 mm.

Engravers:

RF: c.1:1,000,000

Scales: 34 mm. = 20 Milles d'Italie; 6 1/2 Lieues d'une heure

Comments:

Book:

AP: 134
AU:
NAP:

Location:

333 **Savoy** 1751 RV

Duché de SAVOYE / qui comprend / Le Chablais, Le Fossigny, Le Génevois, / La Savoye Propre, La Tarentaise et / La Maurienne; / dressé d'après la Grande Carte / de Piemont de Tomaso Borgomo et autres, / Par le Sr. Robert de Vaugondy. / Géographe ord? du Roy / Avec Privilege 1751.

br: 483 x 517 mm.
pl: 502 x 538 mm.

Engravers:

RF: c.1:325,000

Scales: 85 mm. = 15 Mille Pas geometriques de 60 au Degré; 12 1/2 Milles de Piedmont de 50 au Degré; 5 Lieues de France et d'une heure de 20 au Degré

Comments
State 1: 1751: title: "La Morienne dressé d'après la carte manuscrite de la Lombardie de Mr Honstein"; no "géographe ordinaire du roi"; no frame around scale; no "DAUPHINÉ (lower left); no cardinal points

State 2: 1751: title: "...La Maurienne dressé d'après la Grande Carte de Piedmont de Tomaso Borgonio [sic] et autres"; "géographe ordinaire du roi";

333	**Savoy** (continued)

State 3: 1751: Title: "...Tomaso Borgomo"; "DAUPHINÉ" added; cardinal points and frame around scale added

Book:

AP:
AU: 82
NAP:

Location:

334	**Lombardy, east**	1750	R

Partie Orientale de / La LOMBARDIE / qui comprend / la République de Venise, / le Mantouan, le Cremonese, / l'Evêché de Trente, / et les Confins / du Duché de Toscane, / et de l'Etat de l'Eglise. / Par Le Sr. Robert, / Géographe Ordinaire du Roi, / Avec Privilége 1750.

br: 476 x 550 mm.
pl: 495 x 626 mm.

Engravers:

RF: c.1:725,000

Scales: 77 mm. = 30 Milles Communes d'Italie de 60 au Degré; 10 Lieues d'une heure de 20 au Degré

Comments:

Book:

AP:
AU: 81
NAP:

Location:

335	**Lombardy, west**	1750	R

Partie Occidentale / de la LOMBARDIE / et pays circonvoisins, où sont / les Etats de / Savoye, Piémont, Milan, / Gênes, Plaisance &c. / Par Le Sr. Robert, / Géographe ordinaire du Roy. / Avec Privilége. / 1750.

br: 478 x 556 mm.
pl: 502 x 634 mm.

Engravers: (map) Guill? Delahaye Sculp.

RF: c.1:725,000

Scales: 74 mm. = 30 Milles d'Italie de 60 au degré; 25 Milles de Piémont de 50 au degré; 33 1/2 Milles de Gênes et Milan de 67 au degré; 10 Lieues d'une heure de 20 au degré

Comments: n.b.: scale has no frame

Book:

AP:
AU: 80
NAP:

Location:

336	**Genoa**	1748	RV

Carte Générale / de / l'Etat de GÊNES / Par le Sr. Robert de Vaugondy Fils de Mr. Robert / Géog. ord. du Roi. Avec Privilége. / 1748.

br: 164 x 307 mm.
pl: 172 x 337 mm.

Engravers: (map) Gravé par Guill? Delahaye.

RF: c.1:860,000

Scales: 42 mm. = 20 Milles d'Italie

Comments: Finder map for the following nine large scale maps from the AP [Cat. 337-345]

In title : "Avertissement: Les carreaux marqués par des traits forts et chifres [sic], represente[n]t autant de cartes particulieres et des plus détaillées"

336	**Genoa** (continued)

Book:

AP: 143
AU:
NAP:

Location:

337	**Genoa 1**	1748	RV

I. Carte de l'Etat de Genes où se trouvent NICE, VINTIMILLE, S. REMO &c. / Par le Sr. Robert de Vaugondy. Avec Priv. 1748.

Inset: *Environs de NICE et de VILLE FRANCHE.*

br: 152 x 175 mm.
pl: 167 x 182 mm.

Engravers: (map) Gravé par / Guill. Delahaye

RF: c.1:290,000

Scales: 38 mm. = 2 Lieues d'une heure de Chemin;
Inset: 27 mm. = Echelle de 600 Toises

Comments:

Book:

AP: 144
AU:
NAP:

Location:

338	**Genoa 2**	1748	RV

II. Carte des Confins de l'Étât de Gênes où se trouve une partie du Cté. de NICE, / Par le Sr. Robert de Vaugondy. avec Priv? 1748.

br: 160 x 150 mm.
pl: 175 x 158 mm.

Engravers:

RF: c.1:290,000

Scales: 38 mm. = 2 Lieues d'une heure de chemin

Comments:

Book:

AP: 145
AU:
NAP:

Location:

339	**Genoa 3**	1748	RV

[III] Carte de l'Étât de Gênes / où se trouvent / la Principauté d'ONEILLE, les enviro[n]s / d'ALBINGA, / le Marquisat de FINALE &c. / Par le Sr. Robert de Vaugondy, Fils de Mr. Robert / Geog? ordin? du Roi. Avec Priv. 1748.

br: 162 x 199 mm.
pl: 168 x 203 mm.

Engravers:

RF: c.1:290,000

Scales: 59 mm. = 3 Lieues d'une heure.

Comments: III outside border, top left

Book:

AP: 146
AU:
NAP:

Location:

340 **Genoa 4** 1748 RV

IV. Carte de l'Etat de Gênes ou se trouvent les environs de SAVONE partie / du Marquisat de Finale du PIÉMONT. Par le Sr. Robert de Vaugondy. AvecPriv. 1748.

br: 153 x 200 mm.
pl: 169 x 209 mm.

Engravers:

RF: c.1:290,000

Scales: 39 mm. = 2 Lieues d'une heure de Chemin

Comments:

Book:

AP: 147
AU:
NAP:

Location:

341 **Genoa 5** 1748 RV

V. Carte des Environs de Genes où se trouvent ARENZANO, NERVI, RAPALLO, la BOCHETTA &c. / Par le Sr. Robert de Vaugondy. avec Privilege 1748.

br: 153 x 198 mm.
pl: 170 x 210 mm.

Engravers: (map) Gravé par Guill. Delahaye

RF: c.1:290,000

Scales: 77 mm. = 12 Milles d'Italie de 60 au Degré.; 13 Milles de Gênes de 65 au Degré.; 4 Lieues d'une heure de Chemin de 20 au Degré.

Comments:

Book:

AP: 148
AU:
NAP:

Location:

342 **Genoa 6** 1748 RV

VI. Carte des Confins de l'État de Gênes où sont NOVI, GAVI, et OTTAGIO; avec les envi- / rons d'ALEXANDRIE, TORTONE, et VALENCE. Par le Sr. Robert de Vaugondy. 1748. avec Privilege.

br: 155 x 163 mm.
pl: 168 x 171 mm.

Engravers:

RF: c.1:290,000

Scales: 58 mm. = 3 Lieues d'une heure de Chemin

Comments: n.b.: scale inserted between border and neat line

Book:

AP: 149
AU:
NAP:

Location:

343 **Genoa 7** 1748 RV

VII. Carte des Confins de l'Étât de Gênes où se trouve une partie des Fiefs Imperiaux, / du MILANEZ et du PARMESAN. Par le Sr. Robert de Vaugondy. Avec Pr? 1748.

br: 154 x 188 mm.
pl: 174 x 195 mm.

Engravers: (map) Gravé par Guil. / Delahaye.

RF: c.1:290,000

Scales: 38 mm. = 2 Lieues d'une heure de Chemin

Comments:

343 **Genoa 7** (continued)

Book:

AP: 150
AU:
NAP:

Location:

344 **Genoa 8** 1748 RV

VIII. Carte de l'Etat de Genes ou se trouvent les environs et la VALLÉE DE LAVAGNA; et les / Confins du PARMESAN, et de la TOSCANE. Par le Sr. Robert de Vaugondy. Avec Privilege 1748.

br: 156 x 199 mm.
pl: 172 x 208 mm.

Engravers:

RF: c.1:290,000

Scales: 39 mm. = 2 Lieues d'une heure de Chemin

Comments:

Book:

AP: 151
AU:
NAP:

Location:

345 **Genoa 9** 1748 RV

IX. Carte de l'Étât de Gênes ou se trouvent SESTRI -di Levante, SPETIA et SARZANA. Par le Sr. Robert de Vaugondy. 1748. Avec Priv?

br: 151 x 207 mm.
pl: 172 x 239 mm.

Engravers:

RF: c.1:290,000

Scales: 58 mm. = 3 Lieues d'une heure de Chemin.

Comments:

Book:

AP: 152
AU:
NAP:

Location:

346 **Venice, Republic of** 1748 RV

Etat de la Seigneurie et / Republique de VENISE / avec l'Evêché de TRENTE / Par le Sr. Robert de Vaugondy, fils de M. Robert / Geog? ord? du Roi. Avec Privil? 1748.

br: 166 x 376 mm.
pl: 176 x 392 mm.

Engravers: (map) Guill? Delahaye

RF: c.1:1,150,000

Scales: 45 mm. = 30 Mille pas Geometriques; 10 Lieues d'une heure

Comments:

Book:

AP: 154
AU:
NAP:

Location:

347	**Po River**	1742	R

[La Carte du COURS DU PO, depuis Pavie jusqu'à Ferrare]

Engravers:

RF: c.1:

Scales:

Comments: not seen in any location.
L-D, (1745): "Cette Carte est trés-grande & trés-detaillée, elle comprend partie du Milanois, du Bergamsc, du Bressan, du Vernois, du Vicentin, le Cremasc, le Cremonois, le Mantouan, Partie des Duchés de Plaisance, de Parme, de Modene, etc."

Book:

AP:
AU:
NAP:

Location: L-D (1745), p. 246; Robert, Intro. à la Géographie (1743); RV Catalogue (1777)

348	**Po River 1**	1748	RV

Carte des SOURCES DU PO où se trouve la Partie Meridle. de la Printé. de Piémont Par le Sr. Robert de Vaugondy. A. Pr? du Roi. 1748.

br: 162 x 209 mm.
pl: 166 x 252 mm.

Engravers:

RF: c.1:480,000

Scales: 23 mm. = 2 Lieues d'une heure

Comments: map number engraved as 130 instead of 136

Book:

AP: 136
AU:
NAP:

Location:

349	**Po River 2**	1748	R

II. Cours du PO et partie du PIÉMONT où sont les environs de TURIN / SUZE, PIGNEROL &c Par le Sr. Robert, Géog? ord? du Roi. A.Priv. 1748.

br: 172 x 145 mm.
pl: 176 x 246 mm.

Engravers:

RF: c.1:480,000

Scales: 27 mm. = 2 Lieues d'une heure

Comments:

Book:

AP: 137
AU:
NAP:

Location:

350	**Po River 3**	[1748]	[R]

III. COURS DU PO où sont les environs de Chivazzo, Verrue, Casale Valence, Asti Alexandrie Tortone, Invree, Verceil Novarre &c.

br: 161 x 182 mm.
pl: 174 x 245 mm.

Engravers:

RF: c.1:480,000

Scales:

Comments:

350	**Po River 3** (continued)

Book:

AP: 138
AU:
NAP:

Location:

351	**Po River 4**	[1748]	RV

IV. COURS DU PO où sont les environs de Pavie, Plaisance, Cremone, Picighitone, Lodi, Milan. Par le S. Robert de Vaugondy.

br: 171 x 184 mm.
pl: 176 x 246 mm.

Engravers:

RF: c.1:480,000

Scales: 23 mm. = 2 Lieues d'une heure

Comments:

Book:

AP: 139
AU:
NAP:

Location:

352	**Po River 5**	1748	RV

V. COURS DU PO, où sont les environs de Parme, Guastalle, Mantoue &c. Par le Sr. Robert, de Vaugondy APDR. 1748.

br: 163 x 182 mm.
pl: 166 x 253 mm.

Engravers:

RF: c.1:480,000

Scales: 23 mm. = 2 Lieues d'une heure

Comments:

Book:

AP: 140
AU:
NAP:

Location:

353	**Po River 6**	1748	RV

VI. COURS DU PO, ou sont les environs de la Mirandole, partie du Veronese, Padouan, / Ferrarese, Bolonois, &c. Par le Sr. Robert, de Vaugondy, avec Priv. 1748.

br: 170 x 183 mm.
pl: 176 x 245 mm.

Engravers:

RF: c.1:480,000

Scales: 34 mm. = 3 Lieues d'une heure

Comments:

Book:

AP: 141
AU:
NAP:

Location:

354 **Po River 7 (mouth)** 1748 RV

VII. EMBOUCHURE DU PO, où sont les environs de Rovigo, d'Adria Ariano &c.

br: 170 x 185 mm.
pl: 175 x 245 mm.

Engravers:

RF: c.1:480,000

Scales: 47 mm. = [12] Mille Pas Geometriques; 10 Milles de Piemont de 58 au Deg?; 13 Milles de Genes de 65 au Deg?; 4 Lieues d'une heure de 20 au Deg?

Comments: a second title in cartouche, at right: COURS DU PO / depuis ses Sources, / jusqu'à son Embouchure. / Par le Sr. Robert de Vaugondy. / fils de M. Robert, Geog? ord? du Roi. / Avec Privilege. / 1748.

Book:

AP: 142
AU:
NAP:

Location:

355 **Tuscany, Papal States** 1748 R

ETAT DE L'EGLISE / et Duché / de TOSCANE / Par le Sr. Robert Géog? ordin? / du Roi / 1748.

br: 160 x 198 mm.
pl: 168 x 209 mm.

Engravers:

RF: c.1:3,000,000

Scales: 37 mm. = 20 Lieues d'une heure de chemin.

Comments:

Book:

AP: 155
AU:
NAP:

Location:

356 **Tuscany, Papal States, Corsica** 1750 R

ETAT DE L'EGLISE, / GRANDE DUCHÉ DE TOSCANE, / et / ISLE DE CORSE. / Par le Sr. Robert Géographe ordin. du Roy. / Avec Privilege. / 1750.

br: 484 x 554 mm.
pl: 499 x 635 mm.

Engravers:

RF: c.1:930,000

Scales: 120 mm. = 60 Milles d'Italie de 60 au Degré; 20 Lieues d'une heure de 20 au Degré

Comments: R. Almagià, Documenti Cartografici dello Stato Pontificato, (Vatican Library, 1960), p. 46: "questa carta es per quanto reguarda la figura la migliore che sia stata composta nel periodo, pre-geodetico, ma di fatto essa esce gia da questo periodo, perche, stando a quanto l'autore ci dice utilizza alcuni dati della carta Maire e Boscovich: in che consista questa utilizzasioni non puo per altro dirsi con sicurezza."

Book:

AP:
AU: 83
NAP:

Location:

357 **Tuscany, Papal States, Corsica** 1762 RV

LA LOMBARDIE, / LE DUCHÉ DE TOSCANE / L'ETAT DU ST. SIEGE, / et L'ISLE DE CORSE / Par le S. Robert de Vaugondy / Géog? ordinaire du Roy &c. / Avec Privilege 1762

A Paris chés l'Auteur Quai de l'Horloge près le Pont neuf

br: 243 X 285 mm.

Engravers: (map) Gravé par E. Dussy

RF: c.1:2,500,000

Scales: 44 mm. = 60 Milles Italie de 60 au degré

Comments:
State 1: 1762

State 2: [1778]: privilege, date and imprint removed; Fortin publication

Book:

AP:
AU:
NAP: 31

Location:

358 **Naples, kingdom of** 1762 RV

Royaume de / NAPLES / Par le S. Robert de Vaugondy / Avec Privilege 1762.

Inset: *[Sicile]*

A Paris chès l'Auteur Quai de l'Horloge près le Pont neuf.

br: 244 x 285 mm.
inset: 112 x 163 mm.

Engravers: (map) Gravé par E. Dussy
(cartouche) Arrivet inv. & sculp.

RF: c.1:2,500,000

Scales: 45 mm. = 60 Milles d'Italie de 60 au degré

Comments:
State 1: 1762:

State 2: 1776: title changed: "Royame de / Naples / et de Sicile / Par le Sr. Robert de Vaugondy / Géog? ord. du Roi et / Censeur royal / 1776"; Groux signs cartouche; imprint removed

Book:

AP:
AU:
NAP: 32

Location:

359 **Naples, Kingdom of, north** 1748 R

Partie Septentrionale / du / ROYAUME DE NAPLES. / Par le Sr. Robert Geog? ord? du Roi. / 1748.

br: 160 x 197 mm.
pl: 167 x 207 mm.

Engravers: (map) Gravé par G. Delahaye

RF: c.1:2,600,000

Scales: 43 mm. = 20 Lieues d'une heure de Chemin.

Comments: graticule shown

Book:

AP: 156
AU:
NAP:

Location:

360 **Naples, kingdom of, north** 1750 R

Partie Septentrionale du / ROYAUME / DE NAPLES, / Par le Sr. Robert Geographe ordin. du Roy. / Avec Privilege. / 1750.

br: 478 x 589 mm.
pl: 495 x 624 mm.

Engravers:

RF: c.1:950,000

Scales: 115 mm. = 60 Mille Pas geométriques ou Milles d'Italie; 20 Lieues d'une heure de 20 au degré

Comments:

Book:

AP:
AU: 84
NAP:

Location:

361 **Naples, Kingdom of, south** 1748 R

Partie Méridionale / du Royaume de / NAPLES où se trouvent / la Calabre et la Sicile / Par le Sr. Robert Geog? du Roi / 1748.

br: 158 x 197 mm.
pl: 169 x 208 mm.

Engravers: (map) Gravé par Guill? Delahaye

RF: c.1:2,600,000

Scales: 43 mm. = 20 Lieues d'une heure de Chemin.

Comments: graticule of long. and lat. visible

Book:

AP: 157
AU:
NAP:

Location:

362 **Naples, kingdom of, south** 1750 R

Partie Méridionale / du ROYAUME DE NAPLES, / où se trouvent / La Calabre et l'Isle / et Royaume de SICILE. / Par le Sr. Robert / Géographe ordinaire du Roi / Avec Privilege. / 1750.

Inset: *Supplément / pour / l'ISLE DE MALTE*

br: 478 x 595 mm.
pl: 495 x 628 mm.
Inset: 150 x 150 mm.

Engravers:

RF: c.1:950,000

Scales: 115 mm. = 60 Mille Pas Geometriques ou Milles d'Italie de 60 au degré; 20 Lieues d'une heure de 20 au degré

Comments:

Book:

AP:
AU: 85
NAP:

Location:

363 **Sardinia** 1748 RV

Isle et Royaume / de SARDAIGNE / Par le Sr. Robert de Vaugondy / fils de Mr. Robert Géographe / ord. du Roi / avec Privilege 1748.

br: 209 x 163 mm.
pl: c.232 x 176 mm.

Engravers:

RF: c.1:1,480,000

363 **Sardinia** (continued)

Scales: 38 mm. = 10 Lieue [sic] d'une heure

Comments:

Book:

AP: 135
AU:
NAP:

Location:

364 **Sicily, ancient** 1743 R

Tabula Geographica SICILIAE VETERIS quae Sicania Trinacria eademque Solis Insula et Cyclopum Terra / variis dicitur auctoribus in qua notantur Itineraria Romana, conatibus Geographicis AEg. Robert Geographi Regii /

Inset: *Veterum SYRACUSARUM Typus Ex Cluvero*

Parisiis Apud Auctorem in ripa Sequanae vulgo quay de l'horloge dictâ, 1743, Cum privilegio Regis

br: 472 x 640 mm.

Engravers:

RF: c.1:815,000

Scales:

Comments: Roman roads and distances in Roman numerals
WLCL: no scales, engraver, cartouche; no long. or lat. numbers; lettering on guidelines not completely erased; early state?
Julien Cat. (1763): price, 1#

Book:

AP:
AU:
NAP:

Location: WLCL: AF (59); Julien Cat.(1763): 1310; Robert, Intro. à la Géographie (1743)

365 **Malta** 1748 R

Carte des Isles de MALTE et de GOZE. Par le Sr. Robert Geog? ord? du Roi. Avec Priv? / 1748.

Inset: *Isle de Goze*

br: 156 x 212 mm.
pl: 177 x 223 mm.
inset: 38 x 55 mm.

Engravers:

RF: c.1:

Scales: 82 mm. = 10 Milles d'Italie; 4 Lieues communes de France; Inset: 27 mm. = 10 Milles d'Italie.

Comments: Map lacks usual neat line with long./lat. divisions; compass shows directions as map is oriented to northeast

Book:

AP: 158
AU:
NAP:

Location:

366 **Greece, Ottoman** 1748 R

Carte / de la TURQUIE / D'EUROPE / Par le Sr. Robert Géog. du Roi. / avec Privilege / 1748.

br: 161 x 196 mm.
pl: 165 x 252 mm.

Engravers:

RF: c.1:10,500,000

366 **Greece, Ottoman** (continued)

Scales: 41 mm. = 80 Lieues d'une heure

Comments:

Book:

AP: 159
AU:
NAP:

Location:

367 **Greece, Ottoman** 1755 R

TURQUIE / EUROPÉENNE / Par le Sr. Robert / Geographe ordinaire du Roy. / Avec Privilege. 1755

br: 478 x 452 mm.
pl: 515 x 488 mm.

Engravers: (cartouche) Cht. haussard Sculp.

RF: c.1:2,800,000

Scales: 117 mm. = 180 Millespas [sic] geometriques de 60 au degré; 100 Milles de Turquie de 66 2/3 au degré; 40 Milles de Hongrie de 13 1/3 au degré

Comments:

Book:

AP:
AU: 86
NAP:

Location:

368 **Greece, Ottoman** 1762 RV

HONGRIE, / TURQUIE EUROP?/ Par le S. Robert de Vaugondy / Géog? ordinaire du Roy &c. / 1762

A Paris ches l'Auteur quai de l'Horloge près le Pont Neuf

br: 243 x 203 mm.

Engravers: (map) Gravé par E. Dussy
(cartouche) Arrivet inv. & Sculp.

RF: c.1:7,500,000

Scales: 46 mm. = 180 Mille pas géom? de 60 au degré; 60 Lieues d'une heure

Comments:
State 1: 1762

State 2: [1778]: date and imprint removed; Fortin publication

Book:

AP:
AU:
NAP: 33

Location:

369 **Greece, historical** 1752 RV

GRAECIA VETUS / ex schoedis Sansonianis desumpta, / in quâ / MACEDONIA, THESSALIA, EPIRUS, / ACHAIA et PELOPONESUS, / in minores partes seu populos / distinguuntur; / nec non inter adjacentes insulas speciatim / CRETA delineatur Insula, / Autore Robert de Vaugondy filio, / Geographo Regis ordinairio. / Cum Privilegio. / 1752.

br: 480 x 545 mm.
pl: 492 x 556 mm.

Engravers: (cartouche) E. haussard Sculpcit

RF: c.1:1,700,000

Scales: 129 mm. = 150 Milliaria Romana quorum 75 gradum efficiunt; 1200 Stadia Graeca quorum 600 ad gradum unum requiruntur

369 **Greece, historical** (continued)

Comments:
State 1: 1752

State 2: [1768]: Haliacmon Fl. changes course; changes in lettering (italic to Roman and vice versa) e.g., Amphipolis (in Macedonia), Gnossus to Cnossus (on Crete); place name changes e.g. Sermylia to Olynthus (in Macedonia)

Book:

AP:
AU: 10
NAP:

Location:

370 **Greece, Crete** 1770 RV

L'ISLE DE CANDIE / nommée par les Turcs / I CRITI / extraite de la carte venitienne de Boschini / par le Sr. Sanson Geographe ord. du Roi; / Assujettie aux Observations astronomiques / et augmentée d'un supplément pour les principales Isles / de l'Archipel / Par le Sr. Robert de Vaugondy Géog. ord. du Roi; / de feu S.M.Polonoise, Duc de Lorraine et de Bar, / et de l'Academie Royale des Sciences et Belles Lettres de Nanci; / Avec Privilege 1770

A Paris Quai de l'Horloge du Palais près le Pont Neuf

br: 390 x 545 mm.;
pl: 402 x 556 mm.
inset: 151 x 408 mm.

Engravers:

RF: c.1:

Scales: 100 mm. = 30 Mille pas geometriques de 60 au degré; 10 Lieues marines, ou d'une heure de 20 au degré

Comments: Inset: no title; shows principal islands of the Dodecanese
"La carte venitienne de Boschini" = 'Il regno tutto di Candia.' by Marco Boschini, Venice: Francesco Nicolini, 1651.
listed in RV, Catalogue (1777)

Book:

AP:
AU:
NAP:

Location: BL: K.113.112

371 **Asia** [1740] R

l'ASIE / Divisée en ses principales Parties / ou / Les Empires Roiaume / Etats et Peuples / sont distingués les uns des autres / Par le Sieur Sanson / Géographe ordinaire du Roy / Revue et changée en plusieurs endroits / sur les Mémoires les plus recents / et assujetie aux Observations astronomiques / Par le Sr. Robert Géog. ord. du Roy

BL: Amsterdam: Chez Henry de Leth / à l'enseigne du Pecheur

br: 486 x 599 mm.
pl: 491 x 600 mm.

Engravers:

RF: c.1:

Scales: 80 mm. = 900 Mille pas géometriques ou Milles d'Italie; 3250 Li, ou Stades de la Chine dont 250 font un degré de Latitude; 360 Cos, ou Lieues communes du Mogol, dont 24 font un Degré; 300 Persanges, ou Lieues communes de Perse de 20 au degré; 420 Lieues communes du Japon, de 30 au Degré; 24 Diètes ou Journées de chemin

Comments: only Amsterdam imprint seen
Climates along right hand side

Book:

AP:
AU:
NAP:

371 **Asia** (continued)

Location: BL: 46820 (37), Amsterdam, dated 1730?; BN:Ge C. 8036 (Sanson; dated 1739); Robert, Intro. à la Geog. (1743): dated 1740; L-D (1745), pp. 245-247.

372 **Asia** 1748 R

L'ASIE / Divisée selon ses differens Etats. / Par le Sr. Robert. / Géographe ordinaire du Roi. / Avec Privilege 1748.

br: 162 x 198 mm.
pl: 170 x 206 mm.

Engravers: (map) Gravé par Chambon.

RF: c.1:57,600,000

Scales: 37 mm. = 400 Lieues d'une heure.

Comments:
State 1: 1748

State 2: 1754: title changed: 'Par le Sr. Robert de Vaugondy / Géogr. ord. du Roy. / avec Privilege 1754'

Book:

AP: 160
AU:
NAP:

Location: L.C. Smith Collection, USM: State 2 (1754)

373 **Asia** **1749** R/RV

L'ASIE / divisée en ses principaux / Empires et Royaumes / Par les Srs. Sanson Géogr. Ordin? du Roi. / assujettie aux observations astronomiq. / et corrigée pour la Russie / d'après les Cartes de l'Atlas Russien / par le Sr. Robert Geogr? ordin? du Roi / Avec Privilege 1749.

A Paris Ches l'Autheur / Quay de l'horloge près le Pt. Neuf

br: 495 x 652 mm.
pl: 502 x 670 mm.

Engravers: (cartouche) Gobin sculpsit

RF: c.1:20,000,000

Scales: 55 mm. = 600 mille pas geometriques de 60 au Degré; 666 Milles Turques de 66 2/3 au Degré; 222 Parasangs Persiques de 22 2/9 au Degré; 300 Milles Indiens de 30 au Degré; 567 Milles Arabes de 56 2/3 au Degré; 1040 Woertes de Russie de 104 au Degré; 2500 Lys Chinois de 250 au Degré; 200 Lieues d'une heure et de Marine de 20 au D.

Comments: Table to right

State 1: 1749: R. as author; Gobin cartouche; six unlabeled islands west of Archipelago de St. Lazar (25-27° N, 155° W); "Terre vue en 1741 par M. Tchirikow" (52°N, 187°E)

State 2: 1771: as per RV, Notice Raisonnée, (1774), a re-edition of 1749; RV as author; cartouche: Arrivet inv. et sculp.

State 3: 1776: as per RV, Catalogue, (1777)

State 4: 1778: imprint: Chès Fortin Ingr. / Mécanicien du Roy Rue de la Harpe

State 5: 1781: Delamarche imprint

Book:

AP:
AU:
NAP:

Location: BN: Ge.C.8037 (1749); Ge.C.8048 (1776?); Ge.C.8049; Ge.D.20608 (1778); Ge.D.11923 (without table); Ge.D.11924, 11924A (Delamarche); Ge.A.888 (Delamarche); WLCL: Julien, Theatre du Monde, I (5) [1749]

374 **Asia** 1750 RV

ASIE / Par le Sr. Robert. / De Vaugondy. / Avec Privilege 1750.

br: 150 x 158 mm.

374 **Asia** (continued) 1750 RV

Engravers:

RF: c.1:80,000,000

Scales: 32 mm. = 200 lieues d'une heure

Comments: "Pag. 389" outside border, top right

Book: Méthode Abrégée et facile pour apprendre la Géographie

AP:
AU:
NAP:

[l'Abbé Le François] Paris: Libraires Associés, 1770

Location: WLCL: C2 1770 Le

375 **Asia** 1750 RV

Carte de L'ASIE / dressée sur les Relations les plus nouvelles, / principalement / sur les Cartes de Russie, de la Chine, et / de la Tatarie Chinoise; / et divisée en ses Empires et Royaumes, / Par le Sr. Robert de Vaugondy, / fils de Mr. Robert, Géogr. ordin. du Roy, / Avec Privilege 1750.

br: 470 x 525 mm.
pl: 500 x 552 mm.

Engravers: (map) Guill. Delahaye Sculpsit.

RF: c.1:23,000,000

Scales: 67 mm. = 900 Mille pas géometriques de 60 au degré; 1000 Milles Turques de 66 2/3 au degré; 333 Parasangs Persiques de 22 2/3 au degré; 450 Milles Indiennes de 30 au degré; 850 Milles Arabes de 56 2/3 au degré; 1664 Voerstes de Russie de 104 au degré; 3750 Las [sic] Chinois de 250 au degré; 300 Lieues d'une heure et de Marine de 20 au degre

Comments:
State 1: 1750: many islands south of Japan

State 2: 1751: date changed; islands south of Japan removed [some examples without islands are dated 1750, e.g. BN: GeDD 282; Ge.DD 4796 (114)

State 3: 1775: (BN: Ge D. 1171bis)

State 4: 1776: date changed; some islands restored south of Japan, e.g. Gr.Liqueo

Book:

AP:
AU: 87
NAP:

Location:

376 **Asia** 1758 RV

L'ASIE divisée en ses empires et royaumes. Par le Sr. Robert de Vaugondy

Paris: Chez le Delamarche [LC]

four sheets: 980 x 1120 mm. assembled

Engravers: (cartouche) Arrivet inv. & sculp.

RF: c.1:9,000,000

Scales:

Comments:
State 1? 1758 (as per LC description)

State 2? 1760: as per Julien Cat.

State 3? 1775 (BL)

State 4? after 1786: Delamarche imprint

Julien Cat. (1763) no. 37: gives date 1760; price, 4#

Book:

AP:
AU:
NAP:

Location: BL: K.114.20.2.TAB (1775); LC: G7400 1758.R6 Vault

377 **Asia** [1762] RV

L'ASIE / Par le S. Robert de Vaugondy Géog? / ord. du Roi, de S.M. Polon. Duc de / Lorr. et de Bar, et de la Société / royale de Nanci. / 1762

A Paris, ches l'Auteur Quai de l'Horloge prés le pont neuf

br: 242 x 275 mm.

Engravers: (map) Gravé par E. Dussy
(cartouche) Arrivet inv. & Sculp.

RF: c.1:40,500,000

Scales: 50 mm. = 1200 Mille pas géométriques de 60 au degré; 400 Lieues marines ou d'une heure

Comments:
State 1: 1762

State 2: [after 1773]: "et Censeur Royal." added to title; imprint removed

[State 3: after 1778? Groux cartouche?; Fortin publication?]

Book:

AP:
AU:
NAP: 34

Location:

378 **Asia, historical** 1770 RV

ASIA / ex Cornelio Tacito. / Auctor Robert de Vaugondy Geog? Regis / et regiae Academiae Nanceianae / Socius. / 1770

[Book: Paris: L.F.Delatour, rue St. Jacques; London: Paul Vaillant]

br: 237 x 254 mm.
pl: 322 x 336 mm.

Engravers: (map) Guill? DelaHaye

RF: c.1:

Scales: 60 mm. = 525 Milliaria Romana

Comments:

Book: C. Cornelii TACITI OPERA recognovit, emendavit,supplementis explevit, NOTIS, DISSERTATIONIBUS TABULIS GEOGRAPHICIS ILLUSTRAVIT

AP:
AU:
NAP:

Gabriel Brotier (1771)

Location: BL: 53.g.1-4

379 **Black Sea** 1769 RV

Carte des Environs / de la MER-NOIRE / où se trouvent / l'Ukrayne, la Petite Tartarie / la Circassie, la Georgie, / et les confins de la Russie Europeenne, / et de la Turquie / Dediée et présentée / à Monseigneur le Duc de Choiseul / pair de France Ministre et Secretaire d'Etat, / par Robert de Vaugondy Géographe ord. du Roi / du feu Roi de Pologne Duc de Lorr. et de Bar, / et de la Société royale de Nancy / 1769.

Paris: Chez l'Auteur Quai de l'Horloge du Palais pres le Pont Neuf/ avec privilege

br: 520 x 925 mm. (two sheets assembled)

Engravers: (map) E. Dussy Sculpsit
(cartouche) Arrivet inv. & sculp.

RF: c.1:2,500,000

Scales: 101 mm. = 15 Mille pas géométriques de 60 au degré; 5 Lieues marines ou d'une heure de chemin; 90 mm. = 21 Werst ou Milles de Russie

Comments: map printed perhaps after Russian invasion of Crimea in 1769.

Book:

AP:
AU:
NAP:

Location: Bod: Douce Prints E.2.4(19); BL: K.112.106; 35945 (6)

380 **Near East** 1761 RV

TURQUIE D'ASIE, / ARABIE, PERSE, / TARTARIE / INDEPENDANTE / Par le S. Robert de Vaugondy / Géog? ordinaire du Roy &c. / Avec Privilege 1761.

A Paris, chés l'Auteur Quai de l'Horloge près le Pont neuf

br: 243 x 220 mm.

Engravers: (map) Gravé par E. Dussy
(cartouche) Arrivet inv. & Sculp.

RF: c.1:20,000,000

Scales: 52 mm. = 600 Mille pas géométriques de 60 au degré; 200 Lieues d'une heure de 20 au degré

Comments:
State 1: 1761

State 2: [1778]: imprint removed; privilege and date removed

Book:

AP:
AU:
NAP: 35

Location:

381 **Near East, historical** 1737 R

Carte pour servir la lecture / de l'HISTOIRE ANCIENNE des Egyptiens, des Carthaginois, des Assyriens / des Babyloniens, des Mèdes, des Perses, / des Macédoniens, des Grecs. / Par M. Rollin ancien recteur de l'Université de Paris, / Professeur d'Eloquence au College Royale et Associé / à l'Académie Royale des Inscriptions et Belles-Lettres. / Dressée par le Sr. Robert Geographe ordinaire du Roy. / 1737 / Partie Occidentale de la Carte des Anciennes Monarchies.

A Paris rue St. Jacques dans la maison de Mr Delespine, à l'image St. Paul

two sheets:
Partie Occidentale: br: 505 x 656 mm.
Partie Orientale: br: 510 x 657 mm.

Engravers: (map) Gravé par Le Tellier

RF: c.1:7,500,000

Scales: 93 mm. = 450 Milles Anciens de 75 au Degré; 4000 Stades de 700 au Degré; 140 Parasanges de 23 1/2 au Degré et de Stades Chacune; 150 Lieues de 25 au Degré et de 28 Stades Chacune

Comments: Second sheet continues title: *Partie Orientale de la Carte des Anciennes Monarchies* with inset: Supplément pour l'Asie et l'Afrique
No cartouche; title along top
Partie Occidentale has inset: Supplément pour la Grece and Table for ancient empires
Listed in Robert, Intro. à la Géog. (1743)

State 1: 1737

State 2: 1768

State 3: (after 1788) Delamarche imprint

Book:

AP:
AU:
NAP:

Location: WLCL: AF (8,9); BN: Ge C. 8659; Ge C. 8660; Ge C. 8661 (dated 1768); Ge CC.1358 (12,13) (publ. Delamarche)

382 **Near East, historical** 1753 RV

ANTIQUOR? IMPERIORUM / Tabula, / in qua prae ceteris, / MACEDONICUM seu ALEXANDRI / MAGNI IMPERIUM / et EXPEDITIONES / exarantur, / Conatibus D. Robert de Vaugondy / Geographi Regii. / 1753 / Cum Privilegio.

Inset: *[India]*

br: 470 x 620 mm.;
pl: 487 x 624 mm.
inset (lower right): 128 x 145 mm.

Engravers: (cartouche) El. Haussard fecit

382 **Near East, historical** (continued)

RF: c.1:8,000,000

Scales: 69 mm. = 375 Millaria Romana quorum 75 gradui respondent; 1590 Stadia Aegyptiaca quorum 318 in gradu requiruntur; 78 mm. = 6000 Stadia Aristotelica quorum 1111 gradui respondent; Inset: 27 mm. = 5555 Stadia quorum 1111 Gradui respond?

Comments:

State 1: 1753

State 2: [1768]: course of Nile changed between Trups and Cambysi Aerarium; straightened between T and H of Aethiopia; Zames Mtes added in Arabia Felix
Inset: MONITUM. Ut Scopus hujus tabulae possit omnino adimpleri, non tantum quoad Scilicet Sesostris et Alexandri Magni expeditiones, sed et ut omne Graecorum Imperium ad mortem praesertim Seleuci I. representetur INDIAM hic revocandam duximus.
AU 'Préface,' p. 20: this map includes the following ancient empires: 1) Assyria (2288 BC); 2) Sesotrip (1417 BC); 3) Assyria (900 BC) — Medes, Babylo-Chaldean, Lydians; 4) Perses (522 BC) — Darius Hystapes; 5) Macedonian (331 BC) — Alexander; 6) Ptolemy Cassandrae, Lysimachus, Seleucus (306 BC) — Alexander's generals; 7) Parthians (140 BC)

Book:

AP:
AU: 3
NAP:

Location:

383 **Near East, historical** 1762 RV

Carte / des Premiers Ages / du Monde. // Géographie Ecclésiastique / de la TURQUIE d'ASIE, / et de la PERSE. / Par le S. Robert de Vaugondy / Géog? ordinaire du Roy &c.

A Paris, ches l'Auteur Quai de l'Horloge près le pont neuf

br: 246 x 249 mm.

Engravers: (map) Gravé par E. Dussy
(cartouche) Arrivet inv. & sculp.

RF: c.1:11,000,000

Scales: 53 mm. = 75 Milles Romains; 20 Lieues d'une heure

Comments: Used twice in NAP, differentiated by colour.
Colour key: Patriarchat de Constantinople, d'Antioche, de Jerusalem, d'Alexandria
"Pour la Géographie ecclésiastique de l'Europe, de l'Afrique, et de l'Amerique il faut en consulter les Cartes particulières."
Scales along left hand neat line

State 1: 1762: Arrivet cartouche

State 2: 1778: imprint removed; "Groux sculp." cartouche; Fortin publication

Book:

AP:
AU:
NAP: 47

Location:

384 **Turkey** 1748 R

TURQUIE / ASIATIQUE / Par le Sr. Robert Géog. ord? du Roi / avec Privilege 1748.

br: 160 x 197 mm.
pl: 176 x 208 mm.

Engravers:

RF: c.1:11,500,000

Scales: 38 mm. = 80 Lieues d'une heure

Comments:

384 **Turkey** (continued)

Book:

AP: 161
AU:
NAP:

Location:

385 **Turkey** 1750 RV

TURQUIE / EUROPÉENNE / Par le Sr. Robert de / Vaugondy. / Avec Privilege du Roi.

br: 152 x 160 mm.

Engravers:

RF: c.1:10,000,000

Scales:

Comments: "Pag. 371" outside border, top right cardinal points between border and neat line

Book: Méthode Abrégée et facile pour apprendre la Géographie

AP:
AU:
NAP:

[l'Abbé Le François] Paris: Libraires Associés, 1770

Location: WLCL: C2 1770 Le

386 **Turkey, historical** 1756 RV

ASIA MINOR / in suas partes seu provincias / divisa / Auctore Robert de Vaugondy Geog. Regis, / et Academiae Scientiarum ac Litterarum Nanceianae Socio. / Cum privilegio Regis 1756.

br: 480 x 605 mm.
pl: 492 x 626 mm.

Engravers: (cartouche) E. haussard sculp.

RF: c.1:2,250,000

Scales: 96 mm. = 150 Millaria Romana quorum 75 ad gradum unum requiruntur; 1200 Stadia Graeca quorum 600 Gradui uni respondent

Comments:

State 1: 1756

State 2: [1768]: coastline alteration in eastern part of Black Sea (Colchis); river added in Thrace; place name changes and lettering altered: e.g., on Cyprus, Urania to Olympus M.; in Syria, Seleucia (italics) to Seleucia (Roman)
n.b.: Scale has no frame; roads/trade routes shown; Palmyra outside border (bot. rt.)

Book:

AP:
AU: 11
NAP:

Location:

387 **Arabia** 1749 RV

L'ARABIE / Par le Sr. Robert de Vaugondy / fils de Mr. Robert Geog. ord. du Roi. / Avec Privilege. / 1749.

br: 159 x 178 mm.
pl: 180 x 188 mm.

Engravers:

RF: c.1:16,800,000

Scales: 33 mm. = 100 Lieues d'une heure.

Comments:

387 **Arabia** (continued)

Book:

AP: 162
AU:
NAP:

Location:

388 **Persia** 1749 RV

Royaume de / PERSE / Par le Sr. Robert de Vaugondy / fils de Mr. Robert Geog. du Roi / Avec Privilege 1749.

br: 163 x 179 mm.
pl: 180 x 187 mm.

Engravers:

RF: c.1:14,400,000

Scales: 38 mm. = 90 Lieues de Perse de 18 au degré; 100 Lieues d'une heure de 20 au degré

Comments:

Book:

AP: 163
AU:
NAP:

Location:

389 **Persia, Ottoman** 1753 RV

Etats / du Grand-Seigneur / en Asie, / Empire de PERSE, / PAYS DES USBECS, / ARABIE et EGYPTE. / Par le Sr. Robert de Vaugondy fils, / Geographe ordinaire du Roi. / Avec Privilege. / 1753.

br: 480 x 557 mm.
pl: 492 x 567 mm.

Engravers:

RF: c.1:7,500,000

Scales: 143 mm. = 600 Mille pas geometriques de 60 au degré; 660 Lieues de Turquie de 66 au degré; 190 Farsanges ou Lieues de Perse de 19 au degré; 570 Milles Arabes de 57 au degré; 1000 Milles d'Egypte de 100 au degré

Comments:

Book:

AP:
AU: 90
NAP:

Location:

390 **Holy Land** 1743 R

TERRAE SANCTAE / ad Eusebium et D. Hieronymum / descriptae / à Guillaume Sanson Geog. Reg./ altera editio ab AEg. Robert Geog. Regis

Parisiis in ripâ Sequanae ab horlogio dictâ / Cum privilegio Regis / 1743

inset: *TERRAE CHANAAN REGIONES Vicinae ad Eusebium*
inset: *Quadraginta-duae ISRAELITARUM MANSIONES IN DESERTO juxtamentem divi Hieronymi*

br: 638 x 490 mm.
insets: 110 X 148 mm.

Engravers: Gobin Sc.

RF: c.1:580,000

Scales: 63 mm. = 160 Stadia; 20 Mille Passus; 7 Millaria unius horae

Comments:

390 **Holy Land** (continued)

Book:

AP:
AU:
NAP:

Location: WLCL: AF (84)

391 **Holy Land, Old Testament** 1743 R

TERRE DE CHANAAN / ou Terre Promise à Abraham / et à Sa Posterité / ... Cette carte sert pour la lecture du Pentateuque, Elle est dressée sur les manuscrits de G. Sanson / Geog. du Roi par le Sr. Robert Geog. Ord. du Roi /

Inset: *Cartes des Voyages d'Abraham...*

Inset: *Disposition du Camp des Israelites...*

A Paris Quay de l'Horloge du Palais / Avec Privilege / 1743

br: 425 x 563 mm
insets: 256 x 276 mm.; 150 x 147 mm..

Engravers: (map) Gravé par Lattré
(cartouche) Gobin fecit

RF: c.1:1,360,000

Scales: 13 mm. = 10 milles pas geometriques; 14 mm. = 80 Stades

Comments: Julien Cat. (1763): price, 1#

State 1: 1743

State 2: 1745

Book:

AP:
AU:
NAP:

Location: WLCL: AF (76b) (1745); BN:Ge.D.5088 (1743); Ge.D.20615 (1745); Ge.C.4188 (1745); Julien Cat. (1763): no. 1362, (dates as 1743)

392 **Holy Land, Old Testament** 1745 R

Carte / de la TERRE DES HEBREUX OU ISRAELITES / partagee selon l'ordre de Dieu / aux douze tribus descendantes / des douze fils de Jacob/...Dressée pour la lecture de l'Ecriture Sainte / sur les cartes et manuscrits des Srs Sanson / Par le Sr. Robert Geographe ordinaire du Roy. /

Inset: *La Monarchie / des Hebreux / sous Salomon ou le royaume d'Israel distingué en douze gouvernemens ...*

A Paris / chés l'auteur Quai de l'Horloge du Palais / Avec Privilege du Roy 1745.

br: 487 x 690 mm.
pl: 496 x 697 mm.

Engravers: (map) G. Delahaye sculpsit

RF: c.1:720,000

Scales: 77 mm. = 30 Milles Pas Géometriques; 240 Stades; Inset: 50 mm. = 60 mille Pas Geometriques; 480 Stades

Comments: Table with divisions in bottom right corner
Julien Cat. (1763): price, 1#

Book:

AP:
AU:
NAP:

Location: WLCL: AF (85); BN: Ge.DD. 2987 B (10.433); Ge.C.4189; Ge.D.20606; Julien Cat. (1763): no. 1363

393 Holy Land, Old Testament 1750 R

La JUDÉE ou / TERRE SAINTE, / divisée en ses douze Tribus, / Par le Sr. Robert, Géographe du Roi. / Avec Privilege. / 1750.

Inset: *Carte / de la TERRE DE CANAAN / ou Terre Promise, / à Abraham, et à sa Posterité, / dans laquelle on a tracé la Sortie du Peuple / d'Israel hors de l'Egypte, / et ses 42 demeures dans le desert. / Par le Sr. Robert.*

br: 482 x 593 mm.;
pl: 494 x 626 mm.
Inset (top right): 240 x 248 mm.

Engravers:

RF: c.1:850,000

Scales: 63 mm. = 30 Mille Pas Géométriques; 240 Stades

Comments: "Remarques" above scales

State 1: 1750

State 2: [1768] change of shape of Dead Sea (peninsula of Seboim added); removal of aqueduct between Drusus and Beth Semes.

Book:

AP:
AU: 2
NAP:

Location:

394 Holy Land, New Testament 1747 R

Carte des voyages de Notre Seigneur Jesus Christ et ceux des apostres St. Pierre et St. Paul dans l'Asie et dans l'Europe dans laquelle on trouve les noms des Régions et des lieux dont il est parlé dans le Nouveau Testament. Dressée sur les ouvrages des sieurs Sanson par le Sr. Robert.

Inset: *La Judee sous Herode le grande*
Inset: *Plan de la ville Jerusalem*

Paris: ches l'auteur, 1747

br: 450 x 580 mm.

Engravers:

RF: c.1:7,400,000

Scales:

Comments: WLCL AF (89): N. Sanson, 1665, with Robert's imprint;Title in Latin: 'Jesu Christi Salvatoris Nostri et Apostolorum Petri et Pauli, mansiones, Itinera, Peregrinationes, &c...;' two insets: Jerusalem & Jesus' travels; 370 x 562 mm.
Julien Cat. (1763): price, 1#

Book:

AP:
AU:
NAP:

Location: Julien Cat. (1763): 1365; BN: Ge.C.4190; Ge.D.5086; Ge.D.20622

395 Holy Land, historical 1762 RV

JUDÉE / OU / TERRE SAINTE / Par le S. Robert de Vaugondy / Géog? ordinaire du Roy &c.

A Paris ches l'Auteur quai de l'Horloge prés le pont neuf

br: 247 X 219 mm.

Engravers: (map) Gravé par E. Dussy
(cartouche) Arrivet inv. & sculp.

RF: c.1:1,200,000

Scales: 47 mm. = 300 Milles Hebraïques de 600 au degré; 36 mm. = 30 Milles Romains de 75 au degré

Comments: Used three times in NAP, differentiated by colouring:
'1er Division par Tribus';
'2e Division en six pays ou provinces. savoir: la Galilee, la Samarie, la Judee, la Peree, la Trachonite, l'Iturée';

395 Holy Land, historical (continued)

'3e Divisions sous les Romains en trois parties, savoir: Ire Palestine, II Palestine, III Palestine.'

State 1: 1762

State 2: [1778]; "Groux Sculp." to cartouche; Imprint removed; Fortin publication

Book:

AP:
AU:
NAP: 48,49,50

Location:

396 Holy Land, Ottoman 1762 RV

JUDÉE / OU / TERRE SAINTE / Sous les Turcs. / Par le S. Robert de Vaugondy &c. / 1762

br: 248 x 214 mm.

Engravers: (map) Gravé par E. Dussy
(cartouche) Arrivet inv. & Sculp.

RF: c.1:1,200,000

Scales:

Comments: Same geographical outline and scale as NAP 48-50, but fewer place names and divisions; different cartouche.

State 1: 1762

State 2: 1778: date changed; Fortin publication

Book:

AP:
AU:
NAP: 51

Location:

397 India 1749 RV

INDOUSTAN / ou / ETATS DU MOGOL / Par le Sr. Robert de Vaugondy / fils de Mr. Robert Geog. ord. du Roi. / Avec Privilege / 1749.

br: 205 x 161 mm.
pl: 213 x 180 mm.

Engravers:

RF: c.1:17,280,000

Scales: 32 mm. = 100 Lieues Marines et d'une heure

Comments:

Book:

AP: 164
AU:
NAP:

Location:

398 India, Southeast Asia 1761 RV

INDOSTAN, / PRESQU'ISLES DE L'INDE, / CHINE, / TARTARIE INDEPENDANTE. / Par le S. Robert de Vaugondy / Géographe ord? du Roi. &c. / 1761

A Paris ches l'Auteur quai de l'Horloge prés le pont neuf

br: 245 x 286 mm.

Engravers: (map) Gravé par E. Dussy
(cartouche) Arrivet inv. & Sculp.

RF: c.1:20,000,000

Scales: 39 mm. = 450 Milles pas géom? de 20 [sic] au degré; 150 Lieues d'une heure de 20 au deg?

Comments: Scales: Geometric miles usually "60 au degré."

398 **India, Southeast Asia** (continued)

State 1: 1761

State 2: 1778; date changed; imprint removed; Fortin publication

Book:

AP:
AU:
NAP: 37

Location:

399 **Southeast Asia [East Indies]** 1749 RV

PRESQU'ISLE DE L'INDE / au dela du Gange, où sont / les Rmes. d'Ava, de Siam, de / Tong-King, de Cochinchine, / de Camboge &c. / Par le Sr. Robert de Vaugondy / Fils de Mr. Robert Géog. ord. du Roi / Avec Privilege 1749.

br: 201 x 164 mm.
pl: 208 x 180 mm.

Engravers:

RF: c.1:17,280,000

Scales: 32 mm. = 100 Lieues Marines et d'une heure

Comments:

Book:

AP: 165
AU:
NAP:

Location:

400 **Southeast Asia [East Indies]** 1751 R

LES INDES ORIENTALES / où sont distingués / les Empires et Royaumes / qu'elles contiennent, / tirées du Neptune Oriental. / Par Le S. Robert / Géographe ordinaire du Roy. / Avec Privilege / 1751.

br: 482 x 563 mm.
pl: 493 x 580 mm.

Engravers: (map) Guill. Delahaye Sculp.

RF: c.1:8,500,000

Scales: 100 mm. = 450 Mille pas géometriques de 60 au Degré; 150 Lieues Marines Angloises et Françoises de 20 au Degré; 131 1/4 Lieues Marines d'Espagne de 17 1/2 au Degré

Comments:

Book:

AP:
AU: 91
NAP:

Location:

401 **Russia, Asian** 1749 RV

Carte Générale de / l'EMPIRE DES RUSSES / et de la / TATARIE UNIVERSELLE / Par le Sr. Robert de Vaugondy / Fils de M. Robert Geogr. ord. du Roi. / avec Privilege. / 1749.

br: 160 x 206 mm.
pl: 175 x 215 mm.

Engravers:

RF: c.1:43,200,000

Scales: 36 mm. = 300 Lieues d'une heure.; 1572 Verstes de Russie de 105 au Deg.

Comments:

401 **Russia, Asian** (continued)

Book:

AP: 167
AU:
NAP:

Location:

402 **Russia, Asian** 1772 RV

Nouvelle Représentation / Des Côtes Nord et Est de l'ASIE / pour servir d'eclaircissement aux Articles / du Supplément de l'Encyclopédie qui / concernant le Passage aux Indes par le Nord. / Gravée sous la direction de Mr. de Vaugondy / en 1772

[Book: Paris: Panckoucke, Stoupe, Brunet; Amsterdam: M. M. Rey]

br: 287 x 358 mm.
pl: 316 x 385 mm.

Engravers:

RF: c.1:20,000,000

Scales:

Comments: outside border: "Suppl. 3e carte" (top rt.) "162" (bott. rt.) two insets [Supplements] of Kamchatka: Remarques: Suppl. 1 — [tracing of map in Histoire des Tartares, by d'Abulgasi-Bayadur-Chan (Leyden, 1726)]; Suppl. 2 — [reduction of map published in Nuremburg]

Book: ENCYCLOPÉDIE: Suite du Recueil des Planches, sur les Sciences, les Arts Libéraux, et les Arts Mechaniques, avec leurs explication.... Supplément, ed. Panckoucke (1777)

AP:
AU:
NAP:

Location: UM, Rare Books: AE 25 .E53

403 **Russia, China, Japon** 1762 RV

SIBERIE OU RUSSIE ASIATIQUE, TARTARIE CHINOISE, PAYS DES ELUTS ET ISLES DU JAPON, Par le Sr. Robert de Vaugondy &c.

A Paris ches l'Auteur quai de l'Horloge prés le pont neuf 1762

br: 246 x 309 mm.

Engravers: (map) Gravé par E. Dussy

RF: c.1:20,000,000

Scales: 51 mm. = 200 Lieues d'une heure de 20 au degré

Comments State 1: 1762: date next to imprint

State 2: [1778]: imprint removed; Fortin publication

Scale outside border (bot.)

Book:

AP:
AU:
NAP: 36

Location:

404 **Russia, Asian** 1750 R

Partie Orientale / de L'EMPIRE DE RUSSIE EN ASIE, / où se trouvent les Provinces de / Jakuckskoy, Nerckzinskoy, Selinginskoy, / Ilimskoy, Krasnojarskoy, Narimskoy, / Jenisseiskoy, Mangajeiskoy, &c. / et les Confins de / la Tatarie Chinoise. / Dressée, d'après les Cartes de l'Atlas Russien, / Par le Sr. Robert de Vaugondy / fils de Mr. Robert, Géographe ord? du Roy / Avec Privilege. / 1750.

br: 469 x 594 mm.
pl: 495 x 640 mm.

Engravers:

RF: c.1:8.000,000

404 **Russia, Asian** (continued)

Scales: "Avertissement: Comme la projection de cette Carte ne peut pas souffrir d'Echelle, il suffit d'etre averti que le degré de la Terre contient 104 Woerstes ou Milles Russiens, et 250 Lis Chinois"

Comments: Philippe Buache criticized Vaugondy for claiming that this projection would not support a scale (See Chapter Three)

Book:

AP:
AU: 89
NAP:

Location:

405 **Russia, Asian** 1750 RV

Partie Occidentale / de L'EMPIRE DE RUSSIE, / où se trouve distinguée / la Russie Européenne, / Divisée par ses Gouvernemens en Provinces de / Dwinskoy, Oloneckoy, Bielozerskoy, Archangel, / Nowogorod, Wolockoy, Ustinskoy, Moscow, Kiowie, / Bielogorod, Woroneskoy, et la Petite Tatarie. / La Partie Asiatique / contient les Provinces de / Berezowskoy, Magajeiskoy, Wiatka, Solkamskoy, Pelymskoy, / Usimskoy, Tobolskoy, Surgutskoy, Jenifeiskoy, / Narimskoy, Kuzneckoy, Tomskoy, Krasnojarskoy &c. / le Royaume de Kasan, / le Gouvernement d'Astrachan et les Confins du Royaume des / Kalmaks Eleuths. / Dressée d'après les Cartes de l'Atlas Russien / Par Le Sr. Robert de Vaugondy / fils de Mr. Robert Géographe ordin? du Roi / Avec Privilege. / 1750.

br: 477 x 595 mm.;
pl: 502 x 632 mm.
Inset (top right): 71 x 97 mm.

Engravers: (cartouche) Gobin fecit

RF: c.1:8,000,000

Scales:

Comments: Inset: Explication of Russian words used on the map
From AU 'Préface,' p. 23a: reference to criticisms in the Journal Oeconomique, Decembre 1753, (made by Buache) regarding the Robert map of Russia and the Russian Atlas (1745), the main geographic problems being a) area around Kamchatka and the mouth of the Amur River, b) the land of Jeso, and c) the scale (map has none)

Book:

AP:
AU: 88
NAP:

Location:

406 **Russia, Asian** 1772 RV

*Carte des parties Nord et est de l'Asie / qui comprend / les Côtes de la RUSSIE ASIATIQUE / Le Kamschatka, le Jesso, et les Isles du Japon / dressée en 1764 par M*** / Nouvelle édition réduite par M. de Vaugondy / 1772*

[Book: Paris: Panckoucke, Stoupe, Brunet; Amsterdam: M. M. Rey]

br: 292 x 368 mm.
pl: 325 x 374 mm.

Engravers:

RF: c.1:20,000,000

Scales:

Comments: outside border: "Suppl. 2e Carte" (top rt.) "161" (bot. rt)
Four insets: "Ces quatre Supplemens ont rapport à la carte de l'Amerique indiquée 1re."

Book: ENCYCLOPÉDIE: Suite du Recueil des Planches, sur les Sciences, les Arts Liberaux, et les Arts Mechaniques, avec leurs explication.... Supplément, ed. Panckoucke (1777)

AP:
AU:
NAP:

Location: UM, Rare Books: AE 25 .E53

407 **China** 1749 RV

Empire de / la CHINE / Par le Sr. Robert de Vaugondy / Fils de Mr. Robert Geogr. ord. du Roi / Avec Privilege / 1749.

br: 161 x 198 mm.
pl: 177 x 208 mm.

Engravers:

RF: c.1:17,280,000

Scales: 38 mm. = 120 Lieues d'une heure; 1500 Lys Chinois de 250 au Degré

Comments: Title frame slightly scalloped, not the usual rectangle

Book:

AP: 166
AU:
NAP:

Location:

408 **China** 1751 RV

L'EMPIRE DE LA / CHINE / dressé d'après les Cartes / de l'Atlas Chinois. / Par Le Sr. Robert de Vaugondy. / Geographe ordinaire du Roi. / Avec Privilege. / 1751.

br: 482 x 527 mm.
pl: 495 x 548 mm.

Engravers:

RF: c.1:12,000,000

Scales: 71 mm. = 240 Mille Pas Géométriques de 60 au Degré; 1000 Lys ou Stades Chinois de 250 au Degré; 100 Pus de 25 au Degré; 80 Lieues Marines et d'une heure de 20 au Degré

Comments:

Book:

AP:
AU: 93
NAP:

Location:

409 **Japan** 1749 RV

Le JAPON / Par le Sr. Robert de Vaugondy / Fils de Mr. Robert Géog. du Roi. / Avec Privilege / 1749.

br: 161 x 161 mm.
pl: 181 x 171 mm.

Engravers:

RF: c.1:15,800,000

Scales: 36 mm. = 100 Lieues d'une heure

Comments: Note island "Fatsisio ou l'Empereur envoye en exil les Grands" [repeated on Atlas Universel map, Cat. 410]

Book:

AP: 168
AU:
NAP:

Location:

410 **Japan** 1750 R

L'EMPIRE / DU JAPON, / divisé en sept principales parties, / et subdivisé en soixante et six Royaumes. / Par Le Sr. Robert Géog. ord. du Roi. / Avec Privilege / 1750.

br: 481 x 535 mm.
pl: 495 x 560 mm.

Engravers:

RF: c.1:3,000,000

Scales: 72 mm. = 40 Lieues d'une heure de 20 au Degré; 80 Lieues Japonoises de 40 au Degré

410	**Japan** (continued)		

Comments: Note: island of "Fatsisio ou Isle malheureuse dans laquelle on exile les Grands du Japon"

Book:

AP:
AU: 94
NAP:

Location:

411	**Philippines**	1749	RV

ISLES PHILIPPINES / et MOLUQUES / Par le Sr Robert de Vaugondy / fils de Mr. Robert Géogr. ord. du Roi / Avec Privilege 1749.

br: 207 x 160 mm.
pl: 215 x 180 mm.

Engravers:

RF: c.1:17,280,000

Scales: 32 mm. = 100 Lieues Marines et d'une heure

Comments:

Book:

AP: 169
AU:
NAP:

Location:

412	**Indonesia**	1749	RV

Les ISLES DE / LA SONDE / Par le Sr. Robert de Vaugondy / Fils de Mr. Robert Geog. du Roi. / Avec Privilege / 1749.

br: 162 x 196 mm.
pl: 181 x 207 mm.

Engravers:

RF: c.1:13,400,000

Scales: 41 mm. = 100 Lieues marines et d'une heure.

Comments:

Book:

AP: 170
AU:
NAP:

Location:

413	**Indonesia**	1750	R

ARCHIPEL / DES INDES ORIENTALES, / qui comprend / les Isles de la Sonde, / Moluques et Philippines, / tirées des Cartes du Neptune Oriental, / Par Le Sr. Robert / Géographe ordinaire du Roy. / 1750.

br: 482 x 595 mm.
pl: 502 x 635 mm.

Engravers: (map) Guill? Delahaye Sculp.

RF: c.1:8,000,000

Scales: 96 mm. = 450 Mille pas Geometriques de 60 au Degré; 150 Lieues Marines Angloises et Françoises de 20 au Degré; 131 1/4 Lieues Marines d'Espagne de 17 1/2 au Degré

Comments:

Book:

AP:
AU: 92
NAP:

Location:

414	**Indonesia**	1759	RV

[Les Indes Orientales]

two sheets

Engravers:

RF: c.1:

Scales:

Comments: known only from RV catalogues: Tablettes Parisiennes (1760); Catalogue (1777)

Book:

AP:
AU:
NAP:

Location: RV: Tablettes Parisiennes (1760); Catalogue (1777)

415	**Indonesia**	[1762]	RV

LES ISLES / DE LA SONDE, MOLUQUES, / PHILIPPINES, CAROLINES, / ET MARIANNES. / Par le S. Robert de Vaugondy &c.

A Paris ches l'Auteur quai de l'Horloge près le pont neuf 1762

br: 244 x 267 mm.

Engravers: (map) Gravé par E. Dussy
(cartouche) Arrivet inv. & Sculp.

RF: c.1:20,000,000

Scales: 53 mm. = 600 Mille pas géométriques de 60 au degré; 200 Lieues Marines ou d'une heure

Comments:
State 1: 1762
State 2: [1778]: imprint removed; Fortin publication

Book:

AP:
AU:
NAP: 38

Location:

416	**Pacific Ocean**	1774	RV

CARTE DES DÉCOUVERTES / faites en 1765, 67, et 69 dans / LA MER DU SUD // Dressée par le Sr. de Vaugondy Geog? du Roi / du feu Roi de Pologne Duc de Lorraine et de Bar / de l'Academie Royale des Sciences et Belles-Lettres de Nanci / et Censeur Royal. / 1774

[Book: Paris: De Hansy le jeune, Libraire, rue St. Jacques]

br: 238 x 538 mm.
pl: 258 x 551 mm.

Engravers: (map) Bourgoin Sculp.

RF: c.1:

Scales:

Comments: Approbation of the book given by Robert de Vaugondy as Censeur Royal, 29 Nov. 1773
Map shows routes of Cook, Wallis, Carteret, Byron and Monat

Book: HYDROGRAPHIE DE LA MER DU SUD: Histoire des Nouvelles Decouvertes faites dans la Mer du Sud en 1767, 1768, 1769 & 1770,...les derniers relations M. de Fréville (1774)

AP:
AU:
NAP:

Location: BN: Impr. 8° Pz.20 (1&2)

417 **Australasia** 1756 RV

Carte Réduite / de l'AUSTRALASIE, / Pour servir à la lecture de / l'Histoire des Terres Australes, / par le Sr. Robert de Vaugondy Géog? ord? du Roi, / de l'Academie Royale des Sciences et Belles-Lettres / de Nancy. 1756.

[Book: Paris, Durand, rue du Foin, au Griffon]

br: 230 x 283 mm. (reprint)

Engravers: (map) G. De-la-Haye

RF: c.1:40,000,000

Scales: Lieues Marines ou d'une heure de Chemin [along left hand neat line]:

Comments:

Book: Histoire des Navigations aux Terres Australes.

AP:
AU:
NAP:

Président Charles de Brosses (1756)

Location: CUL: D.18.35; UM: G 160 .B9 1967 (reprint)

418 **Caroline Islands** 1756 RV

Carte / des ISLES CAROLINES, / dressée / Pour l'Histoire des Terres Australes. / par le Sr. Robert de Vaugondy / Géog? ord? du Roi, de l'Académie royale des / Sciences et Belles-Lettres de Nancy. / 1756.

[Book: Paris: Durand, rue du Foin, au Griffon]

br: 200 x 381 mm. (reprint)

Engravers: (map) G. DelaHaye.

RF: c.1:8,500,000

Scales:

Comments:

Book: Histoire des Navigations aux Terres Australes.

AP:
AU:
NAP:

Président Charles de Brosses (1756)

Location: CUL: D.18.35; UM: G 160 .B9 1967 (reprint)

419 **Papua islands** 1756 RV

Carte / des ISLES DES PAPOUS, / pour l'Histoire des Terres Australes; / dessinée d'après celle / de Mr. Is. Tirion. / Par le Sr. Robert de Vaugondy Géog. ordin. du Roi, / de l'Académie royale des Sciences et Belles-Lettres / de Nancy. 1756.

[Book: Paris: Durand, rue du Foin, au Griffon]

br: 286 x 359 mm. (reprint)

Engravers: (map) G. De-la-Haye.

RF: c.1:1,500,000

Scales: 74 mm. = Milles d'Allemagne de 15 au degré

Comments:

Book: Histoire des Navigations aux Terres Australes.

AP:
AU:
NAP:

Président Charles de Brosses (1756)

Location: CUL: D.18.35; UM: G160 .B9 1967 (reprint)

420 **Papua, New Guinea** 1756 RV

Partie de l'AUSTRALASIE / qui comprend / la TERRE DES PAPOUS ou / NOUVLE. GUINÉE, / et la Nouvle. Bretagne. / Pour servir à la lecture de / l'Histre. des Terres Australes. / Par le Sr. Robert de Vaugondy Géog? ord? du Roi. / de l'Académie Royale des Sciences et Belles-Lettres / de Nancy. 1756.

[Book: Paris: Durand, rue du Foin, au Griffon]

br: 218 x 346 mm. (reprint)

Engravers: (map) G. De-la-Haye

RF: c.1:9,500,000

Scales: Lieues marines ou d'une heure de chemin [along right hand neat line]

Comments:

Book: Histoire des Navigations aux Terres Australes.

AP:
AU:
NAP:

Président Charles de Brosses (1756)

Location: CUL: D.18.35; UM: G 160 .B9 1967 (reprint)

421 **Polynesia** 1756 RV

Carte Réduite / de la POLYNÉSIE AUSTRALE, / ou Des Isles / de la Mer du Sud / Dressée pour l'histoire des Terres Australes / par le S. Robert de Vaugondy Géog? ord? du Roi. / de l'Académie royale des Sciences et Belles-Lettres de / Nancy. 1756.

[Book: Paris: Durand, rue du Foin, au Griffon]

br: 233 x 240 mm. (reprint)

Engravers: (map) [G. Delahaye?]

RF: c.1:50,000,000

Scales: Lieues marines de France et d'Angleterre [along left neat line]

Comments: Delahaye conjectured as engraver since he engraved the other maps for this book

Book: Histoire des Navigations aux Terres Australes.

AP:
AU:
NAP:

Président Charles de Brosses (1756)

Location: CUL: D.18.35; UM: G.160.B9 1967 (reprint)

422 **Africa** 1740 R

[Nouvelle Carte de l'Afrique]

Engravers:

RF: c.1:

Scales:

Comments: not seen in any location; known only from catalogues, but parallel to Asia, Europe, and Western Hemisphere maps based on Sanson by Robert and known with De Leth of Amsterdam imprints

Book:

AP:
AU:
NAP:

Location: L-D (1745), p. 245-247; Robert, Intro. à la Géog. (1743); BN: Ge.C.7695 (dated 1741; ed. Sanson)

423 **Africa** 1748 R

L'AFRIQUE / divisée en ses / principales parties. / Par le Sr. Robert Geog. ordinre. / du Roi 1748.

br: 159 x 196 mm.
pl: 167 x 203 mm.

Engravers: (map) Gravé par / G. Delahaye.

RF: c.1:57,600,000

Scales:

Comments:

Book:

AP: 171
AU:
NAP:

Location:

424 **Africa** **1749** R/RV

L'AFRIQUE / divisée en ses principaux / Empires et Royaumes / par les Srs. Sanson Géogr. ordin? du Roi / assujettie aux Observations astronomiques / par le Sr. Robert Géogr. ord. du Roi / avec Privilege / 1749

A Paris chéz l'autheur / Quay de l'Horloge près le Pont Neuf

br: 489 x 660 mm.
pl: 496 x 671 mm.

Engravers: (map) Delahaye l'ainé Sculpsit
(cartouche) Gobin sculpsit

RF: c.1:37,000,000

Scales: 65 mm. = 600 mille Pas géometriques de 60 au Degré; 250 Lieues communes de France de 25 au Degré; 175 Lieues marines d'Espagne de 17 1/2 au Degré; 200 Lieues marines de France et d'Angletterre; [sic] de 20 au D.; 200 Lieues d'une heure et grande lieue de France de 20 au Degré; 567 Milles Arabes de 56 2/3 au degré

Comments: Geographical and historical table to right of map
Julien Cat. (1763): price, 1#;

State 1: 1749: Robert author; Gobin cartouche

State 2: 1771: according to RV, Notice Raisonnée, (1774), the 1749 map was re-edited in 1771, parallel with other continent maps of 1749; title changed slightly: "...Royaumes. / Par le Sr. Robert de Vaugondy Géographe / Ord? du Roi, du feu Roi de Pologne Duc / de Lorraine et de Bar, et de l'Académie royal des Sciences et / belles lettres de Nancy, 1771"; "Arrivet inv. & sculp." on cartouche

State 3: 1776; as per RV, Catalogue, (1777)

State 4: 1778; imprint: chès Fortin Ingr. Mécanicien du Roy Rue de la Harpe

State 5: [after 1784]; imprint: chez le Sr. Delamarche Geog? Rue du Foin St Jacques, au collège de Mtre Gervais

Book:

AP:
AU:
NAP:

Location: BN: Ge.C.7696 (1749; called ed. Sanson); Ge.C.7704, Ge.D.20607 (1778); Ge.D.11446 (Delamarche, 1778); Ge.C.7705 (s.d., Delamarche), Ge.C.7697 (1795, Delamarche) WLCL: Julien, Theatre du Monde I (6)

425 **Africa** 1750 RV

AFRIQUE / Par le Sr. Robert de Vaugondy / Avec Privilege 1750.

br: 151 x 155 mm.

Engravers:

RF: c.1:80,000,000

Scales: 30 mm. = 300 lieues d'une heure

Comments: "Pag. 447" outside border, top right

425 **Africa** (continued)

Book: Méthode Abrégée et facile pour apprendre la Géographie

AP:
AU:
NAP:

[l'Abbé Le François] Paris: Libraires Associés, 1770

Location: WLCL: C2 1770 Le

426 **Africa** 1756 RV

L'AFRIQUE / dressée sur les relations les plus recentes, / et assujettie aux observations astronomiques / Par le Sr. Robert de Vaugondy Geog? ord? du Roy / de Sa. M. Polon? Duc de Lorraine et Bar. et Associé / de l'Academie Rle. des Scien? et B. Lettres de Nancy. / Avec Privilege 1756.

br: 466 x 591 mm.
pl: 495 x 633 mm.

Engravers: (map) Guill? Delahaye Sculpsit.
(cartouche) Gobin fecit

RF: c.1:20,000,000

Scales: 83 mm. = 900 Mille pas geometriques, de 60 au Degré.; 300 Lieues d'une heure de chemin de 20 au Degré; 22 1/2 Dietes où journées de 40 Milles chacunes

Comments: From AU 'Préface,' p. 31a: Problem of longitude of Cape of Good Hope: 2° more west than on earlier maps, thanks to work of d'Après de Manevillette. Problem of sources of the Nile: P. Pays, Jesuit, made discoveries in 1618 which G. Delisle showed on his map; neither N. Sanson nor d'Anville regards Pays source as that of true Nile; d'Anville puts two lakes at 6° N, north of Mts. of the Moon, for source of Nile, following Edrisi and Abulfeda; we have followed G. Delisle: "It is better to follow a received opinion than to forge a chimera in order to give something new in geographic material."

Book:

AP:
AU: 95
NAP:

Location:

427 **Africa** 1761 RV

L'AFRIQUE / Par le S. Robert de Vaugondy / Géog? ord? du Roi, et de S.M. Pol? / Duc de Lorr? et de Bar, et / de la Société royale de Nanci / 1761

A Paris ches l'Auteur quai de l'Horloge près le pont neuf

br: 243 x 284 mm.

Engravers: (map) Gravé par E. Dussy
(cartouche) Arrivet inv. & Sculp.

RF: c.1:50,000,000

Scales: 46 mm. = 900 Mille pas géométriques de 60 au dég?; 300 Lieues d'une heure de 20 au dégre

Comments:

State 1: 1761

State 2: [1778]: removes date and imprint, adds "et Censeur Royal" to title; Fortin publication

State 3: [1790]: note about sources of Nile and d'Anville added; Delamarche issue [RGS 4.c.9]

n.b.: sources of Nile 12°N, 50°E at L. Dambea on State 1

Book:

AP:
AU:
NAP: 39

Location:

428 **Africa** 1767 RV

L'AFRIQUE / divisée en ses / Empires, Royaumes, / et Républiques. / par le Sr. Robert de Vaugondy Géographe ord. du Roi / de S.M. Polon? Duc de Lorr. et de Bar, et de la Societé / royale des Sciences et Belles Lettres de Nancy /

A Paris / ches l'Auteur Quai de l'horloge au Palais / près le Pont-neuf / avec Privilege 1767

four sheets: 977 x 1107 mm. assembled

Engravers: (map) Gravé par E. Dussy
(cartouche) Arrivet inv. & Sculp.

RF: c.1:9,000,000

Scales: 129 mm. = 600 Mille pas géométriques de 60 au degré.; 154 mm. = 210 Lieues marines d'Espagne de 17 au degré.; 169 mm. = 260 Lieues marines de France et d'Angleterre de 20 au degré.; 159 mm. = 840 Lieues de Turquie de 66 au degré; 134 mm. = 600 Milles Arabes de 56 au degré.; 150 mm. = Milles d'Egypte de 100 au degré.

Comments:

[State 1: 1760? (as per Julien Cat. (1763); price 1#]

State 2: 1767 [described in Tooley, The Maps of Africa, MCC 48]
RV lists a World and four continents, each in four sheets in Notice Raisonnée (1774)

Book:

AP:
AU:
NAP:

Location: BL: K.117.20.2 Tab; RV, Notice Raisonnée, (1774); Julien Cat. (1763), no. 46.

429 **Africa, historical** [1755] RV

Carte / de la Navigation d'Hannon, / sur les Côtes d'Afrique, / dressée pour le Mémoire de Mr. de Bougainville, / par le Sr. Robert de Vaugondy / Géographe ord? du Roi

[Book: Paris: Mémoires de l'Academie des Inscriptions, Tom. 26 [1759], 28 [1761?]

br: 231 x 185 mm.
pl: 240 x 190 mm.

Engravers: (map) Gravé par Guill? De-la-Haye

RF: c.1:20,000,000

Scales: 26 mm. = 100 Lieues marines de 20 au degré

Comments: The Memoire discusses the Periplus of Hanno, Carthiginian Admiral, and the differences of opinion concerning its veracity.

Book: Mémoire sur les Découvertes et Les Établissemens faits le long des Côtes d'Afrique par Hannon, Amiral de Carthage

AP:
AU:
NAP:

M. de Bougainville: Acad. des Inscriptions, Sept. 1754; Dec. 1757; Jan. 1758

Location: UM: G88.H25 B7; BN: Ge.DD.2987 (10216) [map only]

430 **Egypt** 1743 R

VETERIS / AEGYPTI / TABULA / ex variis auctoribus / decerpta, / atque Itinerariis adornata / conatibus geographicis / Robert Geographi Regii / Parisiis / Apud auctorem in ripâ Sequanae dictâ / quai de l'horloge. / cum privilegio Regis. 1743

Paris: in ripa Sequanae dictâ quai de l'horloge

br: 612 x 480 mm.

Engravers: (cartouche) Gobin

RF: c.1:1,760,000

Scales:

Comments: Geographical table above cartouche

430 **Egypt** (continued)

Book:

AP:
AU:
NAP:

Location: WLCL: AF (5); BL: Maps. 625. (8); Robert, Intro. à la Géog., (1743); Julien Cat. (1763): no. 1395

431 **Egypt** 1753 RV

Carte de / L'EGYPTE, ANCIENNE / ET MODERNE, / dressée sur celle du R. P. Sicard / et autres, / assujetties aux observations astronomiques / Par le Sr. Robert de Vaugondy / Géographe ordinaire du Roy. / Avec Privilege / 1753.

br: 640 x 465 mm.
pl: 665 x 496 mm.

Engravers: (cartouche) haussard

RF: c.1:650,000

Scales: 99 mm. = 112 1/2 Milles Romains de 75 au degré.; 1008 Stades d'Herodote de 671 1/2 au degré.; 90 Mille pas géométriques de 60 au degré.

Comments: From AU 'Préface,' p. 31a-b: problem of coordinates of Delta: used the observed positions of Alexandra, Peluse, Cairo; Suez at 29°45' N.; Berenice is further south than on previous maps; tried to make the width of the Nile proportional to its true dimensions; as to Lake Moeris: many assume it to be the Calig-Josef (Canal de Joseph), but for RV it is Birk-el Karoon and Bahr Bellomah (Lake without Water), w. of Memphis

Book:

AP:
AU: 96
NAP:

Location:

432 **Africa, north** 1748 R

Partie Occidentale de / La TURQUIE D'AFRIQUE / où l'on trouve les / Re. d'Alger, de Tunis et part. de Tripoli / Par le Sr. Robert Geog? du Roi / Avec Privilege 1748.

Partie Orientale de la TURQUIE D'AFRIQUE / où l'on trouve / une partie du Re. de Tripoli / et l'Egypte.

br: 160 x 158 mm.
pl: 176 x 169 mm.

Engravers:

RF: c.1:14,400,000

Scales: 38 mm. = 300 Mille Pas Geometriques; 100 Lieues d'une heure

Comments: two maps on one sheet

Book:

AP: 172
AU:
NAP:

Location:

433 **Africa, north** [1762] RV

BARBARIE, EGYPTE, / NIGRITIE, GUINÉE, / NUBIE, ABISSINIE. / Par le S. Robert de Vaugondy Géog? ord? / du Roi, de S.M. Pol? Duc de Lorr? et de Bar, / et de la Société royale de Nanci / Avec Privilege.

A Paris ches l'Auteur quai de l'Horloge près le pont neuf 1762

br: 244 x 374 mm.

Engravers: (map) Gravé par E. Dussy
(cartouche) Arrivet inv. & Sculp.

RF: c.1:20,000,000

433 **Africa, north** (continued)

Scales: 60 mm. = 600 Mille pas géometriques de 60 au degré; 200 Lieues Marines ou d'une Heure

Comments:
State 1: 1762

State 2: [1778]: imprint and date removed; Fortin publication

Book:

AP:
AU:
NAP: 40

Location:

434 **Africa, west** 1749 RV

Cote Occidentale d'Afrique / depuis le / Det? de Gibraltar jusqu'au C. Blanc / òu se trouvent les / ETATS DU ROI DE MAROC / Par le Sr. Robert de Vaugondy / Fils de Mr. Robert Géogr. ord. du Roi / Avec Privilege 1749.

br: 165 x 196 mm.
pl: 178 x 206 mm.

Engravers:

RF: c.1:12,000,000

Scales: 46 mm. = 100 Lieues d'une heure

Comments: n.b.: Prime Meridian (I. de Fer) and Tropic of Cancer are only long. and lat. shown

Book:

AP: 173
AU:
NAP:

Location:

435 **Africa, west** 1749 RV

Partie Occidentale de / la NIGRITIE et de / la HAUTE GUINÉE / Par le Sr. Robert de Vaugondy / fils de Mr. Robert Géog. du Roi. / Avec Privilege 1749.

br: 163 x 192 mm.
pl: 178 x 202 mm.

Engravers:

RF: c.1:13,000,000

Scales: 43 mm. = 100 Lieues d'une heure

Comments:

Book:

AP: 174
AU:
NAP:

Location:

436 **Senegal River** 1749 RV

Cours de la RIVIERE / DE SENEGAL / depuis le Rocher Gouinea jusqu'a / la Mer. / Par le Sr. Robert de Vaugondy / Fils de Mr. Robert Géog. du Roi / Avec Privilege / 1749.

Embouchure du Senegal

br: 161 x 169 mm.
pl: 164 x 282 mm.

Engravers:

RF: c.1:3,700,000

Scales: 54 mm. = 35 Lieues marines et d'une heure a 20 au Degré; Embouchure du Senegal: 56 mm. = Echelle de dix Lieues

Comments: Three maps in one. RF. is for 'Cours de la Riviere'; 'Embouchure du Senegal' has no long/lat

436 **Senegal River** (continued)

Book:

AP: 175
AU:
NAP:

Location:

437 **Africa: Guinea** 1749 RV

Partie Orientale de / la HAUTE GUINÉE / Par le Sr. Robert de Vaugondy / Fils de Mr. Robert Géog. du Roi / Avec Privilége / 1749.

br: 162 x 178 mm.
pl: 174 x 187 mm.

Engravers:

RF: c.1:13,400,000

Scales: 33 mm. = 80 Lieues d'une heure

Comments:

Book:

AP: 176
AU:
NAP:

Location:

438 **Africa: Congo** 1749 RV

CONGO ou / BASSE GUINÉE / Par le Sr. Robert de Vaugondy / fils de Mr. Robert Geog. du Roi / Avec Privilege / 1749.

br: 163 x 172 mm.
pl: 178 x 180 mm.

Engravers:

RF: c.1:13,400,000

Scales: 43 mm. = 100 Lieues d'une heure

Comments:

Book:

AP: 177
AU:
NAP:

Location:

439 **Africa: Congo** 1762 RV

CONGO, / CAFRERIE. / Par le S. Robert de Vaugondy / Géog? ordinaire du Roy &c. / 1762

A Paris ches l'Auteur quai de l'Horloge près le pont neuf

br: 243 x 283 mm.

Engravers: (map) Gravé par E. Dussy
(cartouche) Arrivet inv. & Sculp.

RF: c.1:20,000,000

Scales: 43 mm. = 450 Mille pas géometriques de 60 au degré; 150 Lieues marines ou d'une heure

Comments:
State 1: 1762

State 2: [1778]; date and imprint removed; Fortin publication

Book:

AP:
AU:
NAP: 41

Location:

440	**Africa, south**	1749	RV

PAYS DES CAFRES. / Par le Sr. Robert de Vaugondy / fils de Mr. Robert Geog. du Roi / Avec Privilege / 1749.

br: 163 x 231 mm.
pl: 173 x 237 mm.

Engravers:

RF: c.1:13,000,000

Scales: 43 mm. = 100 Lieues d'une heure.

Comments:

Book:

AP: 178
AU:
NAP:

Location:

441	**Africa: Zanzibar**	1749	RV

Cote de / ZANGUEBAR. / Par le Sr. Robert de Vaugondy / fils de Mr. Robert Geog. du Roi / Avec Privilege 1749.

br: 190 x 163 mm.
pl: 196 x 173 mm.

Engravers:

RF: c.1:13,000,000

Scales: 34 mm. = 80 Lieues d'une heure

Comments:

Book:

AP: 179
AU:
NAP:

Location:

442	**Africa: Nubia**	1749	RV

NUBIE / ABISSINIE et / COTE D'AJAN. / Par le Sr. Robert de Vaugondy / Fils de Mr. Robert Geog. du Roi A Priv 1749.

br: 164 x 184 mm.
pl: 177 x 194 mm.

Engravers:

RF: c.1:18,240,000

Scales: 31 mm. = 100 Lieues d'une heure.

Comments: Sources of the Nile at 12° N., 50° E. at mts. south of L. Dambee

Book:

AP: 181
AU:
NAP:

Location:

443	**Africa: Madagascar**	1749	RV

ISLE DE MADAGASCAR / ou de St. Laurent / et Isles circonvoisines / Par le Sr. Robert de Vaugondy / Fils de Mr. Robert Géog. du Roi APr. 1749.

br: 163 x 193 mm.
pl: 178 x 202 mm.

Engravers:

RF: c.1:17,280,000

Scales: 32 mm. = 100 Lieues Marines de 20 au Degré

Comments:

443	**Africa: Madagascar** (continued)

Book:

AP: 180
AU:
NAP:

Location:

444	**Canary Islands**	1749	RV

LES / ISLES CANARIES / Par le Sr. Robert de Vaugondy / Fils de Mr. Robert Geog. ord. du Roi / Avec Privilege 1749.

Inset: *Isles de MADERE et PORTO-SANTO*

br: 162 x 192 mm.
pl: 175 x 202 mm.

Engravers:

RF: c.1:3,000,000

Scales: 36 mm. = 20 Lieues d'une heure; Inset: 36 mm. = 15 Lieues d'une heure.

Comments: n.b.: "Premier Merdien fixe à l'Isle de Fer par la Declaration de Louis XIII en 1634" runs through I. de Fer

Book:

AP: 182
AU:
NAP:

Location:

445	**Cape Verde Islands**	1749	RV

LES ISLES / DU CAP-VERD / Par le Sr. Robert de Vaugondy / Fils de Mr. Robert Géog. ord. du Roi / Avec Privilege 1749.

br: 162 x 192 mm.
pl: 175 x 201 mm.

Engravers:

RF: c.1:2,000,000

Scales: 43 mm. = 15 Lieues marines et d'une heure

Comments:

Book:

AP: 183
AU:
NAP:

Location:

446	**North America**	1748	R

AMERIQUE / SEPTENTRIONALE. / Par le Sr. Robert Géographe / ordinaire du Roi. / 1748.

br: 158 x 197 mm.
pl: 167 x 205 mm.

Engravers:

RF: c.1:50,400,000

Scales: 23 mm. = 200 Lieues d'une heure de Chemin

Comments: Ohio R. shown as R. des Ouabaches; NW left blank

Book:

AP: 184
AU:
NAP:

Location:

Septentrion
PARTIE NORD-OUEST DE L'AMERIQUE
Stachtan Nitada
Grande Terre
Lac Bernarda
Michinipi ou la Grande eau de 600 lieues de tour
Tetes Plates
Assinipoels
Cristinaux
Bergi
Anian
Det d'Anian
Fou sang
Quivira
Nouvelle Albion
Tahu
Teguayo ou Tolm
CALIFORNIE
TERRES
BAYE DE BAFFIN
Det de Baffin
James Island
Cumberland
Welcome
BAYE D'HUDSON
DETROIT D'HUDSON
NOUVELLE BRETAGNE
TERRE DE LABRADOR
ESQUIMAUX
Cristinaux
CANADA
Quebec
Algonquins
Outa-ouaes
Renards
Outagamis
Mascoutens
Miamis
PAYS DES OSAGES
LOUISIANE
NOUVEAU MEXIQUE
APACHES
NOUVELLE NAVARRE
MER VERMEILLE
CALIFORNIE
VIRGINIE
CAROLINE
ETATS UNIS
FLORIDE
GEORGIE
New York
Philadelphie
Charles Town
George Town
Bermudes
GOLFE DU MEXIQUE
MER DU SUD ou PACIFIQUE
CUBA
GRANDES ANTILLES
Jamaique
HISPANIOLA ou S.DOMINGO
ISLES LUCAYES
Havana
MEXIQUE
Acapulco
Honduras
TERRE FERME
Carthagene
AMERIQUE M
Occident
Midi
ECHELLES
Mille pas géometriques, de 60 au Dégré.
Lieues Marines de france, et d'Angleterre, de 20 au D.
Lieues d'Espagne, de 17½ au Dégré.

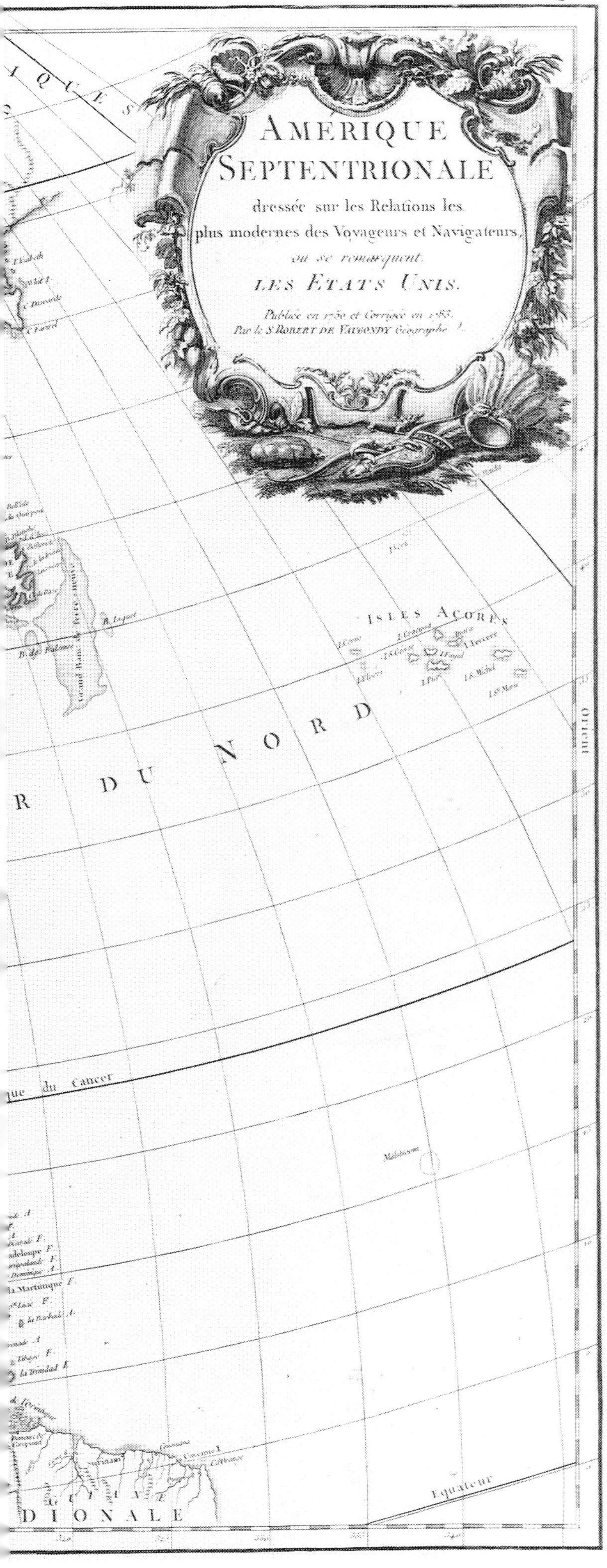

"Amérique Septentrionale," 1783, showing the new United States of America. (Courtesy of the William L. Clements Library, University of Michigan)

447	North America	1750	RV

AMERIQUE / SEPTENTRIONALE / Par le Sr. Robert de Vaugondy / Avec Privilege 1750.

br: 150 x 150 mm.

Engravers:

RF: c.1:60,000,000

Scales: 35 mm. = 300 Lieues d'une heure

Comments: "Pag. 482" outside border, top right; cardinal points: Septentrion & Midy outside br.; Occident & Orient betw. br. & nt. ln.

Book: Méthode Abrégée et facile pour apprendre la Géographie

AP:
AU:
NAP:

[l'Abbé Le François] Paris: Libraires Associés, 1770

Location: WLCL: C2 1770 Le

448	North America	1750	RV

AMÉRIQUE / SEPTENTRIONALE, / dressée, sur les Relations les / plus modernes des Voyageurs et Navigateurs, / et divisée suivant les differentes / possessions des Européens. / Par Le Sr. Robert de Vaugondy, / fils de Mr. Robert Géographe ordin? du Roy. / Avec Privilege. / 1750.

br: 480 x 590 mm.
pl: 500 x 632 mm.

Engravers:

RF: c.1:15,000,000

Scales: 90 mm. = 840 Mille pas géometriques de 60 au Degré; 280 Lieues marines de france et d'Angleterre de 20 au Degré; 245 Lieues d'Espagne de 17 1/2 au Degré

Comments:

State 1: 1750: no frame around scales; lakes north of L. Sup. will change; many place names not present in Canada; no islands at mouth of Miss. R.; no dotted line borders; Is. Bermudes at 37°N, 315°W.

State 2: 1750; frame around scales; Canadian place names added; dotted borders showing French claims before 1763: S. of St. Lawrence to Bay des Chaleurs; islands added at mouth of Miss. R.; diff. geography for lakes N. of L. Superior, e.g., Lac des Assinipoils

State 3: [1768]: 'Is. Bermudes ou Sommer' removed; dotted border of province of Quebec (Treaty of Paris, 1763); dotted border (Treaty of Ft. Stanwix) through Niagara and L. Erie; extension of borders of Va.,Carolina,Ga.,Fla.Occ. west to Miss.R.; "Ma-ri-land" [sic]; geography of lakes N. of L. Superior changed, e.g., Lac des Sioux

State 4: 1775: title altered: "...Européens / Publiée en 1750 et Corrigée en 1775 / Par ...Géog. ord. du Roi et / du feu Roi de Pologne Duc de Lorr? et de Bar, de la Societé / Royale de Nanci et Censeur Royal."; cartouche moved top right; scales bot. left; inset top left: PARTIE NORD-OUEST DE L'AMERIQUE (de la Fuente geography); Baye de Baffin geography changed; Assinipoels moved far north.

State 5: 1783: see Cat. 450

Book:

AP:
AU: 97
NAP:

Location: BL: C. 38. f. 4(41) (state 1); WLCL: 2-I-9b (state 2), 2-I-21 (1775)

449	North America	1762	RV

CANADA, / LOUISIANE, / POSSESSIONS ANGL? / Par le S. Robert De Vaugondy Géog? ord? / du Roi, de S.M. Polonoise Duc de / Lorr? et de Bar, et de la Société royale / de Nanci. / 1762.

Inset: *[Northwest America with De Fonte discoveries]*

A Paris ches l'Auteur quai de l'Horloge près le pont neuf

449	North America (continued)

br: 245 x 293 mm.
inset: 143 x 124 mm.

Engravers: (map) Gravé par E. Dussy
(cartouche) Arrivet inv. & Sculp.

RF: c.1:20,000,000

Scales: 55 mm. = 600 Mille pas géometriques de 60 au degré; 200 Lieues marines ou d'une heure
"Le Supplement [inset] est de moitié d'échelle"

Comments: Below inset: "Les Terres pointillées dans ce Supplément sont les Découvertes attribuées à un Amiral Espagnol nomme De Fonte ou De Fuente. Nous les avons tracées d'après la Carte de M. Joseph Delisle publiée en 1753 sans cependant les adopter, à moins que dans la suite elles ne devieñent authentiques."

State 1: 1762

State 2: 1778: date changed; Fortin publication

State 3: An 3: new cartouche (simple frame); 'corrigés par le Cen. Lamarche son successeur' added to title

Book:

AP:
AU:
NAP: 43

Location: NAC: NMC 118952 (1762); 6740 (An 3)

450	North America	1783	RV

AMÉRIQUE / SEPTENTRIONALE / Dressée sur les Relations les / plus modernes des Voyageurs et Navigateurs, / ou se remarquent. / LES ETATS UNIS. / Publiée en 1750 et Corrigée en 1783 / Par le S. Robert de Vaugondy Géographe / [ordinaire du Roi.]

br: 480 x 590 mm.
pl: 500 x 632 mm.

Engravers:

RF: c.1:15,000,000

Scales: 90 mm. = 840 Mille pas géometriques de 60 au Degré; 280 Lieues marines de france et d'Angleterre de 20 au Degré; 245 Lieues d'Espagne de 17 au Degré

Comments: State 5 of RV's Amérique Septentrionale, 1750, from AU, [Cat. 448] as State 4 but adds I. Shelrock on the Tropic of Cancer, 261°E.; configuration of Great Lakes as d'Anville (not Bellin as previous states); Is. Bermudes added at 27°N, 318° E. adds Indian tribes around Ohio R., e.g., Les Six Nations, Renards, Miamis; adds place names along east coast, e.g., Newport, N York, De Laware; "dont la source est inconnue" at N end of Miss. R.; Inset: "suivant Drake" added to Nouvelle Albion

State 6: after 1786?: removes "ordinaire du Roi" from title; Delamarche?

Book:

AP:
AU:
NAP:

Location: BL: c. 44. f. 8 (102); WLCL: 2-J-3 (both state 6)

451	North America, northwest	1768	RV

*CARTE SYSTEMATIQUE DES PAYS SEPTENTRIONAUX DE L'ASIE ET DE L'AMÉRIQUE Dressée Par le Sr. Robert de Vaugondy / d'après les Memoires et Observations Géographiques et Critiques de M*** publiés à Lausanne, format in-4° de 268 pages, 1765 / A Paris / Ches l'Auteur Quay de / l'Horloge du Palais / proche le Pont Neuf / 1768 / Multa renascentur quae iam cecidere.*

Paris: Ches l'auteur Quay de / l'Horloge du Palais / proche le Pont neuf

br: 347 x 237 mm.

Engravers:

RF: c.1:

451	North America, northwest (continued)		

Scales:

Comments: 180 longitude written 160. a re-editing of S. Engel's map, with notes in the accompanying mémoire explaining the differences: Asia terminates at 185° (Engel: 175°); Cote d'Anian at 190° (Engel: 180°)

Book: Lettre de M. Robert de Vaugondy à M*** au sujet d'une Carte Systèmatique des Pays Septentrionaux de l'Asie et de l'Amerique. Didier Robert de Vaugondy, Paris: A. Boudet, 1768 — AP: AU: NAP:

Location: JCB: M1D, bound with Engel, Mémoires et Observations Critiques

452	North America, northwest	[1772]	RV

CARTE DES NOUVELLES DÉCOUVERTES / dressée par Phil. Buache Pr. Géogr? du Roi / presentée à l'Acad. des Sciences le 9 Aout 1752 / et approuvée dans son assemblée / du 6 Septembre suivant

Extrait d'une Carte Japonoise / de l'Univers / apportée en Europe par Kaempfer / et deposée dans le cabinet de / feu Mr. Han-Sloane / président de la Société / royale de Londres

[Book: Paris: Panckoucke, Stoupe, Brunet; Amsterdam: M. M. Rey]

br: 292 x 378 mm.
pl: 316 x 399 mm.

Engravers:

RF: c.1:50,000,000

Scales:

Comments: two maps in one

outside border: "Suppl. 6e Carte" (top rt.); 165 (bot. rt.)

Book: ENCYCLOPÉDIE; Suite du Recueil des Planches, sur Les Sciences, les Arts Libéraux, et les Arts Mechaniques, avec leurs explications.... Supplément, ed. Panckoucke (1777) — AP: AU: NAP:

Location: UM, Rare Books: AE 25 .E53; NAC: NMC 9844, 9980, 9981

453	North America, northwest	1772	RV

*CARTE DES PARTIES NORD ET OUEST DE L'AMÉRIQUE / dressée d'après les relations les plus authentiques / par M*** en 1764 / Nouvelle edition réduite par M. de Vaugondy / en 1772*

[Book: Paris: Panckoucke, Stoupe, Brunet Amsterdam: M. M. Rey]

br: 290 x 368 mm.
pl: 312 x 401 mm.

Engravers:

RF: c.1:20,000,000

Scales:

Comments: outside border: "Suppl. 1re Carte" (top right); 160 (bot. rt.)

Book: ENCYCLOPÉDIE: Suite du Recueil des Planches, sur les Sciences, les Arts Libéraux, et les Arts Mechaniques, avec leurs explications... Supplément, ed. Panckoucke (1777) — AP: AU: NAP:

Location UM, Rare Books: AE 25 .E53; NAC: NMC 6687

454	North America, northwest	[1772]	RV

CARTE GÉNÉRALE DES DÉCOUVERTES / de AMIRAL DE FONTE / et autres Navigateurs Espagnols, Anglois et Russes / pour la recherche du Passage à la Mer du Sud. / Par M. De l'Isle de l'Académie royale des Sciences &c / Publiée à Paris en Septembre 1752

Inset: *Carte dressée sur la lettre de l'Amiral de Fonte / par l'Ecrivain de la Californie*

[Book: Paris: Panckoucke, Stoupe, Brunet; Amsterdam: M.M. Rey]

br: 288 x 378 mm.;
pl: 314 x 397 mm.
Inset: 98 x 92 mm.

Engravers:

RF: c.1:20,000,000

Scales:

Comments: outside border: "Suppl. 7e Carte" (top rt.); 166 (bot. rt.)

Book: ENCYCLOPÉDIE: Suite du Recueil des Planches, sur les Sciences, Les Arts Libéraux, et Les Arts Mechaniques, avec leur explications — AP: AU: NAP:

Supplément, ed. Panckoucke (1777)

Location: UM, Rare Books: AE 25 .E53; NAC: NMC 6912, 17082, 9982, 6913

455	North America, northwest	1772	RV

CARTE GÉNÉRALE / DES DÉCOUVERTES DE / L'AMIRAL DE FONTE / representant / la grande probabilité d'une / Passage au Nord Ouest / Par / Thomas Jefferys / Geographe du Roi / à Londres 1768.

[Book: Paris: Panckoucke, Stoupe, Brunet; Amsterdam: M. M. Rey]

br: 287 x 354 mm.
pl: 326 x 374 mm.

Engravers:

RF: c.1:30,000,000

Scales: 60 mm. = 240 Lieues Marines de 20 au degré

Comments: outside border, bot. left: "Cette Carte a été traduite par M. de Vaugondy et gravée à Paris en 1772"; outside border: "Suppl. 8e Carte" (top rt.); 167 (bot. rt.)

Book: ENCYCLOPÉDIE: Suite du Recueil des Planches, sur les Sciences, les Arts Libéraux, et les Arts Mechaniques, avec leurs explications... Supplément, ed. Panckoucke (1777) — AP: AU: NAP:

Location: UM, Rare Books: AE 25 .E53; NAC: NMC 9979

456	North America, northwest	1774	RV

NOUVEAU SYSTEME GÉOGRAPHIQUE / Par lequel on concilie les anciennes connoissances sur / les Pays nord-ouest de l'Amérique / avec les nouvelles découvertes des Russes / au nord de la Mer du sud. / Par Mr. de Vaugondy / 1774.

A Paris chés l'Auteur Quai de l'Horloge du Palais / près le pont neuf.

br: 286 x 366 mm.
pl: 297 x 378 mm.

Engravers:

RF: c.1:20,000,000

Scales:

Comments: shows DeFonte discoveries; accompanies book

Book: Mémoire sur les pays de l'Asie et de l'Amérique, situés au nord de la mer du Sud... — AP: AU: NAP:

Didier Robert de Vaugondy, Paris: RV & A. Boudet, 1774

Location: BN: Ge.D.12870; WLCL: C 1774 Ro

457 **Canada** 1749 RV

LES LACS DU CANADA / et / NOUVELLE ANGLETERRE / Par le Sr. Robert de Vaugondy / Fils de Mr. Robert Géog. ordin. du Roi / Avec Privilege / 1749

br: 163 x 216 mm.
pl: 179 x 244 mm.

Engravers:

RF: c.1:9,500,000

Scales: 46 mm. = 80 Lieues d'une heure.

Comments: Great Lakes as per Bellin (1744)

Book:

AP: 193
AU:
NAP:

Location:

458 **Canada** 1753 RV

CARTE DES PAYS / CONNUS SOUS LE NOM DE / CANADA / dans laquelle sont distinguées / les possessions françoises; & angl? / Dédiée et Présentée / A Monseigneur le Comte d'Argenson / Ministre Sécrétaire d'Etat / pour le département de la guerre, / par le Sr. Robert de Vaugondy fils / Géographe ordinaire du Roi.

Inset: *Supplément / pour l'Isle de / Terre-Neuve /*

A Paris Chès l'Auteur, Quai de l'horloge du Palais proche le pont-neuf. Avec Privilege 1753

br: 479 x 667 mm.
pl: 497 x 685 mm.

Engravers:

RF: c.1:4,000,000

Scales: 111 mm. = 240 Mille Pas géometriques de 60 au degré; 100 Lieues communes de France de 25 au degré.; 80 Lieues marines de France et d'Angleterre de 20 au degré; Inset: "l'Echelle est la même que / celle de la carte."

Comments:
State 1: 1753: area south of Lake Erie shows Ohio River running east to west, no forts

State 2: [1755]; Ohio River follows Bellin (1755), i.e., N-E to S-W; English and French forts added, e.g. F. Duquesne

State 3: 1778: new title: *CARTE DU / CANADA / et des ETATS-UNIS / de l'Amérique Septentrionale / par le Sr. Robert de Vaugondy Geog.ord.du Roi / du Feu Roi de Pologne Duc de Lorraine et de Bar,de la / Société Royale de Nancy et Censeur Royal. / A Paris / Chez Fortin Ingr. Mécanicien du Roi pour les Globes et / Spheres; Rue de la Harpe pres celle du Foin / 1778.* Coat of arms removed from cartouche; crown of Indian feathers added; many place name additions, e.g. Villiamsburg, Ticonderoga, Great Dismal Swamp

See Chapter Four re: controversy on first publication

Book:

AP:
AU:
NAP:

Location: NAC: NMC 24548 [1753], 52294 [1755], 24639 [1778]; WLCL: 4-B-31 [1755]; Arkway: 1778; Julien Cat. (1763): no. 921, price 1# 5s.

459 **Canada** 1755 RV

Partie de / l'Amérique Septent? / qui comprend / LA NOUVELLE FRANCE / ou le CANADA, / Par le Sr. Robert de Vaugondy Géog? / Ordinaire du Roy. / Avec Privilege 1755.

Inset: *Supplément / pour / Les Lacs du Canada*

br: 477 x 605 mm.
pl: 495 x 612 mm.
Inset: 220 x 375 mm.

459 **Canada** (continued) 1755 RV

Engravers: (cartouche) C. haussard sculp.;

Inset: C. haussard fecit

RF: c.1:4,000,000

Scales: 83 mm. = 180 Mille pas géometriques de 60 au dégré; 60 Lieues Marines ou d'une heure de 20 au dégré

Comments:
State 1: 1755: no I. Charlton in James Bay; F. de la Presque Isle written perpendicular to shore of L. Erie; Ouabache has no R. (for River)

State 2: 1755: I. Charlton added; F. de la Presque Isle written parallel to shore of Erie; R. added to Ouabache; dotted border south of L. du S. Sacrément and L Champlain to 45th parallel; Jamés Bay (with accent)

State 3: [1768]: Treaty of Paris 1763 border dotted line, province of Quebec; NOUVELLE ECOSSE added; Jamess [sic] Bay; B. de Korck changed to B. de Cork (Newfoundland); N boundary of NY removed; boundary added S of L. Ont. (Treaty of Ft. Stanwix)

State 4: [1778]; "C. haussard fecit" and "C. haussard sculp." removed from cartouches; "Groux" added to cartouche

State 5: [after 1784]; Delamarche imprint; "ordinaire du roi/Avec privilege" removed; borders of Etats-Unis added

Book:

AP:
AU: 98
NAP:

Location:

460 **Canada** [1773] RV

Partie de la Carte / du CAPITAINE CLUNY / Auteur d'une ouvrage anglois intitulé / American Traveller / publié à Londres en 1769

[Book: Paris: Panckoucke, Stoupe, Brunet; Amsterdam: M. M. Rey]

br: 290 x 495 mm.
pl: 316 x 505 mm.

Engravers:

RF: c.1:50,000,000

Scales:

Comments: outside border: "Suppl. Carte 10" top rt.; 169 bot.rt.

Book: ENCYCLOPÉDIE: Suite du Recueil des Planches, sur les Sciences, les Arts Libéraux, et les Arts Mechaniques, avec leurs explication...

AP:
AU:
NAP:

Supplément, ed. Panckoucke (1777)

Location: UM, Rare Books: AE 25 .E53; NAC: NMC 8059, 8060, 8267.

461 **Canada: Hudson Bay** 1749 RV

BAYES / D'HUDSON et de BAFFINS, / et TERRE DE LABRADOR. / Par le Sr. Robert de Vaugondy / fils de Mr. Robert Geogr. du Roi / Avec Privilege 1749.

br: 163 x 161 mm.
pl: 173 x 171 mm.

Engravers:

RF: c.1:22,080,000

Scales: 41 mm. = 160 Lieues d'une heure.

Comments: note: "Septentrion" written outside border (congested long. numbers on conic projection)

Book:

AP: 200
AU:
NAP:

Location:

462	Canada: Hudson Bay	1773	RV

Carte qui représente les différentes connoissances que l'on a eues des TERRES ARCTIQUES depuis 1650 jusq'en 1747 ausquelles il faut comparer la Carte suivante Ce. 10 Par M. De Vaugondy 1773.

[Book: Paris: Panckoucke, Stoupe, Brunet; Amsterdam: M.M. Rey]

br: 294 x 347 mm.
pl: 313 x 375 mm.

Engravers:

RF: c.1:

Scales: various

Comments: outside border "Suppl. Carte 9e." (top rt); 168 (bot.rt.)
four maps on one sheet: Selon Sanson en 1650; Selon Delisle en 1700; Selon Delisle en 1703; Carte de la BAYE D'HUDSON parcourue en 1746 et 1747 par Henri Ellis / pour la recherche du passage par le Nord-Ouest.

Book: ENCYCLOPÉDIE: Suite du Recueil des Planches, sur les Sciences, les Arts Libéraux, et les Arts Mechaniques, avec leurs explications... Supplément, ed. Panckoucke (1777)
AP:
AU:
NAP:

Location: UM, Rare Books: AE 25 .E53; NAC: NMC 6173, 6274, 6951

463	Canada: Cape Breton Island	1749	RV

ISLE ROYALE / Par le Sr. Robert de Vaugondy / Fils de Mr. Robert Géog. ord. du Roi / Avec Privilege / 1749.

br: 163 x 194 mm.
pl: 179 x 206 mm.

Engravers:

RF: c.1:1,350,000

Scales: 42 mm. = 10 Lieues mar. Francoises Angloises et d'une heure

Comments:

Book:
AP: 198
AU:
NAP:

Location:

464	Canada: Newfoundland	1749	RV

GOLFE DE ST. LAURENT, / ISLE et BANCS / de TERRE NEUVE. / Par le Sr. Robert de Vaugondy / Fils de Mr. Robert Geogr. ord. du Roi / Avec Privilege 1749.

br: 161 x 181 mm.
pl: 176 x 198 mm.

Engravers:

RF: c.1:7,700,000

Scales: 35 mm. = 50 Lieues d'une heure.

Comments:

Book:
AP: 197
AU:
NAP:

Location:

465	Canada: Newfoundland	1749	RV

ISLE DE / TERRE-NEUVE / Par le Sr. Robert de Vaugondy / Fils de Mr. Robert Geog. ord. du Roi / Avec Privilege / 1749.

br: 170 x 191 mm.
pl: 178 x 202 mm.

Engravers:

465 Canada: Newfoundland (continued)

RF: c.1:3,200,000

Scales: 35 mm. = 20 Lieues d'une heure

Comments: In center of island: "L'interieur du Pays est entierement inconnu"

Book:
AP: 199
AU:
NAP:

Location:

466	Canada: Acadia	1749	RV

L'ACADIE. / Par le Sr. Robert de Vaugondy / Fils de Mr. Robert Géogr. ord. du Roi / Avec Privilege / 1749.

br: 161 x 203 mm.
pl: 180 x 214 mm.

Engravers:

RF: c.1:2,700,000

Scales: 41 mm. = 20 Lieues Marines et d'une heure

Comments:

Book:
AP: 196
AU:
NAP:

Location:

467	Canada: St. Lawrence River	1749	RV

Partie du Canada / où se trouvent / le FLEUVE ST. LAURENT et / LA NOUVELLE ECOSSE / Par le Sr. Robert de Vaugondy / Fils de Mr. Robert Geogr. ord. du Roi / Avec Privilege 1749

br: 164 x 218 mm.
pl: 178 x 244 mm.

Engravers:

RF: c.1:5,800,000

Scales: 38 mm. = 40 Lieues marines d'une heure

Comments:

Book:
AP: 195
AU:
NAP:

Location:

468	New England	1749	RV

NOUVELLE ANGLETERRE / NLLE. YORK, NLLE. JERSEY. / PENSILVANIE MARILAND / et VIRGINIE. / Par le Sr. Robert de Vaugondy Fils / de Mr. Robert Geog. ordin. du Roi / Avec Privilege 1749.

br: 162 x 193 mm.
pl: 176 x 203 mm.

Engravers:

RF: c.1:5,800,000

Scales: 38 mm. = 40 Lieues d'une heure.

Comments: "198" outside border; should be "194"

Book:
AP: 194
AU:
NAP:

Location:

469	**Ohio River, environs**	1755	RV

Partie / de l'Amerique Septentrionale, / qui comprend / LE COURS DE L'OHIO, / LA NLLE. ANGLETERRE, LA NLLE YORK, / LE NEW JERSEY, LA PENSYLVANIE, / LE MARYLAND LA VIRGINIE, / LA CAROLINE / Par le Sr. Robert de Vaugondy Géographe / ordinaire du Roi. / Avec Privilége 1755.

Inset: *Supplément Pour / La Caroline*

br: 480 x 620 mm.;
pl: 494 x 628 mm.
Inset: 180 x 216 mm.

Engravers: (cartouche) M.C. haussard fecit

RF: c.1:2,900,000

Scales: 75 mm. = 120 Mille pas géometriques, de 60 au degré; 40 Lieues Marines, d'une heure, et de 20 au degré; 138 Milles d'Angleterre, de 69 au degré

Comments:
State I: 1755: dotted border is pre-Treaty of Paris, 1763, from 45th parallel south along the Appalachians; "LOUISIANE" in western frontier

State 2: [1768]; dotted border of Treaty of Paris, 1763, with Virginie and Caroline Sept. et Merid. extended west to Mississippi; Penn. extended west to and into L. Erie; 45th parallel border removed and province of Quebec border added; "LOUISIANE" removed; "New-Hamp-Shire" and "Prov. de Sagadahock" added

Book:
AP:
AU: 99
NAP:

Location:

470	**Virginia, Maryland**	1755	RV

Carte / de la VIRGINIE et / du MARYLAND / Dressée sur la grande carte Angloise de Mrs. Josué / Fry et Pierre Jefferson, / Par le Sr. Robert de Vaugondy Géographe / ordinaire du Roi. / Avec Privilege / 1755.

br: 483 x 642 mm.
pl: 492 x 666 mm.

Engravers: (cartouche) Gravé E. haussard

RF: c.1:870,000

Scales: 96 mm. = 45 Mille pas géométriques de 60 au degré; 15 Lieues Marines d'Angleterre de 20 au degré; 51 1/4 Milles de 69 au degré

Comments
State 1: 1755: western border of Penn. to 297°30'; "Limites de Lord Fairfax" run NW-SE across Appal. Mts.; Haussard cartouche

State 2: [1776?]; west border of Penn. extended to left edge of map; "Limites de Lord Fairfax" removed; dated by C. Verner 1776a? (L. Fairfax's refusal to declare allegiance to new state after Decl. of Ind.), but is in edition accompanying Dictionnaire Géog. (1768)

State 3: [1778]; "Gravé E. haussard" removed; "Groux" added to cartouche; Verner dates 1776b?, but probably Fortin publication (i.e. after 1778)

State 4: [after 1784]; Delamarche: "du roi" and "avec privilege" removed from title; Verner: 1793a?

State 5: [1793?] "ordinaire" and date removed from title; Verner: 1793b?

Bibliography: C. Verner, *IM* XXI (1967);

Book:
AP:
AU: 100
NAP:

Location:

471	**Louisiana**	1749	RV

COURS DU MISSISSIPI / et / LA LOUISIANE / Par le Sr. Robert de Vaugondy. / fils de Mr. Robert Geog. du Roi / Avec Privilege / 1749.

br: 163 x 216 mm.
pl: 173 x 237 mm.

Engravers:

RF: c.1:9,600,000

Scales: 46 mm. = 80 Lieues d'une heure.

Comments: n.b.: 31°30'N, 283° E: "Ici fut tué Mr. de la Salle en 1686."

Book:
AP: 188
AU:
NAP:

Location:

472	**Florida**	1749	RV

LA FLORIDE / divisée en / Floride et Caroline / Par le Sr. Robert de Vaugondy / Fils de Mr. Robert Geogr. Ordin. du Roi / Avec Privilege / 1749

br: 164 x 174 mm.
pl: 180 x 181 mm.

Engravers:

RF: c.1:9,600,000

Scales: 35 mm. = 60 Lieues d'une heure

Comments:

Book:
AP: 189
AU:
NAP:

Location:

473	**California**	1772	RV

Carte / de la CALIFORNIE / et des Pays Nord-Ouest / separés de l'Asie par / le détroit d'Anian / extraite de deux cartes publiées / au commencement du 17e Siécle / Par le S. Robert de Vaugondy Géog. ord. du Roi / du feu Roi de Pologne Duc de Lorraine et de Bar / et de l'Academie Royale des Sciences et Belles Lettres / de Nanci et Censeur royal 1772.

[Book: Paris: Panckoucke, Stoupe, Brunet; Amsterdam: M. M. Rey]

br: 289 x 322 mm.
pl: 318 x 394 mm.

Engravers:

RF: c.1:20,000,000

Scales: 58 mm. = 40 Lieues marines de 20 au degré

Comments: two maps on one sheet
outside border: "Suppl. 4e Carte" (top rt.); 163 (bot. rt.)
maps are reductions from 4 sheet Visscher map of 1612 and from Mappemonde of Pierre Plantius dedicated to Louis XIII, 1641 (published by Visscher)

Book: ENCYCLOPÉDIE: Suite du Recueil des Planches, sur les Sciences, les Arts Libéraux, et les Arts Mechaniques, avec leurs explications... Supplément, ed. Panckoucke (1777)
AP:
AU:
NAP:

Location: UM, Rare Books: AE 25 .E53

474 **California** [1772] RV

Carte de la CALIFORNIE / Suivant / I la Carte manuscrite de l'Amérique de Mathieu / Neron Pecciolen dressé à Florence en 1604 / II Sanson 1656 / III De l'Isle Amérique Sept. 1700 / IV le Pere Kino Jesuite en 1705 / V la Société de Jésuites en 1767 / La Côte orientale depuis de C. des Vierges jusqu'à l'embou- / chure / du R. Colorado est extraite de la Carte du P. Ferdinande / Gonsaque dressée en 1746.

[Book: Paris: Panckoucke, Stoupe, Brunet; Amsterdam: M. M. Rey]

br: 293 x 367 mm.
pl: 316 x 399 mm.

Engravers:

RF: c.1:

Scales: various

Comments: outside border: "Supple 5e Carte" (top rt.); 164 (bot. rt.)
five maps on one sheet, showing history of mapping of California

Book: ENCYCLOPÉDIE: Suite du Recueil des Planches, sur les Sciences, les Arts Libéraux, et les Arts Mechaniques, avec leurs explications
Supplément, ed. Panckoucke (1777)

AP:
AU:
NAP:

Location: UM, Rare Books: AE 25 .E53

475 **West Indies: Antilles** 1749 RV

LES ISLES ANTILLES / Par le Sr. Robert de Vaugondy / Fils de Mr. Robert Géogr. ord. du Roi / avec Privilege / 1749.

br: 162 x 228 mm.
pl: 175 x 241 mm.

Engravers:

RF: c.1:17,800,000

Scales: 43 mm. = 100 Lieues marines de 20 au Degre.

Comments:

Book:

AP: 190
AU:
NAP:

Location:

476 **West Indies: Antilles** 1750 R

Partie / DE LA MER DU NORD, / où se trouvent / LES GRANDES ET PETITES ISLES / ANTILLES, / et Les Isles Lucayes. / Par Le Sr. Robert / Géographe ordinaire du Roy. / Avec Privilege / 1750.

br: 484 x 592 mm.
pl: 496 x 625 mm.

Engravers: (map) Guill? Delahaye Sculpsit

RF: c.1:5,000,000

Scales: 92 mm. = 48 Mille Pas Géometriques; 16 Lieues Marines de 20 au Degré.

Comments:
State 1: 1750: has no cardinal points between neat line and border

State 2: 1750: cardinal points engraved between neat line and border

Book:

AP:
AU: 102
NAP:

Location: WLCL: 7-G-1750 (States 1 & 2); NAC: G 1100 .C656 1775 H1 Vault

477 **West Indies: Martinique** 1749 RV

L'Isle de / LA MARTINIQUE. / Par le Sr. Robert de Vaugondy / Fils de Mr. Robert Geogr. ord. du Roi / Avec Privilege / 1749.

br: 162 x 192 mm.
pl: 177 x 201 mm.

Engravers:

RF: c.1:4,300,000

Scales: 41 mm. = 3 Lieues d'une heure.

Comments:

Book:

AP: 192
AU:
NAP:

Location:

478 **West Indies: Santo Domingo** 1749 R

Carte de l'Isle / de SAINT-DOMINGUE / dressée d'après la Carte originale / de Mr. Frezier / par le Sr. Robert Géographe ordin? du Roi / Avec Privilege 1749.

br: 462 x 598 mm.
pl: 470 x 608 mm.

Engravers:

RF: c.1:1,350,000

Scales: 121 mm. = 90 mille pas géometriques de 60 au degré; 30 grandes lieues de France, Marine, et d'une heure de 20 au degré.; 26 1/4 Lieues Espagnols de 17 1/2 au degré

Comments: Frézier's map of St. Domingo from 1722 (Tooley)
n.b.: islands on E. cut into neat line border
Julien Cat. (1763): price, 1# 5s.

Book:

AP:
AU:
NAP:

Location: WLCL: 8-H-20; Julien Cat. (1763): no. 943; RV, Catalogue (1777)

479 **West Indies: Santo Domingo** 1749 RV

L'Isle / ST. DOMINGUE / Par le Sr. Robert de Vaugondy / Fils de Mr. Robert Geogr. ord. du Roi / Avec Privilege 1749.

br: 161 x 214 mm.
pl: 179 x 223 mm.

Engravers:

RF: c.1:3,800,000

Scales: 44 mm. = 30 Lieues d'une heure.

Comments:

Book:

AP: 191
AU:
NAP:

Location:

480 **West Indies: Santo Domingo** 1750 R

ISLES DE / SAINT DOMINGUE / ou Hispaniola, / et de la MARTINIQUE. / Par le Sr. Robert Geographe ord. du Roy. / Avec Privilege. / 1750.

Inset: *ISLE DE / LA MARTINIQUE*

br: 482 x 517 mm.;
pl: 497 x 540 mm.
Inset: 161 x 191 mm.

Engravers:

480	**West Indies: Santo Domingo** (continued)

RF: c.1:1,700,000

Scales: 98 mm. = 30 Lieues Marines et d'une heure de 20 au Degré; 26 Lieues d'Espagne de 17 au Degré; Inset: 37 mm. = 3 Lieues d'une heure

Comments:

Book:

AP:
AU: 103
NAP:

Location:

481	**Central America**	[1762]	RV

NOUVELLE ESPAGNE, / NOUVEAU MEXIQUE, / ISLES ANTILLES / Par le S. Robert de Vaugondy / Géog? ordinaire du Roy &c.

Inset: *[ISLES ACORES OU TERCERES]*

A Paris ches l'Auteur quai de l'Horloge près le pont neuf

br: 245 x 313 mm.
inset: 42 x 70 mm.

Engravers: (map) Gravé par E. Dussy
(cartouche) Arrivet inv. & Sculp.

RF: c.1:20,000,000

Scales: 55 mm. = 600 Mille pas géometriques de 60 au degré; 200 Lieues marines de 20 au degré

Comments: inside neat line, under scales: "Les Audiences de Guadalajara, de Mexico, et de Guatimala, forment Le Mexique ou la Nouvelle Espagne."

State 1: [1762]: no date

State 2: [1778]: imprint removed; Fortin publication

Book:

AP:
AU:
NAP: 44

Location:

482	**Central America: Mexico**	1749	RV

Partie du MEXIQUE ôu de / la Nouvle. Espagne ou se trouve / l'Audce. de Guadalajara, / Nouveau Mexique, / Nouvelle Navarre, / Californie &c / Par le Sr. Robert de Vaugondy / Fils de Mr. Robert Géog. du Roi A.Pr. 1749.

br: 161 x 199 mm.
pl: 180 x 207 mm.

Engravers:

RF: c.1:14,400,000

Scales: 40 mm. = 100 Lieues marines de 20 au Deg? et d'une heure

Comments:

Book:

AP: 187
AU:
NAP:

Location:

483	**Central America: Mexico**	1749	RV

Partie du MEXIQUE / ou de la Nouvlle. Espagne / ôu se trouve / l'Audience du Mexique / Par le Sr. Robert de Vaugondy / Fils de Mr. Robert Géogr? ordin? du Roi. / Avec Privilege 1749.

br: 162 x 240 mm.
pl: 177 x 245 mm.

Engravers:

483	**Central America: Mexico** (continued)

RF: c.1:8,200,000

Scales: 55 mm. = 80 Lieues mar. Francoises, Angloises et d'une heure; 70 Lieues Espagnoles de 17 au Degré.

Comments: Note: number engraved as 138 instead of 186

Book:

AP: 186
AU:
NAP:

Location:

484	**Central America: southern**	1749	RV

Partie du MEXIQUE ou de / la Nouvelle Espagne / oû se trouve / l'Audience de GUATIMALA / Par le Sr. Robert de Vaugondy / Fils de Mr. Robert Géogr. ord. du Roi / Avec Privilege. / 1749.

br: 162 x 240 mm.
pl: 180 x 243 mm.

Engravers:

RF: c.1:8,200,000

Scales: 55 mm. = 80 Lieues mar. Francoises Angloises et d'une heure; 70 Lieues Espagnoles de 17 1/2 au Degré

Comments:

Book:

AP: 185
AU:
NAP:

Location:

485	**South America**	1748	R

AMERIQUE / MERIDIONALE. / Par le Sr. Robert Géog? ord. du Roi / 1748.

br: 158 x 197 mm.
pl: 168 x 203 mm.

Engravers: (map) Gravé par Guill? Delahaye

RF: c.1:52,800,000

Scales: 23 mm. = 200 Lieues d'une heure de Chemin.

Comments:

Book:

AP: 201
AU:
NAP:

Location:

486	**South America**	1750	RV

AMERIQUE / MÉRIDIONALE. / Par le Sr. Robert de Vaugondy / Avec Privilege / 1750.

br: 150 x 153 mm.

Engravers:

RF: c.1:60,000,000

Scales: 33 mm. = 300 lieues d'une heure

Comments: "Pag. 487" outside border, top right

Book: Méthode Abrégée et facile pour apprendre la Géographie

AP:
AU:
NAP:

[l'Abbé Le François] Paris: Libraires Associés, 1770

Location: WLCL: C2 1770 Le

487 **South America** 1750 RV

AMÉRIQUE / MÉRIDIONALE. / Dressée, sur les Mémoires les plus récents / et assujétie aux observations astronomiques / Par le Sr. Robert de Vaugondy / fils de Mr. Robert Géographe ordin. du Roy. / Avec Privilege. / 1750.

br: 480 x 593 mm.
pl: 498 x 635 mm.

Engravers: (cartouche) Gobin fecit [see Comments]

RF: c.1:20,000,000

Scales: 90 mm. = 840 Mille Pas géometriques, de 60 au Degré.; 280 Lieues Marines ou d'une heure, de Degré.; 245 Lieues d'Espagne, de 17 1/2 au degré.

Comments:
State 1: 1750: cartouche: "Gobin fecit"; bottom portion of cartouche is very simple scrollwork and one alligator; perhaps this is unfinished?

State 2: 1750: cartouche unsigned; top portion as in State 1; below, alligator reworked, foliage, fruits, and waterfalls added; "ISLES GALAPES" (name), I. de Martin Vaz (name), and I. de Fernand de Noronha (name and island with rocks) added; Arouba changed to Oruba;

State 3: 1776: date changed; "Censeur Royal" added to title; note top half of cartouche more faded than bottom;

Book:

AP:
AU: 101
NAP:

Location: BL: maps. c.f.4. (1750, State 1); BN: Ge.D.11797 (1776); JCB: AU [1776]; NAC: G 1100 .C656 1775 H1 Vault

488 **South America, north** 1749 RV

Partie Occidentale de / la TERRE FERME / Par le Sr. Robert de Vaugondy / Fils de Mr. Robert Geog. ord. du Roi / Avec Privilege / 1749.

br: 160 x 237 mm.
pl: 177 x 243 mm.

Engravers:

RF: c.1:10,800,000

Scales: 42 mm. = 80 Lieues d'une heure

Comments:

Book:

AP: 202
AU:
NAP:

Location:

489 **South America, north** 1749 RV

Partie Orientale / de la TERRE FERME ou / l'on trouve la GUIANE.
Partie Orientale du Fl. Maragnon / ou AMAZONE / tirée de la Carte de Mr. de la Condamine. / Par le Sr. Robert de Vaugondy. / Fils de Mr. Robert Geog. ord. du Roi A. Pr. 1749

br: 160 x 223 mm.
pl: 173 x 238 mm.

Engravers:

RF: c.1:11,000,000

Scales: 40 mm. = 80 Lieues Marines de 20 au Deg?

Comments:

Book:

AP: 203
AU:
NAP:

Location:

490 **Peru, north** 1749 RV

Partie Septentrionale du / PEROU / Où l'on trouve la partie occidentale / du Fleuve Maragnon ou AMAZONE / tirée de la Carte de / Mr. de la Condamine / Par le Sr. Robert de Vaugondy / Fils de Mr. Robert Geogr. ordin. du Roi / avec Privil. 1749.

br: 162 x 192 mm.
pl: 175 x 202 mm.

Engravers:

RF: c.1:11,000,000

Scales: 40 mm. = 80 Lieues marines de 20 au Degré

Comments:

Book:

AP: 204
AU:
NAP:

Location:

491 **Peru, south** 1749 RV

Partie Meridionale / du PEROU / Par le Sr. Robert de Vaugondy / Fils de Mr. Robert Geogr. ord. du Roi / Avec Privilege / 1749.

br: 161 x 192 mm.
pl: 173 x 203 mm.

Engravers:

RF: c.1:11,000,000

Scales: 51 mm. = 100 Lieues marines ou d'une heure

Comments:

Book:

AP: 205
AU:
NAP:

Location:

492 **Chile** 1749 RV

LE CHILI. / Par le Sr. Robert de Vaugondy / Fils de Mr. Robert Géogr. ord. du Roy / Avec Privilege / 1749.

br: 190 x 163 mm.
pl: 203 x 174 mm.

Engravers:

RF: c.1:12,000,000

Scales: 46 mm. = 100 Lieues marines et d'une heure.

Comments:

Book:

AP: 206
AU:
NAP:

Location:

493 **South America, south** 1762 RV

PARAGUAY, / CHILI, / TERRE MAGELLAN? / Par le S. Robert de Vaugondy / Géog? ordinaire du Roy &c. / 1762.

A Paris ches l'Auteur quai de l'Horloge près le pont neuf

br: 244 x 217 mm.

Engravers: (map) Gravé par E. Dussy
(cartouche) Arrivet inv. & Sculp.

RF: c.1:20,000,000

Scales: 41 mm. = 90 Mille pas géom? de 60 au degré; 30 Lieues marines ou d'une heure

493 **South America, south** (continued)

Comments:
State 1: 1762

State 2: [1778]: imprint and date removed; Fortin publication

Book:

AP:
AU:
NAP: 46

Location:

494 **South America: Straits of Magellan** 1749 RV

LES TERRES / MAGELLANIQUES / Par le Sr. Robert de Vaugondy / Fils de Mr. Robert Geogr. ord. du Roi / avec Privilege / 1749.

br: 192 x 163 mm.
pl: 198 x 177 mm.

Engravers:

RF: c.1:12,000,000

Scales: 46 mm. = 100 Lieues marines ou d'une heure

Comments: n.b.: Falkland Is. as "I. Nouvelles peu connus"

Book:

AP: 207
AU:
NAP:

Location:

495 **South America:Straits of Magellan** 1756 RV

Carte / du DÉTROIT DE MAGELLAN / et des Isles Malouines / dressée pour la lecture / de l'Histoire des Terres Australes / par le Sr. Robert de Vaugondy Géog? ord? du Roi. / de l'Académie royale des Sciences et Belles-Lettres de Nancy. / 1756.

[Book: Paris: Durand, rue du Foin, au Griffon]

br: 232 x 445 mm. (reprint)

Engravers: (map) G.De-la-Haye.

RF: c.1:2,500,000

Scales: 87 mm. = 40 Lieues marines ou d'une heure de 20 au Degré.

Comments: "I. des Etats nommé Albemarle par Barthelmi Sharp" at 310°E long.

Book: Histoire des Navigations aux Terres Australes.

AP:
AU:
NAP:

Président Charles de Brosses (1756)

Location: CUL: D.18.35; UM: G 160 .B9 1967 (reprint)

496 **Paraguay** 1749 RV

LE PARAGUAY / Par le Sr. Robert de Vaugondy / Fils de Mr. Robert Geog. du Roi / Avec Privilege / 1749.

br: 160 x 188 mm.
pl: 172 x 198 mm.

Engravers:

RF: c.1:14,400,000

Scales: 38 mm. = 100 Lieues marines ou d'une heure

Comments:

Book:

AP: 208
AU:
NAP:

Location:

497 **Brazil** 1749 RV

Terre de / Sainte Croix, / nommée à present / BRÉSIL. / Par le Sr. Robert de Vaugondy / Fils de Mr. Robert Geog. du Roi A.Pr. / 1749.

br: 193 x 163 mm.
pl: 202 x 175 mm.

Engravers:

RF: c.1:14,400,000

Scales: 37 mm. = 100 Lieues marines ou d'une heure

Comments:

Book:

AP: 209
AU:
NAP:

Location:

498 **Brazil** 1762 RV

TERRE-FERME / PEROU, BRÉSIL, / PAYS DE L'AMAZONE / Par le S. Robert de Vaugondy / Géog? ordinaire du Roi &c. / 1762

A Paris ches l'Auteur quai de l'Horloge près le pont neuf

br: 243 x 284 mm.

Engravers: (map) Gravé par E. Dussy
(cartouche) Arrivet inv. & Sculp.

RF: c.1:20,000,000

Scales: 41 mm. = 90 Mille pas géom? de 60 au degré; 30 Lieues Marines ou d'une heure

Comments:
State 1: 1762

State 2: [1778]: date and imprint removed; Fortin publication

Book:

AP:
AU:
NAP: 45

Location:

499 **Brazil** 1776 RV

Partie de la Carte de l'AMÉRIQUE MÉRIDIONALE de M. Green / publiée à Londres par acte de Parlement du 19 fevrier 1753 Chès Jefferys. / Gravée sous l'inspection de Robert de Vaugondy Géographe ord? du Roi et Censeur royal 1776.

[Book: Paris: Boudet 1776]

br: 292 x 218 mm.
pl: 301 x 233 mm.

Engravers: (map) Gravé par E. Dussy.

RF: c.1:40,000,000

Scales:

Comments: has table of long. & lat. comparing observations from Bellin's map (1744) and d'Anville's map; a note describes why Mr. Green's map is preferable to M. Bellin's

Book: Dissertation historique et géographique sur le méridien de démarcation entre les domaines d'Espagne et de Portugal.

AP:
AU:
NAP:

Don George Juan & Don Antoine de Ulloa, 1776.

Location: WLCL: C2 1776 Ju

500 **Practice map** [1748] [R]

INTRODUCTION à la / coñoissance et à l'usage / DES CARTES.

br: 160 x 198 mm.
pl: 175 x 247 mm.

Engravers:

RF: c.1:625,000

Scales: 44 mm. = 15 Mille pas Géométriques ou minutes de degrés de latitude; 5 Lieues d'une heure de Chemin

Comments: an introduction to map-reading with representative symbols on the map

cardinal points outside the border

Book:

AP: 1
AU:
NAP:

Location:

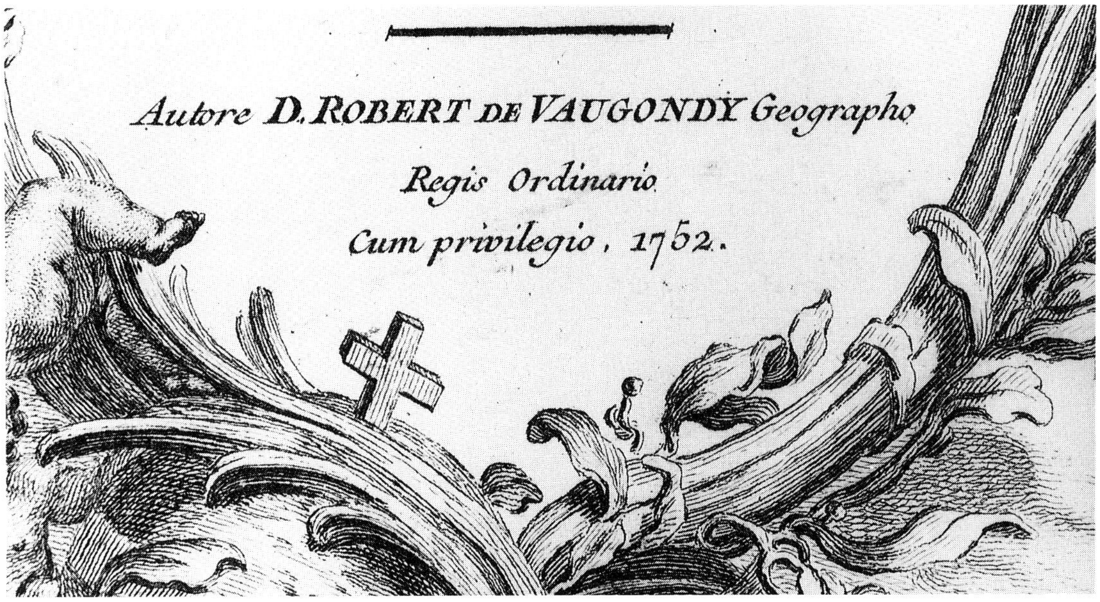

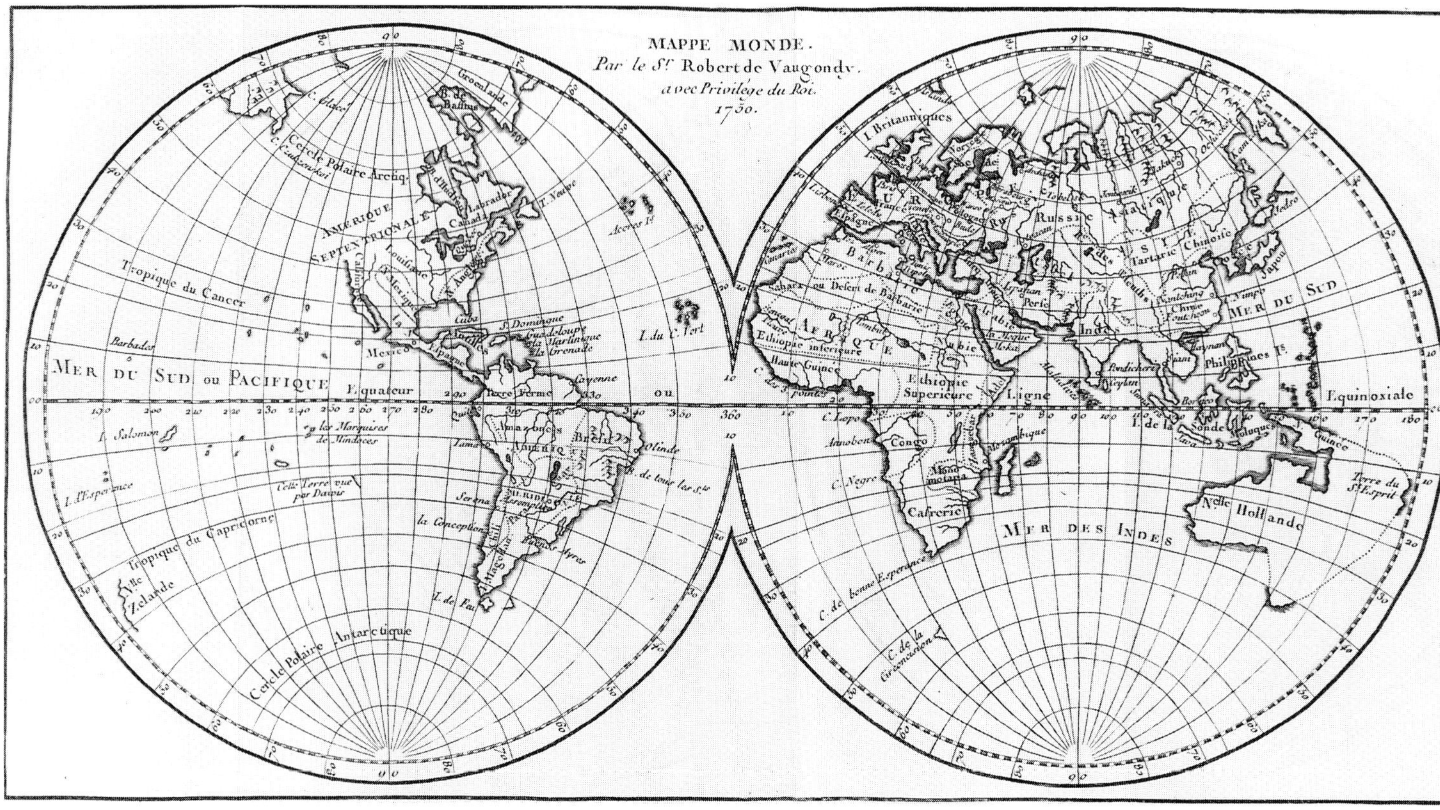

Mappemonde [Cat. 7] from **Méthode abrégée et facile pour apprendre la Géographie** *(Photo courtesy William L. Clements Library)*

Atlas Portatif Universel et Militaire

ATLAS PORTATIF UNIVERSEL ET MILITAIRE / composé d'après les meilleures Cartes, tant gravées que manuscrites / des plus célèbres Géographes et Ingenieurs. / Par M. Robert, Géographe ordinaire du Roi. / A Paris / chez l'Auteur, Quay de l'Horloge. / Durand Libraire, rue St. Jacques, au Griffon, / Pissot Fils, Libraire, Quay des August? à la Sagesse. / Avec Privilege du Roi / 1748.

Description:
Engraved title page: "Maison-neuve sculpsit, "Guill. Delahaye scripsit";
Introduction: "Introduction à la connoissance et à l'usage des Cartes"
Table of Contents: "Ordre des Cartes de l'Atlas Portatif";
209 maps

Editions and Locations

1748	BN: Estampes xf.23 136 maps; lacks detail maps of North and South America, China, and India
1748/49	LC:Philips 5981 80 maps; table of contents entitled "Liste des cartes du supplément à l'Atlas Portatif"
[1749]	BN: Ge.FF.9869 LC: Philips 608 NAC: G 1015.R61 1749 209 maps, t.p., intro., t. of c. Variants: BOD: 4 Delta 302: no intro. or t. of c. CUL: Atlas 7.74.14: interleaved with Mercator *Atlas Minor* Harvard, Houghton Library: MA.18.48.S: 207 maps in four vols. NL: Greenlee 4891.R65.1784 : dated 1784 but no map later than 1749; no t.p.; "du Roi" crossed out with ink in every instance

new title-page:

ATLAS PORTATIF, / UNIVERSEL, / Composè d'après les meilleures Cartes, tant gravées que / manuscrites des plus célèbres Géographes et Ingenieurs / Par Robert, Géographe ordinaire du Roi. / A Falaise Chez Pîtel-Préfontaine Libraire; et A Paris / Chez Panckouke Libraire, Rue et à côté de la Comédie Française.

1769?	BN: Ge.FF.2658 209 maps; no date on title-page; title-page signature: "Baisiez Scr." maps set far to the right on oblong paper
1769	University of Southern Maine: Smith Collection 209 maps; date added to title page; Mappemonde dated 1768, engraved by Dussy; Asie dated 1754
[1799?]	Nancy, Bibliothèque Municipale: 263.500 113 maps title page imprint: "Au Dépôt de Géographie, Rue Geoffroy Langevin, No. 328//Se trouve A Paris au Bureau de l'Atlas / National, Rue de la Harpe, No. 26. Et chez tous les Marchands de Géographie" "géographe du roi" not removed; many maps dated 1769; Mappemonde 1774, with Russian discoveries but no de la Fuente geography; an edition after death of Panckoucke (1798)?

SHORT TITLES OF MAPS IN THE *ATLAS UNIVERSEL PORTATIF ET MILITAIRE*

Numerals in bold type refer to number of map catalogue entry.

1 **500** INTRODUCTION à la / connoissance et à l'usage / DES CARTES. 1748 R

2 **5** MAPPEMONDE / ou Description du / Globe Terrestre 1748 R

3 **329** L'EUROPE / divisée / suivant ses principx. Etats 1748 R

4 **41** CARTE / DES / ISLES BRITANNIQUES 1748 R

5 **47** ROYME. D'ANGLETERRE / divisé par Provinces 1748 R

6 **51** ROYAUME / D'ECOSSE / divisé en Provinces 1748 R

7 **55** Isle et Royaume / d'IRLANDE 1748 R

8 **67** Royaumes de / SUEDE et DE NORWEGE 1748 R

9 **71** Royaume de / DANEMARCK / 1748 R

10 **74** RUSSIE BLANCHE / ou MOSCOVIE 1748 R

11 **95** FRANCE / ECCLESIASTIQUE 1748 R

12 **82** La France / divisée / par Gouvernemens généraux. 1749 RV

13 **96** Gouvernemens / Generaux de / PICARDIE ET ARTOIS 1748 R

14 **100** Gouvernement / General / de NORMANDIE 1748 R

15 **104** Gouvernement / General / de BRETAGNE / 1748 R

16 **107** Gouvernemens / Generaux / du Maine et Perche, 1748 R

17 **111** Gouvernement General/d'ORLEANOIS 1748 R

18 **114** Gouvernemens / Generaux, / de l'ISLE DE FRANCE, 1748 R

19 **115** Banlieue / et Environs de / PARIS. 1748 R

20 **128** LORRAINE / et ALSACE1748 R

21 **132** Gouvernemens Generaux / de POITOU de SAINTONGE 1748 R

22 **136** Gouvernemens Generaux / de la MARCHE / et du LIMOSIN 1748 R

23 **138** Gouvernemens Generaux / du BERRY, / du BOURBONNOIS, / et du NIVERNOIS. 1748 R

24 **144** Gouvernemens Generaux / de BOURGOGNE, et, /de FRANCHE-COMTÉ 1748 R

25 **147** Gouvernemens Generaux / du LYONNOIS et / de l'AUVERGNE 1748 R

26 **151** Gouvernemens Generaux / de la GUIENNE PROPRE 1748 R

27 **153** Gouvernemens / Généraux / de Foix et de Roussillon, 1748 R

28 **154** Gouvernement / General / du LANGUEDOC. 1748 R

29 **158** Gouvernement / General / du DAUPHINÉ 1748 R

30 **160** Gouvernement / General / de PROVENCE. 1748 R

31 **165** Carte / des PAYS BAS / CATHOLOQUES 1748 R

32 **169** Carte Generale / du Comté de / FLANDRE. 1748 R

33 **171** I. Carte du Comté de Flandre ou sont les envs. de l'ECLUSE, 1748 R

34 **172** II. Carte de la Flandre òu sont les environs d'AXEL, 1748 R

35 **173** III. Carte de la Flandre où sont les Environs de GRAVELINE, 1748 R

36 **174** IV. Carte du Comté de Flandre où sont les environs de BRUGES 1748 R

37 **175** V. Carte / de la Flandre / où sont les environs de / GAND, 1748 R

38 **176** VI. Carte du Cté. de Flandres où sont les environs de St. OMER, 1748 R

39 **177** VII. Carte de la Flandre où sont les environs d'YPRES 1748 R

40 **178** VIII. Carte de Flandre ou sont les environs d'OUDENARDE, 1748 R

41 **179** IX. Carte du Comté de Flandre òu sont les environs de la Bassée / de DOUAY, 1748 R

42 **180** Carte Generale / du Duché / de BRABANT 1748 R

43 **181** I. Carte du Brabant où sont les environs de BERG-OP-ZOOM, 1748 R

44 **182** II. Carte du Brabant où sont les environs de BRÉDA, et de BOSLEDUC. 1748 R

45 **183** III. Carte du Brabant ou se trouve[n]t le MAESLAND, 1748 R

46 **184** IV. Carte du Brabant / où se trouve[n]t la Mairie de SANTHO / VEN, 1748 R

47 **185** V. Carte du Brabant où se trouve[n]t les Mair[ie]s. d'HERENTALS, 1748 R

48 **186** VI. Carte du Brabant ou se trouv[en]t les envir[on]s. d'EYNDHOVEN 1748 R

49 **187** VII. Carte du Brabant / ou sont les environs / de BRUXELLES, 1748 R

50 **188** VIII. Carte du Brabant / ou sont les envs. de / LOUVAIN, 1748 R

51 **189** IX. Carte du Brabant / ou sont les Mairies de / ROO, 1748 R

52 **190** X. Carte du Brabant / où sont les environs / de TILLEMEONT, 1748 R

53 **194** Carte Générale des / Comtés / de HAYNAUT, de NAMUR, et / de CAMBRESIS. 1748 R

54 **195** I. Carte du Comté du Haynaut où sont les environs d'AHT, 1748 R

55 **196** II. Carte du Cté. de Haynaut où sont les environs de HALLE / D'ENGHIEN, 1748 R

56 **197** Carte du Cté de Haynaut et des Frontieres du Cté. de Flandre / òu sont DOUAY 1748 R

57 **198** IV. Carte du Cté. de Haynaut ou sont les environs de MONS, 1748 R

58 **199** V. Carte des Ctes. de Haynaut et de Namur, et partie du Lyegeois, 1748 R

59 **200** VI. Carte du Cté. de Namur òu sont les environs de NAMUR; 1748 R

60 **201** VII. Carte du Cté de Haynaut òu sont les environs de CAMBRAI 1748 R

61 **202** VIII. Carte du Cté. de Haynaut où sont les environs de LANDRECHIES, 1748 R

62 **203** IX. Carte du Cté. de Haynaut où sont les environs de PHILIPPEVILLE, 1748 R

63 **204** X. Carte du Comté du Namur où se trouvent les environs de DINANT. 1748 R

64 **205** Carte Générale / de la Seigneurie / de LYEGE;/et du Duché / de LIMBOURG. 1748 R

65 **206** I. Carte de l'Etat de Lyege ou se trouvent la TERRE de LOEN, 1748 R

66 **207** II. Carte de L'Etat de Lyege ou se trouvent le Cté. de HORN 1748 R

67 **208** III. Carte de l'Etat de Liege où sont les environs de MAESTRICHT, 1748 R

68 **209** IV. Carte du Duché de Limbourg où sont les environs de FAUQUEMONT, 1748 R

69 **210** V. Carte de l'Etat de Lyege ou se trouvent LYEGE, 1748 R

70 **211** VI. Carte du Duché de Limbourg ou sont les environs de LIMBOURG. 1748 R

71 **213** Carte Générale / du Duché / de LUXEMBOURG 1748 R

72 **214** I. Carte du Dché de Luxembourg ou sont les Frontieres du NAMUROIS / et du LYEGEOIS. 1748 R

73 **215** II. Carte du Dé. de Luxembourg où se trouvt. la Mairie de la MARCHE, 1748 R

74 **216** III. Carte du Dhé. de Luxembourg où sont les Ctés. de ST. VEIT, 1748 R

75 **217** IV. Carte du Duché de Luxembourg où sont les Seignries. de LOMPRÉ 1748 R

76 **218** V. Carte du Dhé. de Luxembourg où sont les Ctés. de LA ROCHE, 1748 R

77 **219** VI. Carte du Dché. de Luxembourg òu sont les Prevôté de BIDBURG, 1748 R

78 **220** VII. Carte des Frontieres du Luxembourg où sont les environs de BOUILLON, 1748 R

79 **221** VIII. Carte du Dché. de Luxembourg où sont les Prevotés de NEUFCHATEL, 1748 R

80 **222** IX. Carte du Dché. du Luxembourg ou sont les environs de LUXEMBG. 1748 R

81 **223** X. Carte du Dché. de Luxembourg òu se trouvent, MOUSON, 1748 R

82 **224** XI. Carte du Dché. de Luxembourg et des Confins de la LORRAINE 1748 R

83 **226** Carte / des Provinces-Unies / des PAYS-BAS. 1748 R

84 **229** Carte du Comté / de ZÉLANDE 1748 R

85 **230** Carte Generale des IX / de la HOLLANDE; 1748 R

86 **231** I. Carte du Comté de Hollande où se trouve l'ISLE DE TEXEL. 1748 R

87 **232** Partie du Comté / de Hollande / òu sont les environs / d'ALCMAAR, 1748 R

88 **233** Carte de la Hollande / où se trouve une partie de la VELUWE, 1748 R

89 **234** Partie du Comté / de Hollande, / où sont les environs / de HAARLEM, 1748 R

90 **235** V. Carte de la Hollande et de la Seigneurie, d'UTRECHT, où sont les environs d'AMSTELDAM 1748 R

91 **236** Partie du Duche /de GUELDRE / où se trouve / la VELUWE. 1748 R

92 **237** VII. Carte de la Hollande ou sont les envs. de ROTTERDAM, 1748 R

93 **238** VIII. Carte de la Hollande òu son [sic] les environs de SCHOONHOVE, 1748 R

94 **239** IX. Carte de la Hollande òu sont les environs de NIMEGUE, 1748 R

95 **241** L'Empire / d'ALLEMAGNE / divisé par Cercles 1748 R

96 **257** Cercle / de WESTPHALIE, / divisé suivant ses / differentes Principautés 1749 RV

97 **260** Cercle de la / BASSE SAXE, / divisée par ses Principautés 1749 RV

98 **263** Partie Septle. du Cercle / de la HAUTE SAXE 1749 RV

99 **262** Partie Meridionale / du Cercle de la / HAUTE SAXE 1749 RV

100 **266** Cercle des Quatre / Electeurs 1749 RV

101 **268** Cercle / de FRANCONIE / divisé en ses princip[a]ux. Etats 1749 RV

102 **271** Cercle de SOUABE / où sont distingué 1749 RV

103 **273** Cercle de BAVIERE / où se trouvent le / Palatinat de Baviere 1748 R

104 **275** L'Archiduché / d'AUTRICHE / et les Duchés de / Stirie, Carinthie et Carniole 1748 R

105 **279** Comte de TIROL / et / Evêché de Brixen. 1748 R

106 **281** Les Suisses / les Alliés de Suisses / et leurs Sujets. 1748 R

107 **284** Royaume de BOHÉME, / Marquisat de MORAVIE, / et SILESIE 1748 R

108 **287** Royaume de / HONGRIE 1749 RV

109 **250** I. Carte du COURS DU RHIN ou sont les environs de WALTSHUT, 1748 R

110 **251** II. COURS DU RHIN ou sont les environs de BASLE, 1748 R

111 **252** III. COURS DU RHIN où sont les environs de BRISACH, 1748 R

112 **253** IV. COURS DU RHIN où sont les environs de RHINAUW, 1748 R

113 **254** V. Carte du COURS DU RHIN où sont les Environs de STRASBOURG. 1748 R

114 **255** VI. Carte du COURS DU RHIN où sont les environs du FORT, 1748 R

115 **256** VII. COURS DU RHIN où sont les environs de PHILISBOURG, 1748 R

116 **289** Etats de la Couron[n]e de / POLOGNE / ou sont les Royaues 1748 R

117 **295** Royaume de PRUSSE / et / Prusse Rle. ou Polonoise. 1748 R

118 **290** Royaume de POLOGNE / divisé en / Haute et Basse Pologne 1749 RV

119 **297** Grand Duché / de LITHUANIE / divisé par Palatinats. 1749 RV

120 **298** Curlande, Livonie, Ingrie/et Partie Meridionale/de Finlande. 1749 RV

121 **75** Partie Occidentale de / la Russie Polonoise 1749 RV

122 **76** Partie Orientale / de la Russie Polonoise 1749 RV

123 **299** Royaumes / d'ESPAGNE, et de / PORTUGAL. 1748 R

124 **307** Partie Septentrionale / des Etats de CASTILLE 1749 RV

125 **308** Partie Septle. des Etats de / CASTILLE 1749 RV

126 **310** Partie Meridion[a]le des Etats de / CASTILLE 1749 RV

127 **313** Royaume / de NAVARRE / divisé en six Mérindades. 1749 RV

128 **315** Partie Septentle. de la / Couronne d'ARAGON 1749 RV

129 **316** Partie Méridionale de la / Couronne d'ARAGON 1749 RV

130 **311** Partie Meridionale des / ETATS DE CASTILLE 1749 RV

131 **318** Partie Septentrionale / du Royaume / de PORTUGAL 1749 RV

132 **320** Partie Meridionale / du Royaume / de PORTUGAL 1749 RV

133 **323** L'ITALIE / et ses Isles 1748 R

134 **332** Duché de / SAVOYE 1748 R

135 **363** Isle et Royaume / de SARDAIGNE 1748 RV

136 **348** Carte des SOURCES DU PO où se trouve 1748 RV

137 **349** II. Cours du PO et partie du PIÉMONT 1748 R

138 **350** III. COURS DU PO où sont les environs 1748 R

139 **351** IV. COURS DU PO où sont les environs de Pavie 1748 RV

140 **352** V. COURS DU PO, où sont les environs de Parme 1748 RV

141 **353** VI. COURS DU PO, ou sont les environs de la Mirandole 1748 RV

142 **354** VII. EMBOUCHURE DU PO, où sont les environs de Rovigo 1748 RV

143 **336** Carte Générale / de / l'Etat de GÊNES. 1748 RV

144 **337** I. Carte de l'Etat de Genes où se trouvent NICE 1748 RV

145 **338** II. Carte des Confins de l'Étât de Gênes où se trouve une partie...de NICE. 1748 RV

146 **339** Carte de l'Étât de Gênes / où se trouvent...d'ONEILLE, 1748 RV

147 **340** IV. Carte de l'Etat de Gênes ou se trouvent les environs de SAVONE 1748 RV

148 **341** V. Carte des Environs de Genes où se trouvent ARENZANO 1748 RV

149 **342** VI. Carte des Confins de l'État de Gênes où sont NOVI 1748 RV

150 **343** VII. Carte des Confins de l'État de Gênes où se trouve...du MILANEZ et du PARMESAN 1748 RV

151 **344** VIII. Carte de l'Etat de Genes ou se trouvent les environs et la VALLÉE DE LAVAGNA 1748 RV

152 **345** IX. Carte de l'Étât de Gênes ou se trouvent SESTRI-di Levante 1748 RV

153 **162** Carte de l'Isle de / CORSE 1748 RV

154 **346** Etat de la Seigneurie et / Republique de VENISE 1748 RV

155 **355** ETAT DE L'EGLISE / et Duché / de TOSCANE 1748 R

156 **359** Partie Septentrionale / du / ROYAUME DE NAPLES. 1748 R

157 **361** Partie Méridionale / du Royaume de / NAPLES 1748 R

158 **365** Carte des Isles de MALTE et de GOZE. 1748 R

159 **366** Carte / de la TURQUIE / D'EUROPE 1748 R

160 **372** L'ASIE / Divisée selon ses differens Etats. 1748 R

161 **384** TURQUIE / ASIATIQUE 1748 R

162 **387** L'ARABIE 1749 RV

163 **388** Royaume de / PERSE 1749 RV

164 **397** INDOUSTAN / ou / ETATS DU MOGOL 1749 RV

165 **399** PRESQU'ISLE DE L'INDE / au dela du Gange 1749 RV

166 **407** Empire de / la CHINE 1749 RV

167 **401** Carte Generale de / l'EMPIRE DES RUSSES 1749 RV

168 **409** Le JAPON 1749 RV

169 **411** ISLES PHILIPPINES / et MOLUQUES 1749 RV

170 **412** Les ISLES DE / LA SONDE 1749 RV

171 **423** L'AFRIQUE / divisée en ses / principales parties. 1748 R

172 **432** Partie Occidentale de / La TURQUIE D'AFRIQUE 1748 R

173 **434** Cote Occidentale d'Afrique 1749 RV

174 **435** Partie Occidentale de / la NIGRITIE 1749 RV

175 **436** Cours de la RIVIERE DE SENEGAL 1749 RV

176 **437** Partie Orientale de / la HAUTE GUINÉE 1749 RV

177 **438** CONGO ou / BASSE GUINÉE 1749 RV

178 **440** PAYS DES CAFRES. 1749 RV

179 **441** Cote de / ZANGUEBAR. 1749 RV

180 **443** ISLE DE MADAGASCAR 1749 RV

181 **442** NUBIE / ABISSINIE et / COTE D'AJAN 1749 RV

182 **444** LES / ISLES CANARIES 1749 RV

183 **445** LES ISLES / DU CAP-VERD 1749 RV

184 **446** AMERIQUE / SEPTENTRIONALE. 1748 R

185 **484** Partie du MEXIQUE ou de / la Nouvelle Espagne...de GUATIMALA. 1749 RV

186 **483** Partie du MEXIQUE / ou de la Nouvlle. Espagne...Mexique 1749 RV

187 **482** Partie du MEXIQUE òu de / la Nouvle. Espagne...de Guadalajara 1749 RV

188 **471** COURS DU MISSISSIPI / et LA LOUISIANE 1749 RV

189 **472** LA FLORIDE / divisée en / Floride et Caroline 1749 RV

190 **475** LES ISLES ANTILLES 1749 RV

191 **479** L'Isle / ST. DOMINGUE 1749 RV

192 **477** L'Isle de / LA MARTINIQUE. 1749 RV

193 **457** LES LACS DU CANADA / et / NOUVELLE ANGLETERRE 1749 RV

194 **468** NOUVELLE ANGLETERRE / NLLE. YORK, NLLE. JERSEY. / PENSILVANIE MARILAND / et VIRGINIE. 1749 RV

195 **467** Partie du Canada...le FLEUVE ST. LAURENT et / LA NOUVELLE ECOSSE 1749 RV

196 **466** L'ACADIE. 1749 RV

197 **464** GOLFE DE ST. LAURENT, / ISLE et BANCS / de TERRE NEUVE. 1749 RV

198 **463** ISLE ROYALE 1749 RV

199 **465** ISLE DE / TERRE-NEUVE 1749 RV

200 **461** BAYES / D'HUDSON et de BAFFINS. / et TERRE DE LABRADOR. 1749 RV

201 **485** AMERIQUE / MERIDIONALE. 1748 R

202 **488** Partie Occidentale de / la TERRE FERME 1749 RV

203 **489** Partie Orientale / de la TERRE FERME ou / l'on trouve la GUIANE 1749 RV

204 **490** Partie Septentrionale du / PEROU / Où l'on trouve la...Fleuve Maragnon ou AMAZONE 1749 RV

205 **491** Partie Meridionale / du PEROU 1749 RV

206 **492** LE CHILI 1749 RV

207 **294** LES TERRES / MAGELLANIQUES 1749 RV

208 **496** LE PARAGUAY 1749 RV

209 **497** Terre de / Sainte Croix, / nommée à present / BRÉSIL. 1749 RV

ATLAS
UNIVERSEL,
Par M. ROBERT Géographe ordinaire du Roy,
ET
Par M. ROBERT DE VAUGONDY son fils Géographe ord. du Roy, et de
S. M. Polonoise Duc de Lorraine et de Bar, et Associé de
l'Academie Royale des Sciences et belles Lettres de Nancy.
AVEC PRIVILEGE DU ROY.
1757.
A PARIS
Chez Les AUTEURS Quay de l'Horloge du Palais.
BOUDET Libraire Imprimeur du Roi, rue St. Jacques.

Atlas Universel

ATLAS/UNIVERSEL / Par M. Robert, Géographe ordinaire du Roy / et / Par M. Robert de Vaugondy son fils Géographe ord. du Roy, et de / S.M. Polonoise Duc de Lorraine et de Bar, et Associé de / l'Académie Royale des Sciences et belles Lettres de Nancy. / Avec Privilege du Roy / 1757. / A Paris / Chèz Les Auteurs Quay de l'Horloge du Palais / Boudet Libraire Imprimeur du Roi rue St. Jacques.

Description: 103/108 maps;
Engraved title page: "Ch. Baquoy sculpsit" "J. Oger Scripsit";
Avertissement," r°-v°
"Préface Historique..." A-R recto
table of contents: "Table des Articles" R verso
subscription list *-***3 verso

Editions and Locations:

[pre-1757] BL: C.38.f.4: 50 maps
K.4.TAB.53,54: 103 maps in 2 vols
BN: Ge.DD.282
Ge.DD.4796 (114)
Estampes Xf.24;
Harvard: MA.18.57 F: 103 maps in 2 vols.
in these, maps are in early states, e.g., unfinished cartouches, incomplete place names; probably early subscription installments

1757/8 BN: Ge.DD.1267 : 103 maps
Nancy, BM: 145.276 : 108 maps in 2 vols. with 8 others tipped in
BL: C.38.f.3 : 108 maps
C.38.f.5 : 108 maps
C.38.f.6 : 108 maps
CUL: Atlas 3.75.2
Biblioteca Vaticana: Mai XI Z.I.12
NAC: G 1015 .R62 1757: 101 maps; missing 24 (**97**) and 66 (**277**)
LC: Philips 619, 620 : 108 maps
Huntington Library: HN 238461: 108 maps (2 v.)
NL: Ayer 135.R6.1757
UM: GA.317.1757
WLCL: W-3-C : 100 maps
USM: Smith Collection

[1768] CUL: Acton aa.45.8
108 maps; many engraving changes on maps; particularly in lettering, changing italic to Roman and vice-versa; borders of North America reflect Treaty of Paris (1763); serves as Vol.VII of Bruzen de la Martinière, *Dictionnaire Géographique* (1768)

[after 1776] JCB: Z.R642.1757. Case 9d
108 maps; Groux signs many cartouches; "Censeur Royal" added to Robert de Vaugondy titles; 'Amérique Septentrionale' dated 1775; 'Mappemonde' dated 1776.
NAC: G 1015 .R62 1783: 107 maps, dated 1749-1783 (missing 101 (**487**))

[1784?] BN: Ge.DD.4520
126 maps, among which many by Jaillot, d'Anville, Rizzi-Zannoni, Buache and Delisle; some Vaugondy re-edited by Delamarche

[1786?] LC: Philips 4292
111 maps; Amérique Septentrionale' dated 1783 and uses "Etats Unis"; Canada: border of 1783; three maps added: Etats-Unis, 1785, La Judée, 1786, Indes Orientales, 1786, all maps published by Boudet, but no mention of Robert de Vaugondy as geographer; perhaps a last edition by Boudet before his death (1787) when Delamarche bought stock?

[1788+] BN: Ge.DD.1467
Ge.DD.1520
BL: C.44.f.8
RGS: 1.c.26
LC: Philips 678, 3524, 5982, 5996
all are Delamarche imprints

Facing page *(Photo courtesy William L. Clements Library)*

SHORT TITLES OF MAPS IN THE *ATLAS UNIVERSEL*

Numerals in bold type refer to number of map catalogue entry.

1 **13** ORBIS VETUS / in utraque continente 1752 RV

2 **393** La JUDÉE ou / TERRE SAINTE, 1750 R

3 **382** ANTIQUOR? IMPERIORUM 1753 RV

4 **38** ROMANI IMPERII 1752 RV

5 **46** BRITANNICAE INSULAE 1750 RV

6 **92** GALLIA ANTIQUA 1750 R

7 **247** GERMANIA ANTIQUA 1756 RV

8 **306** HISPANIA / ANTIQUA 1750 RV

9 **330** ITALIA ANTIQUA 1751 RV

10 **369** GRAECIA VETUS 1752 RV

11 **386** ASIA MINOR 1756 RV

12 **93** IMPERIUM / CAROLI MAGNI 1752 RV

13 **8** MAPPEMONDE 1752 RV

14 **32** L'EUROPE 1751 RV

15 **43** LES ISLES / BRITANNIQUES 1754 R

16 **48** LE ROYAUME / D'ANGLETERRE,1753 R

17 **52** LE ROYAUME / D'ECOSSE 1751 R

18 **56** ROYAUME / D'IRLANDE1750 R

19 **72** ROYAUME DE / DANEMARCK1750 R

20 **68** LES ROYAUMES / DE SUEDE ET DE NORWEGE,1756 RV

21 **79** Partie / Septentrionale / de la / RUSSIE EUROPÉENNE 1753 RV

22 **78** Partie Méridionale / de la / RUSSIE EUROPÉENNE 1752 RV

23 **84** LE ROYAUME DE / FRANCE 1750 R

24 **97** Gouvernement Général de / PICARDIE ET ARTOIS, 1753 R

25 **101** Gouvernement Général de / NORMANDIE 1751 R

26 **105** Gouvernement / General de / BRETAGNE 1751 R

27 **108** Gouvernemens / Generaux / du MAINE ET PERCHE, 1753 R

28 **112** Gouvernement Général / d'ORLEANOIS, 1753 RV

29 **113** Gouvernement Général / de / L'ISLE DE FRANCE 1754 R

30 **116** ENVIRONS / DE PARIS 1753 R

31 **125** Partie Septentrionale /...de / CHAMPAGNE 1752 R

32 **126** Partie Méridionale /...de / CHAMPAGNE 1752 R

33 **129** Carte / de la LORRAINE ET DU BARROIS 1756 RV

34 **127** L'ALSACE 1754 R

35 **133** Gouvernemens Généraux / du POITOU, DU PAYS D'AUNIS, / ET DE / SAINTONGE-ANGOUMOIS 1753 RV

36 **137** Gouvernemens / Généraux / de la MARCHE, DU LIMOSIN, / ET DE L'AUVERGNE. 1753 RV

37 **139** Gouvernements Généraux / du BERRY, DU NIVERNOIS, / ET DU BOURBONOIS. 1753 R

38 **142** Partie / Septentrionale.../ DE BOURGOGNE 1752 R

39 **143** Partie Meridionale /...de / BOURGOGNE 1752 R

40 **146** Partie Septentrionale / du COMTÉ DE BOURGOGNE / OU FRANCHE-COMTÉ 1749 R

41 **145** Partie Meridionale / du COMTÉ DE BOURGOGNE / OU FRANCHE-COMTÉ, 1749 R

42 **149** Partie Septentrionale /.../ de la GUIENNE, 1752 R

43 **148** Partie Orientale /.../de la / GUIENNE, 1753 R

44 **150** Partie Méridile./...de / GUIENNE, 1753 R

45 **155** Gouvernement Général du / LANGUEDOC, 1752 R

46 **159** Gouvernement Général / du DAUPHINÉ., 1751 R

47 **161** Gouvernement Général de / PROVENCE, 1754 RV

48 **166** PAYS-BAS / CATHOLIQUES, 1751 R

49 **170** COMTÉ DE FLANDRE, 1752 RV

50 **191** Partie Septentrionale du / DUCHÉ DE BRABANT, 1752 RV

51 **192** Partie Meridion. du / DUCHÉ DE BRABANT, 1752 RV

52 **225** Carte du Duche de / LUXEMBOURG, 1753 RV

53 **193** Comtés de HAINAUT, / et de CAMBRESIS, 1754 R

54 **212** La Principauté / de LIÉGE / et / le Duché / de LIMBOURG. 1754 RV

55 **228** LES PROVINCES-UNIES / DES PAYS-BAS, 1753 R

56 **243** Carte de / L'EMPIRE D'ALLEMAGNE, 1756 RV

57 **282** Carte de la / REPUBLIQUE DES SUISSES, 1756 RV

58 **258** CERCLE DE / WESTPHALIE, 1751 R

59 **261** Cercle de / BASSE SAXE, 1752 R

60 **265** Partie Meridionale du / Cercle de HAUTE SAXE, 1756 R

61 **264** Partie Septentrionale / du Cercle de HAUTE SAXE, 1751 RV

62 **267** Carte des Cercles / DU HAUT ET DU BAS RHIN, 1753 RV

63 **269** Cercle de FRANCONIE, 1752 R

64 **272** CERCLE DE SOUABE, 1751 R

65 **274** LE CERCLE DE BAVIERE, 1751 R

66 **277** Partie Septentrionale / du / CERCLE D'AUTRICHE, 1752 RV

67 **278** Partie Méridionale / du CERCLE D'AUTRICHE, 1752 RV

68 **280** LE TYROL, 1753 R

69 **285** LE ROYAUME DE BOHEME, 1751 R

70 **288** Royaume / de HONGRIE, 1751 R

71 **292** Le Royaume de / POLOGNE, 1752 RV

72 **296** LA PRUSSE, 1751 R

73 **301** Royaumes / d'ESPAGNE et de / PORTUGAL, 1750 R

74 **309** Partie Septentrionale de / LA COURONNE DE CASTILLE, 1752 RV

75 **312** Partie Meridionale / des ETATS DE CASTILLE, 1751 RV

76 **314** Etats de la Couronne / d'ARAGON, 1752 RV

77 **321** Partie Meridionale / du Royaume de / PORTUGAL. 1751 R

78 **319** Partie Septentrionale / du Royaume de / PORTUGAL. 1751 R

79 **325** L'ITALIE, 1750 R

80 **335** Partie Occidentale / de la LOMBARDIE, 1750 R

81 **334** Partie Orientale de / La LOMBARDIE, 1750 R

82 **333** Duché de SAVOYE, 1751 RV

83 **356** ETAT DE L'EGLISE,/ GRANDE DUCHÉ DE TOSCANE, / et / ISLE DE CORSE. 1750 R

84 **360** Partie Septentrionale du / ROYAUME/DE NAPLES, 1750 R

85 **362** Partie Méridionale / du ROYAUME DE NAPLES, 1750 R

86 **367** TURQUIE / EUROPÉENNE 1755 R

87 **375** Carte de L'ASIE, 1750 RV

88 **405** Partie Occidentale / de L'EMPIRE DE RUSSIE, 1750 RV

89 **404** Partie Orientale / de L'EMPIRE RUSSIE EN ASIE, 1750 R

90 **389** Etats / du Grand-Seigneur / en Asie, / Empire de PERSE, 1753 RV

91 **400** LES INDES ORIENTALES, 1751 R

92 **413** ARCHIPEL / DES INDES ORIENTALES, 1750 R

93 **408** L'EMPIRE DE LA / CHINE, 1751 RV

94 **410** L'EMPIRE / DU JAPON, 1750 R

95 **426** L'AFRIQUE, 1756 RV

96 **431** Carte de / L'EGYPTE, ANCIENNE / ET MODERNE, 1753 RV

97 **448** AMÉRIQUE / SEPTENTRIONALE, 1750 RV

98 **459** Partie de / l'Amérique Septent? / qui comprend / LA NOUVELLE FRANCE / ou le CANADA, 1755 RV

99 **469** Partie / de l'Amerique Septentrionale, / qui comprend / LE COURS DE L'OHIO, 1755 RV

100 **470** Carte / de la VIRGINIE et / du MARYLAND, 1755 RV

101 **487** AMÉRIQUE / MÉRIDIONALE., 1750 RV

102 **476** Partie / DE LA MER DU NORD, 1750 R

103 **480** ISLES DE / SAINT DOMINGUE...et de la MARTINIQUE. 1750 R

104 **327** Carte de / L'ITALIE...les Routes des Postes 1756 R

105 **303** Carte / des Royaumes / d'ESPAGNE et de PORTUGAL...Les Routes des Postes 1757 R

106 **245** Carte / de l'Empire / D'ALLEMAGNE...Les Routes des Postes 1757 R

107 **45** Carte / des GRANDES ROUTES / D'ANGLETERRE, D'ECOSSE, / ET D'IRLANDE. 1757 R

108 **86** Carte / du Royaume de FRANCE...Les Routes des Postes 1758 R

NOUVEL ATLAS
PORTATIF
destiné principalement pour l'instruction
DE LA JEUNESSE,
d'après la Géographie Moderne
de feu l'Abbé Delacroix
Par ROBERT DE VAUGONDY Géographe ;
Revu, Corrigé et augmenté avec la division de la France en
Départemens par le Cit. LAMARCHE, Géographe Successeur
du Cit. FORTIN ingénieur mécanicien pour les Globes et Sphères.
A PARIS
Chez le Cen Lamarche, Géographe,
Rue du Foin St Jacques au Collège de Mtre Gervais.
L'AN III de la République Française,
1795.

Nouvel Atlas Portatif

NOUVEL ATLAS / PORTATIF / destiné principalement pour l'instruction / de la Jeunesse. / d'après La Géographie Moderne / de feu l'Abbé Delacroix / Par le S. Robert de Vaugondy Géographe ord? du Roi, / de S.M.Polonoise Duc de Lorraine et de Bar, / de La Société Royale des Sciences et Belles Lettres de Nancy, / A Paris / Chés le S. Robert Géographe ord. du Roi, / Quai de l'Horloge du Palais proche le Pont Neuf / Avec Privilege 1762.

Description: 52 maps
Engraved title page: "Arrivet delineavit & Sculpcit";
introduction: "Discour sur l'étude de la Géographie" pp. 3-6;
table of contents: "Table des Cartes" pp. 7-9.

Editions and Locations:

1762	BN: Estampes xf.29 Ge.FF.3593 Harvard: MA.18.62.2 LC: Philips 5984 All maps have imprint outside border, lower left
1778	BL: Maps.39.c.3 : most maps in 1762 state LC: Philips 649 NL: Greenlee 4981.R64.1778 Title page altered: "...Nancy, / et Censeur Royal. / A Paris / Chés le Sr. Fortin Ingr. Mécanicien du Roy/ Rue de la Harpe près la rue du Foin"; some maps dated 1778; imprint removed from maps; Groux signs some cartouches [NAP: 2,3,4, 47, 48, 49, 50]; "Censeur Royale" added to titles of most maps; 'Royaume de Naples'(32 (**358**)) dated 1776 in LC and NL copies
1784+	BN: Ge.FF.539 LC: Philips 4298 Title page altered: "Chez le S. Delamarche Géographe / Rue du Foin, St. Jacques, au Collége de Mtre. Gervais."; Title page engraver: "Groux Sculp."; Mappemonde dated 1780
1790/1800	RGS: 4.c.9: Mappemonde dated 1786; map of Africa has note about d'Anville and sources of the Nile added
An XII (1804)	Harvard: MA.19.13.3 maps re-dated with Republican dates: An III - An XII; title page imprint: "chez Delamarche Géographe / Rue du Jardinet No. 13 / vis-à-vis celle de l'Eperon."
1806	LC: Philips 3528 with six other maps tipped in; title page imprint: "chez Delamarche Géographe / Rue du Jardinet No. 13 / vis-à-vis celle de l'Eperon."
1814	Harvard: MA 19.13.2 LC: G 1019.R774.1814 Vault title page altered: "...et adopté pour l'usage de l'Ecole Impériale / Militaire de St. Germain / A Paris / chez Delamarche Géographe / Rue du Jardinet No. 13 / vis-à-vis celle de l'Eperon." maps re-worked by Félix Delamarche and dated 1811-1813.
1817	USM: Smith Collection title page imprint: "chez Fx. Delamarche et Chles. Dien / Rue Jardinet No. 13 / vis-à-vis celle de l'Eperon."

SHORT TITLES OF MAPS IN THE *NOUVEL ATLAS PORTATIF*

Numerals in bold type refer to number of map catalogue entry.

1 **11** MAPPE-MONDE 1762 RV

2,3,4 **15** MAPPEMONDE, 1761 RV

5 **35** L'EUROPE 1762 RV

6 **44** LES ISLES / BRITANNIQUES 1762 RV

7 **49** L'ANGLETERRE 1762 RV

8 **53** L'ECOSSE 1762 RV

9 **57** L'IRLANDE 1762 RV

10 **66** LA SUEDE 1762 RV

11 **70** NORWEGE 1762 RV

12 **81** Partie Septentrionale /.../ RUSSIE EUROPEENNE 1762 RV

13 **80** Partie Meridionale /.../ RUSSIE EUROPEENNE, 1762 RV

14 **85** ROYAUME DE / FRANCE 1762 RV

15 **99** FLANDRE FRANCOISE, / PICARDIE ET ARTOIS, 1762 RV

16 **103** NORMANDIE, / BRETAGNE, / MAINE, ANJOU, 1762 RV

17 **141** BERRI, NIVERNOIS, BOURBONOIS, / LYONOIS, BOURGOGNE, 1762 RV

18 **135** POITOU, AUNIS, / SAINTONGE ET ANGOUMOIS, 1762 RV

19 **152** GUIENNE ET GASCOGNE, / BASSE NAVARRE / et BEARN. 1762 RV

20 **157** LANGUEDOC, / DAUPHINÉ, / PROVENCE. 1762 RV

21 **244** Empire / d'ALLEMAGNE 1762 RV

22 **167** PAYS-BAS, / PROVINCES-UNIES, / WESTPHALIE. 1762 RV

23 **259** Cercles de / HAUTE-SAXE / et de / BASSE-SAXE 1762 RV

24 **270** HAUT ET BAS RHIN, / FRANCONIE, / SOUABE. 1761 RV

25 **276** Cercles de / BAVIERE / et d'AUTRICHE 1762 RV

26 **286** BOHEME SILÉSIE, / MORAVIE, / LUSACE. 1761 RV

27 **283** LES SUISSES / ET / LES GRISONS, 1762 RV

28 **294** Royaume de / POLOGNE 1762 RV

29 **302** Royaumes / d'ESPAGNE, / et de / PORTUGAL. 1762 RV

30 **326** L'ITALIE 1761 RV

31 **357** LA LOMBARDIE, / LE DUCHÉ DE TOSCANE / L'ETAT DU ST.SIEGE, / et L'ISLE DE CORSE 1762 RV

32 **358** Royaume de / NAPLES 1762 RV

33 **368** HONGRIE, / TURQUIE EUROP? 1762 RV

34 **377** L'ASIE 1762 RV

35 **380** TURQUIE D'ASIE, / ARABIE, PERSE, / TARTARIE, 1761 RV

36 **403** SIBERIE OU RUSSIE ASIATIQUE, TARTARIE CHINOISE, PAYS DES ELUTS ET ISLES DU JAPON, 1762 RV

37 **398** INDOSTAN, / PRESQU'ISLES DE L'INDE, / CHINE, / TARTARIE INDEPENDANTE. 1761 RV

38 **415** LES ISLES / DE LA SONDE, MOLUQUES, / PHILIPPINES, CAROLINES, / ET MARIANNES. 1762 RV

39 **427** L'AFRIQUE 1761 RV

40 **433** BARBARIE, EGYPTE, / NIGRITIE, GUINÉE, / NUBIE, ABISSINIE. 1762 RV

41 **439** CONGO, / CAFRERIE. 1762 RV

42 **21** L'AMÉRIQUE 1762 RV

43 **449** CANADA, / LOUISIANE, / POSSESSIONS ANGL? 1762 RV

44 **481** NOUVELLE ESPAGNE, / NOUVEAU MEXIQUE, / ISLES ANTILLES 1762 RV

45 **498** TERRE-FERME / PEROU, BRÉSIL, / PAYS DE L'AMAZONE 1762 RV

46 **493** PARAGUAY, / CHILI, / TERRE MAGELLAN? 1762 RV

47 **383** Carte / des Premiers Ages / du Monde. // Géographie Ecclésiastique, 1762 RV

48,49,50 **395** JUDÉE / OU / TERRE SAINTE 1762 RV

51 **396** JUDÉE / OU / TERRE SAINTE / Sous les Turcs. 1762 RV

52 **94** LA GAULE 1762 RV

Atlas Factice de Géographie Ancienne [WLCL: W-3-D]

In the the William L. Clements Library at the University of Michigan is an atlas entitled *Atlas Factice de Géographie Ancienne*, a collection of 104 seventeenth and eighteenth century maps of ancient history. The atlas was part of the collection formed by Henri Vignaud, an American diplomat resident in Paris at the end of the nineteenth century. All the cartographers represented in the atlas were associated in some way with Nicolas Sanson: his sons, Nicolas and Guillaume Sanson; his nephew, Pierre DuVal; his grandson, Pierre Moullart-Sanson. Two other geographers represented in the atlas, Philippe de La Rue and Pierre Bertius, shared editors with Sanson: Melchoir Tavernier and Pierre Mariette. Over fifty of the maps in the *Atlas Factice* are listed in the inventory of maps sold by Jean Mariette to Gilles Robert in 1733. (See Chapter One)

This group of maps may have been part of Gilles Robert's working collection and show every sign of having been used in this way. Five of the maps have a manuscript grid drawn over them for copying; three maps are printed on both sides as if for proofs; three of the Moullart-Sanson maps are excellent examples of execrable map lettering — upside down placenames; no consistency between Roman and italic — leaving one to think (or hope) that these too are proof states. The earliest map is Nicolas Sanson's 1627 wall map, "Galliae Antiquae Descriptio Geographiae;" the latest is Gilles Robert's 1745 "Carte de la Terre des Hebreux." Robert maps from this collection are designated in the map catalogue as WLCL: AF.

Atlas des Révolutions du Globe [BN:Ge.CC.1370]

In the Bibliothéque Nationale in Paris is an atlas of ancient history attributed to Gilles Robert. Though it has no title page, it has been catalogued as *Atlas des Révolutions du Globe* and consists of 66 examples of the same two-sheet map of Eurasia (each sheet 500 x 725 mm.; RF 1:7,500,000). Each map is hand coloured in a different way to show the boundaries of various political states and kingdoms. The atlas is arranged chronologically, beginning with the Creation and Flood and ending in 1737 with the unification of Lorraine to France. There are manuscript pages between the maps describing the time period, observing the map's differences from its predecessors, and describing the "Etats Nouveaux."

Weiss, Gilles Robert's biographer in Michaud's *Biographie Universelle* (1854, Vol. 36, p.138-9), describes the atlas as Gilles Robert's "most curious and least known work...It was not published and the copper plate no longer existed in 1773, according to a letter from the son of the author to the publisher Boudet. An example of this atlas, perhaps the only one which exists, was bought for sixty francs at the sale of M. Lamy, 13 January 1808 (no. 4440 in the catalogue). It is thought that the work served as a model to Picaud of Nantes for his *Révolutions de l'univers*, which was better engraved but on a much smaller scale (1:1,230,000). That work, published in 1763, was sold for 60 francs by Julien [the geographer and map dealer]."

None of the maps in this atlas are dated nor do they bear engravers' names, making it impossible to date the collection accurately. Didier Robert de Vaugondy does not mention it in his "Préface Historique" to the *Atlas Universel* , but does suggest a source of inspiration: the posthumous work of Johann Matthias Haas, an "atlas historique," edited by Augustus Gottlieb Boehme and described by Vaugondy as "a work in which one finds the revolutions and successions of great empires which have taken place in the world up to our day" [p. 14b]. The *Atlas des Révolutions du Globe* does employ a style used both by the Sansons and the Roberts in which a base map is re-coloured in order to show different themes, in this case the political changes in the old world.

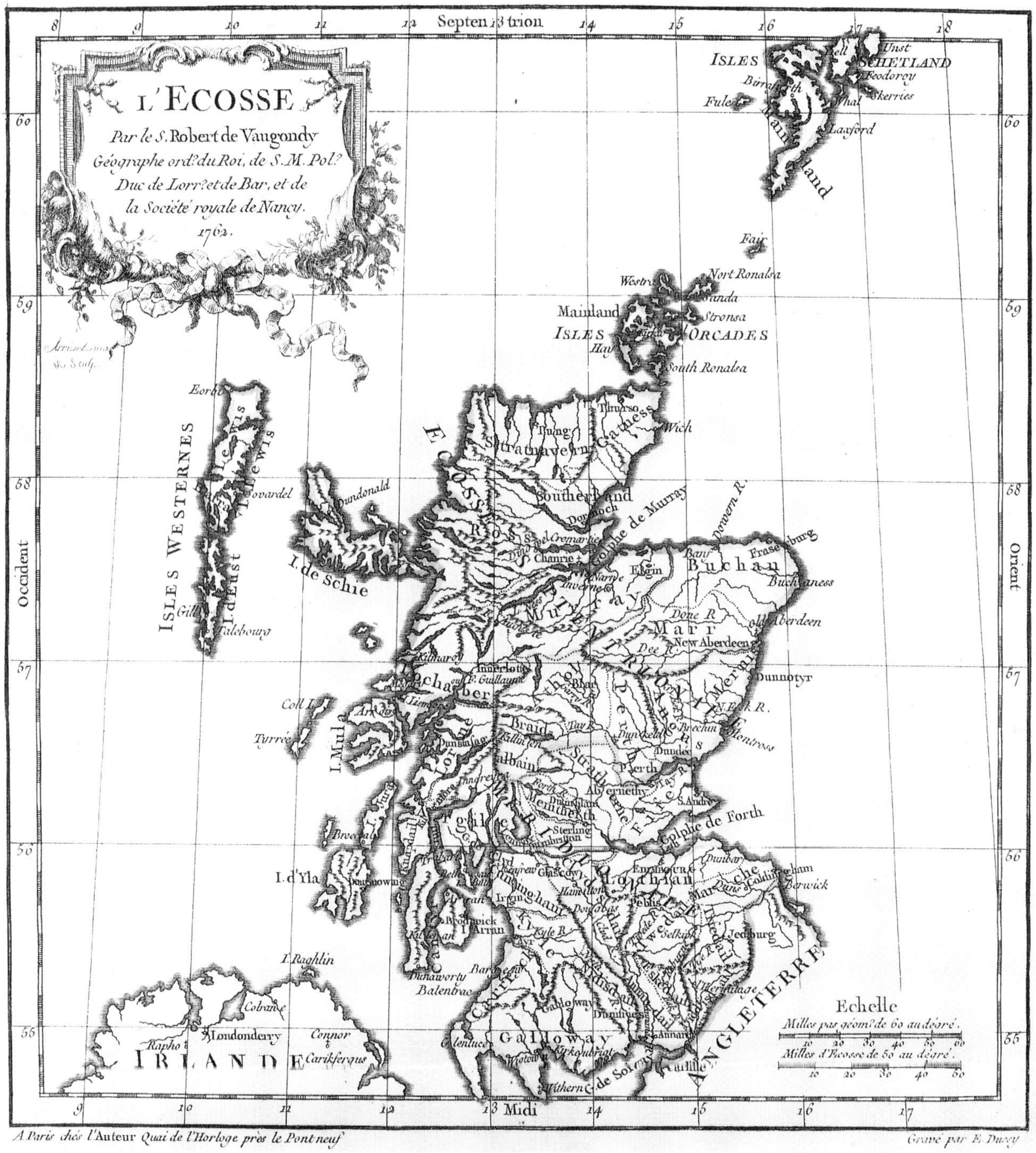

L'Ecosse [Cat. 53] from the **Nouvel Atlas Portatif**, *1762. (Photo courtesy of the William L. Clements Library, University of Michigan)*

INTRODUCTION TO THE GLOBE CATALOGUE

This catalogue is a finding list for globes made by Didier Robert de Vaugondy. In his own catalogues, he lists terrestrial and celestial globes as well as Ptolemaic and Copernican armillary spheres ranging in size from three inches (c. 75 mm.) to eighteen (c. 455 mm.). I have not set out to see all the Vaugondy globes and consequently can only record the existence of most of the globes listed here. There are probably many more, unnoticed, in libraries, private collections, and dealers' showrooms. Very often the distinguishing features of a Vaugondy globe or sphere are the greenish-yellow varnish used on the globe's surface, the zodiac pasted on the wide horizon of the same hue as the globe, and the distinctive red color of the edges of the horizon, meridian, and four supports of the horizon which themselves usually have lists pasted on with the major cities of the world and their coordinates.
The catalogue is laid ut in a similar fashion to the map catalogue, viz.:

Cat. No.	Globe type	Date

Measurements

Title

Imprint

Location

Comments

This listing makes use of two catalogues and one work on globes.

Duprat: Gabrielle Duprat, *Liste des Globes Terrestres et Célestes anciens — antérieurs à 1850 — conservés dans les collections publiques de France*, Paris: CNRS, 1970.
Fiorini: Matteo Fiorini, *Sfere terrestri et celesti in Italia*, Rome, 1899.
Yonge: Ena Yonge, *A Catalogue of Early Globes made prior to 1850 and Conserved in the United States*, New York: American Geographical Society, 1968.

1 Armillary Sphere [1750]

175 mm.

imprint:

Location: Vienna: private collection

Comments: presented by Rudolf Schmidt, VII Internationale Coronelli Symposium 1989

2 Terrestrial 1750

175 mm.

GLOBE TERRESTRE / dedié et presenté / au Roy / par son très humble, très obeissant / et très soumis serviteur, et fidèle suject / Robert de Vaugondy / 1750 / gravé par G. Delahaye

imprint: A Paris, Chès le Sr. Robert / Géographe ordre. du Roi / Quai de l'Horloge

Location: Paris: CNAM F.b.4.3851[2] [Duprat 161]

Comments: Gravé par G. Delahaye

3 Celestial [1751]

150 mm.

imprint:

Location: USA, Newport News, Va.: The Mariners Museum [Yonge p. 54, illustrated]

Comments: etched on glass

4 Terrestrial [1751]

150

imprint:

Location: USA, Newport News, VA: The Mariners Museum; [Yonge p. 54]

Comments: etched on glass; now missing from collection

5 Celestial 1751

455 mm.

imprint:

Location: France, Chartres: Museum [Duprat 163]; Troyes: Musée des Beaux-Arts [Duprat 169]; Nancy: Musée historique lorrain [Duprat 165]

Comments:

6 Celestial 1751

455 mm.

GLOBE CELESTE / dressé par ordre / du Roi, / et calculé pour l'année 1770 / par le Sr. Robert de Vaugondy fils / Avec approbation de l'Académie / Royale des Sciences / Août 1751

imprint: A Paris / Chés l'Auteur sur / le Quay de l'horloge / au Palais

Location: Paris: Ministère des Affaires Etrangères [Duprat 167]

Comments: Gobin sculpsit; G. Delahaye scripsit

7 Terrestrial 1751

455 mm.

imprint:

Location: France, Chartres: Musée [Duprat 162]; Troyes: Musée des Beaux-Arts [Duprat 168]; Nancy: Musée historique lorrain [Duprat 164]

Comments: "Commandés par Mme de Pompadour" [Chartres]

8 Terrestrial 1751

455 mm.

Globe Terrestre / dressé par ordre / du roi / par le Sr. Robert de Vaugondy / avec Approbation de l'Académie / Royale des Sciences / Août 1751

imprint: A Paris / Ches l'Auteur Quay / de l'Horloge / du Palais

Location: France, Paris: Ministère des Affaires Etrangères [Duprat 166]

Comments: cartouche: "dessiné et gravé par Gobin"; " Géographie et la lettre gravées par Guill. Delahaye"

9 Celestial 1754

220 mm.

imprint:

Location: Italy, Parma: Regia Biblioteca Palatina di Parma [Fiorini, p. 418]; Tassarolo: Palazzo Spinola [Fiorini, p. 418]; Venice: Pinacoteca Quirini-Stampalia [Fiorini, p. 418]

Comments:

10 Terrestrial 1754

220 mm.

imprint:

Location: Italy, Parma: Regia Biblioteca Palatina di Parma [Fiorini, p. 418]; Tassarolo: Palazzo Spinola [Fiorini, p. 418]; Venice: Pinacoteca Quirini-Stampalia [Fiorini, p. 418]

Comments:

11 Terrestrial 1756

74 mm.

imprint:

Location: France, Paris: Musée du Louvre [Duprat 170]

Comments: in a case with celestial map on inside

12 Celestial 1764

455 mm.

Le Globe Celeste / dressé par ordre du Roi en 1751 / par le Sr. Robert de Vaugondy / Nouvelle Edition augmentée / des constellations de Mr de La Caille / Avec Privilege 1764

imprint:

Location: France, Paris: Bibliothèque nationale [Duprat 174]; Versailles [Duprat 176]; Clermont-Ferrand: Bibliothèque Municipale [Duprat 172]

Comments:

13 Terrestrial 1764

461 mm.

imprint:

Location: France, Clermont-Ferrand: Bibliothèque Municipale [Duprat 171]

Comments:

14 Celestial 1764

not given

imprint:

Location: Italy, Lucca: Biblioteca Governativa [Fiorini, p. 418]; Venice: Osservatorio Patriarcale [Fiorini, p. 418]

Comments:

15 Terrestrial 1773

455 mm.

GLOBE TERRESTRE / dressé par ordre / du Roi / Par le Sr. Robert du Vaugondy 1751 / Nouvelle edition corrigée et augmentée / par l'Auteur Géographe ord. du Roi / de S.M. Polonaise Duc de Lorraine et Bar / de la Société royale des Sciences et Belles-Lettres / de Nancy 1773 / Avec Priv.

imprint: A Paris / chés L'Auteur Quay / de l'horloge / du Palais

Location: France, Paris: Bibliothèque Nationale [Duprat 173]; Versailles [Duprat 175]; Strasbourg: Musée des Beaux-Arts [Duprat 177]

Comments: cartouche: Arrivet inv. & sculp.

16 Celestial 1773

461 mm.

imprint:

Location: France, Strasbourg: Musée des Beaux-Arts [Duprat 178]

Comments:

17 Terrestrial 1773

not given

imprint:

Location: Italy, Lucca: Biblioteca Reale di Caserta [Fiorini, p. 418]

Comments:

18 Terrestrial 1784

1600 mm.

imprint:

Location: France, Versailles: Musée au Château [Duprat 179]

Comments: called the "Globe Bergevin"

19 Terrestrial 1785

557 mm.

imprint:

Location: France, Angers: Bibliothèque Municipale [Duprat 180]

Comments: with Delamarche

20 Armillary Sphere 1787

240 mm.

imprint:

Location: USA, Portland, Maine: Univ. of Maine, Smith Collection [Yonge p. 97]

Comments: with Delamarche

21 Terrestrial 1787

240 mm.

imprint:

Location: USA, Portland, Maine: Univ. of Maine, Smith Collection [Yonge p. 24]; France, Autun: Bibliothèque Municipale [Duprat 181]

Comments: with Delamarche

22 Terrestrial 1804

322 mm.

imprint:

Location: France, Dôle: Bibliothèque Municipale [Duprat 182]

Comments: with Delamarche

23 Terrestrial 1816

246 mm.

imprint:

Location: France, Autun: Bibliothèque Municipale [Duprat 183]

Comments: with Delamarche & Dien

24 Terrestrial 1821

246 mm.

imprint:

Location: France, Autun: Bibliothèque Municipale [Duprat 184] NAC: G/3160/1821/.G6

Comments: by Félix Delamarche & Charles Dien

25 Celestial 1821

c. 250 mm.

imprint:

Location: NAC: G 3160 1821 .G6

Comments: no imprint; seems to be a pair with **24**

26 Terrestrial 1824

458 mm.

imprint:

Location: France, Albi: Bibliothèque Municipale [Duprat 185]

Comments: with Delamarche

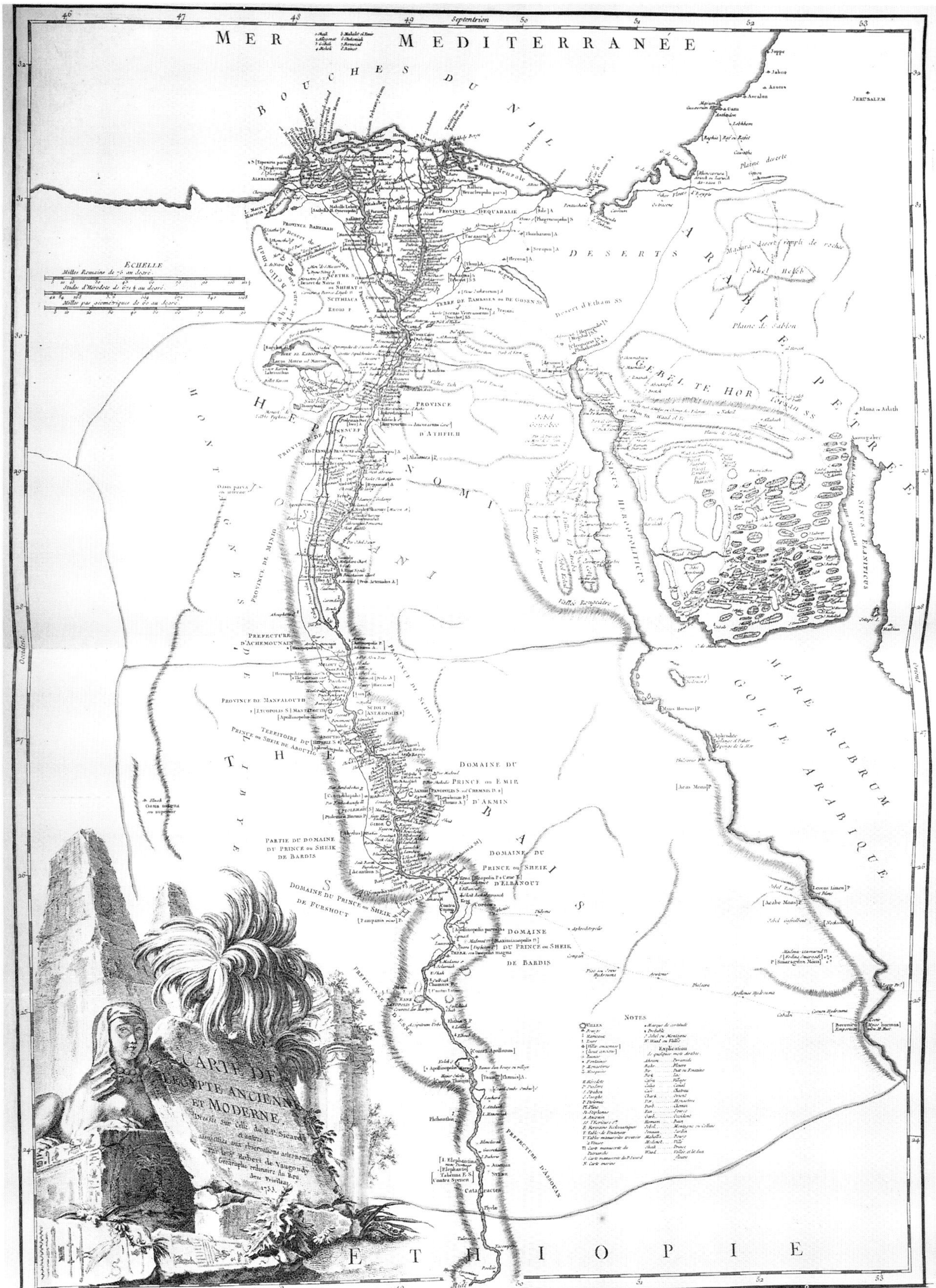

Map of Egypt Ancient and Modern, 1753, from the **Atlas Universel** *[Cat. 431]. (Photo courtesy of the William L. Clements Library, University of Michigan)*

SHORT TITLE MAP INDEX

Numerals in bold type refer to number of map catalogue entry. Page numbers in italics refer to illustrations. [d] = detail. [c] = cartouche. Roman numerals in titles are treated as letters.

INDEX

Numerals in bold type refer to number of map catalogue entry. Page numbers in [italics] refer to illustrations.

AB = 0.2042